Methicillin-Resistant
Staphylococcus aureus
(MRSA) Protocols

METHODS IN MOLECULAR BIOLOGY™

John M. Walker, SERIES EDITOR

436. **Avian Influenza Virus,** edited by *Erica Spackman,* 2008
435. **Chromosomal Mutagenesis,** edited by *Greg Davis and Kevin J. Kayser,* 2008
434. **Gene Therapy Protocols:** *Volume 2: Design and Characterization of Gene Transfer Vectors,* edited by *Joseph M. LeDoux,* 2008
433. **Gene Therapy Protocols:** *Volume 1: Production and In Vivo Applications of Gene Transfer Vectors,* edited by *Joseph M. LeDoux,* 2008
432. **Organelle Proteomics,** edited by *Delphine Pflieger and Jean Rossier,* 2008
431. **Bacterial Pathogenesis:** *Methods and Protocols,* edited by *Frank DeLeo and Michael Otto,* 2008
430. **Hematopoietic Stem Cell Protocols,** edited by *Kevin D. Bunting,* 2008
429. **Molecular Beacons:** *Signalling Nucleic Acid Probes, Methods and Protocols,* edited by *Andreas Marx and Oliver Seitz,* 2008
428. **Clinical Proteomics:** *Methods and Protocols,* edited by *Antonio Vlahou,* 2008
427. **Plant Embryogenesis,** edited by *Maria Fernanda Suarez and Peter Bozhkov,* 2008
426. **Structural Proteomics:** *High-Throughput Methods,* edited by *Bostjan Kobe, Mitchell Guss, and Huber Thomas,* 2008
425. **2D PAGE: Volume 2:** *Applications and Protocols,* edited by *Anton Posch,* 2008
424. **2D PAGE: Volume 1:** *Sample Preparation and Pre-Fractionation,* edited by *Anton Posch,* 2008
423. **Electroporation Protocols,** edited by *Shulin Li,* 2008
422. **Phylogenomics,** edited by *William J. Murphy,* 2008
421. **Affinity Chromatography:** *Methods and Protocols, Second Edition,* edited by *Michael Zachariou,* 2008
420. **Drosophila:** *Methods and Protocols,* edited by *Christian Dahmann,* 2008
419. **Post-Transcriptional Gene Regulation,** edited by *Jeffrey Wilusz,* 2008
418. **Avidin–Biotin Interactions:** *Methods and Applications,* edited by *Robert J. McMahon,* 2008
417. **Tissue Engineering,** *Second Edition,* edited by *Hannsjörg Hauser and Martin Fussenegger,* 2007
416. **Gene Essentiality:** *Protocols and Bioinformatics,* edited by *Andrei L. Osterman,* 2008
415. **Innate Immunity,** edited by *Jonathan Ewbank and Eric Vivier,* 2007
414. **Apoptosis in Cancer:** *Methods and Protocols,* edited by *Gil Mor and Ayesha Alvero,* 2008
413. **Protein Structure Prediction,** *Second Edition,* edited by *Mohammed Zaki and Chris Bystroff,* 2008
412. **Neutrophil Methods and Protocols,** edited by *Mark T. Quinn, Frank R. DeLeo, and Gary M. Bokoch,* 2007
411. **Reporter Genes:** *A Practical Guide,* edited by *Don Anson,* 2007
410. **Environmental Genomics,** edited by *Cristofre C. Martin,* 2007
409. **Immunoinformatics:** *Predicting Immunogenicity In Silico,* edited by *Darren R. Flower,* 2007
408. **Gene Function Analysis,** edited by *Michael Ochs,* 2007

407. **Stem Cell Assays,** edited by *Vemuri C. Mohan,* 2007
406. **Plant Bioinformatics:** *Methods and Protocols,* edited by *David Edwards,* 2007
405. **Telomerase Inhibition:** *Strategies and Protocols,* edited by *Lucy Andrews and Trygve O. Tollefsbol,* 2007
404. **Topics in Biostatistics,** edited by *Walter T. Ambrosius,* 2007
403. **Patch-Clamp Methods and Protocols,** edited by *Peter Molnar and James J. Hickman,* 2007
402. **PCR Primer Design,** edited by *Anton Yuryev,* 2007
401. **Neuroinformatics,** edited by *Chiquito J. Crasto,* 2007
400. **Methods in Membrane Lipids,** edited by *Alex Dopico,* 2007
399. **Neuroprotection Methods and Protocols,** edited by *Tiziana Borsello,* 2007
398. **Lipid Rafts,** edited by *Thomas J. McIntosh,* 2007
397. **Hedgehog Signaling Protocols,** edited by *Jamila I. Horabin,* 2007
396. **Comparative Genomics,** *Volume 2,* edited by *Nicholas H. Bergman,* 2007
395. **Comparative Genomics,** *Volume 1,* edited by *Nicholas H. Bergman,* 2007
394. **Salmonella:** *Methods and Protocols,* edited by *Heide Schatten and Abraham Eisenstark,* 2007
393. **Plant Secondary Metabolites,** edited by *Harinder P. S. Makkar, P. Siddhuraju, and Klaus Becker,* 2007
392. **Molecular Motors:** *Methods and Protocols,* edited by *Ann O. Sperry,* 2007
391. **Methicillin-Resistant Staphylococcus aureus (MRSA) Protocols,** edited by *Yinduo Ji,* 2007
390. **Protein Targeting Protocols,** *Second Edition,* edited by *Mark van der Giezen,* 2007
389. **Pichia Protocols,** *Second Edition,* edited by *James M. Cregg,* 2007
388. **Baculovirus and Insect Cell Expression Protocols,** *Second Edition,* edited by *David W. Murhammer,* 2007
387. **Serial Analysis of Gene Expression (SAGE):** *Digital Gene Expression Profiling,* edited by *Kare Lehmann Nielsen,* 2007
386. **Peptide Characterization and Application Protocols,** edited by *Gregg B. Fields,* 2007
385. **Microchip-Based Assay Systems:** *Methods and Applications,* edited by *Pierre N. Floriano,* 2007
384. **Capillary Electrophoresis:** *Methods and Protocols,* edited by *Philippe Schmitt-Kopplin,* 2007
383. **Cancer Genomics and Proteomics:** *Methods and Protocols,* edited by *Paul B. Fisher,* 2007
382. **Microarrays,** *Second Edition: Volume 2, Applications and Data Analysis,* edited by *Jang B. Rampal,* 2007
381. **Microarrays,** *Second Edition: Volume 1, Synthesis Methods,* edited by *Jang B. Rampal,* 2007
380. **Immunological Tolerance:** *Methods and Protocols,* edited by *Paul J. Fairchild,* 2007
379. **Glycovirology Protocols,** edited by *Richard J. Sugrue,* 2007
378. **Monoclonal Antibodies:** *Methods and Protocols,* edited by *Maher Albitar,* 2007
377. **Microarray Data Analysis:** *Methods and Applications,* edited by *Michael J. Korenberg,* 2007

METHODS IN MOLECULAR BIOLOGY™

Methicillin-Resistant *Staphylococcus aureus* (MRSA) Protocols

Edited by

Yinduo Ji

*Department of Veterinary and Biomedical Sciences,
University of Minnesota, St. Paul, MN*

HUMANA PRESS ✱ TOTOWA, NEW JERSEY

© 2007 Humana Press Inc.
999 Riverview Drive, Suite 208
Totowa, New Jersey 07512

www.humanapress.com

All rights reserved. No part of this book may be reproduced, stored in a retrieval system, or transmitted in any form or by any means, electronic, mechanical, photocopying, microfilming, recording, or otherwise without written permission from the Publisher. Methods in Molecular Biology™ is a trademark of The Humana Press Inc.

All papers, comments, opinions, conclusions, or recommendations are those of the author(s), and do not necessarily reflect the views of the publisher.

This publication is printed on acid-free paper. ∞
ANSI Z39.48-1984 (American Standards Institute)
Permanence of Paper for Printed Library Materials.

Cover illustration: Figure 2, Chapter 1, "Clinical, Epidemiological, and Laboratory Aspects of Methicillin-Resistant *Staphylococcus aureus* (MRSA) Infection," by Elizabeth Palavecino.

Cover design by Nancy Fallatt

Production Editor: Michele Seugling

For additional copies, pricing for bulk purchases, and/or information about other Humana titles, contact Humana at the above address or at any of the following numbers: Tel.: 973-256-1699; Fax: 973-256-8341; E-mail: orders@humanapr.com; or visit our Website: www.humanapress.com

Photocopy Authorization Policy:
Authorization to photocopy items for internal or personal use, or the internal or personal use of specific clients, is granted by Humana Press Inc., provided that the base fee of US $30.00 per copy is paid directly to the Copyright Clearance Center at 222 Rosewood Drive, Danvers, MA 01923. For those organizations that have been granted a photocopy license from the CCC, a separate system of payment has been arranged and is acceptable to Humana Press Inc. The fee code for users of the Transactional Reporting Service is: [978-1-58829-655-9/07 $30.00].

Printed in the United States of America. 10 9 8 7 6 5 4 3 2 1

e-ISBN: 978-1-59745-468-1

Library of Congress Cataloging-in-Publication Data
Methicillin-Resistant *Staphylococcus aureus* (MRSA) protocols/edited by Yinduo Ji.
   p. ; cm. -- (Methods in molecular biology, ISSN 1064-3745; 390)
   Includes bibliographical references and index.
   ISBN 1-58829-655-5 (alk. paper)
   1. *Staphylococcus aureus* infections--Laboratory manuals. 2. Methicillin resistance--Laboratory manuals.
I. Ji, Yinduo. II. Title: MRSA protocols. III. Series: Methods in molecular biology (Clifton, N.J.); v. 390.
   [DNLM: 1. Staphylococcal Infections. 2. Clinical Protocols. 3. Methicillin Resistance. 4. *Staphylococcus aureus*. W1 ME9616J v. 390 2007 / WC 250 M9392 2007]
   QR201.S68M49 2007
   616.9′297--dc22
                                                                   2006032912

Preface

The recent emergence of drug-resistant pathogens—especially multiple-drug-resistant isolates and methicillin-resistant *Staphylococcus aureus* (MRSA)—is causing serious public health concerns because most of the pathogens can lead to fatal infections. The availability of whole-genome sequences and advanced high-throughput technologies allows us to develop a specific and rapid diagnostic method, investigate potential mechanisms of bacterial evolution to antibiotic resistance, and identify novel targets for the development of more effective therapeutic and/or preventive agents.

The aim of *MRSA Protocols* is to provide a comprehensive collection of the most up-to-date techniques for the detection and investigation of MRSA. Each chapter starts with a brief introduction to the method and its purpose and then goes on to provide very detailed protocols for every step of the analysis. Most of the chapters also contain a section with tips on individual steps that are not usually found in a methods book but that may represent the difference between immediate success and lengthy troubleshooting.

MRSA Protocols is an excellent starting point for anyone who wants or needs to set up a new method to study MRSA. Most of the methods are oriented toward routine clinical diagnosis, research, and actual practice for treatment of patients infected by MRSA. Although it mainly focuses on MRSA, it should be a valuable reference for technicians and scientists working on other pathogens.

Yinduo Ji

Contents

Preface v
Contributors ix

1 Clinical, Epidemiological, and Laboratory Aspects of Methicillin-Resistant *Staphylococcus aureus* (MRSA) Infections
 Elizabeth Palavecino 1

2 MRSA Case Studies
 Kurt D. Reed, Mary E. Stemper, and Sanjay K. Shukla 21

3 Minimum Inhibitory Concentration (MIC) Analysis and Susceptibility Testing of MRSA
 German Bou 29

4 Internal Transcribed Spacer (ITS)-PCR Identification of MRSA
 Shin-ichi Fujita 51

5 Pulsed-Field Gel Electrophoresis of MRSA
 Kurt D. Reed, Mary E. Stemper, and Sanjay K. Shukla 59

6 Multilocus Sequence Typing (MLST) of *Staphylococcus aureus*
 Nicholas A. Saunders and Anne Holmes 71

7 Staphylococcal Cassette Chromosome *mec* (SCC*mec*) Analysis of MRSA
 Teruyo Ito, Kyoko Kuwahara, and Keiichi Hiramatsu 87

8 Targeted Gene Disruption for the Analysis of Virulence of *Staphylococcus aureus*
 J. Ross Fitzgerald 103

9 Molecular Analysis of Staphylococcal Superantigens
 Patrick M. Schlievert and Laura C. Case 113

10 Investigation of Biofilm Formation in Clinical Isolates of *Staphylococcus aureus*
 James E. Cassat, Chia Y. Lee, and Mark S. Smeltzer 127

11 Comparative Analysis of Staphylococcal Adhesion and Internalization by Epithelial Cells
 Xudong Liang and Yinduo Ji 145

12 Comparative Analysis of MRSA
 Fumihiko Takeuchi, Tadashi Baba, and Keiichi Hiramatsu ... 153

13	Genomic Analysis of Gene Expression of *Staphylococcus aureus* **Chuanxin Yu, Junsong Sun, Li Zheng, and Yinduo Ji** *169*
14	Proteomic Approach to Investigate MRSA **Patrice Francois, Alexander Scherl, Denis Hochstrasser, and Jacques Schrenzel** *179*
15	Environmental Surveillance for MRSA **J. Scott Weese** *201*
16	Control and Prevention of MRSA Infections **Liangsu Wang and John F. Barrett** *209*
17	Treatment of Infections Caused by Resistant *Staphylococcus aureus* **Gregory M. Anstead, Gabriel Quinones-Nazario, and James S. Lewis II** *227*
	Index 259

Contributors

GREGORY M. ANSTEAD, MD, PhD • *Division of Infectious Diseases, Department of Medicine, University of Texas Health Science Center at San Antonio, San Antonio, TX*
TADASHI BABA, PhD • *Department of Infection Control Science, Juntendo University, Tokyo, Japan*
JOHN F. BARRETT, PhD • *Infectious Disease Research, Merck & Co., Inc., Rahway, NJ*
GERMAN BOU, PhD • *Microbiology Service, Juan Canalejo Hospital, La Coruña, Spain*
LAURA C. CASE, BS • *Department of Microbiology, University of Minnesota Medical School, Minneapolis, MN*
JAMES E. CASSAT, PhD • *Department of Microbiology and Immunology, University of Arkansas for Medical Sciences, Little Rock, AR*
J. ROSS FITZGERALD, PhD • *Centre for Infectious Diseases, Chancellor's Building, New Royal Infirmary, University of Edinburgh, Edinburgh, Scotland, United Kingdom*
PATRICE FRANCOIS, PhD • *Department of Internal Medicine, Service of Infectious Diseases-Genomic Research Laboratory, Geneva, Switzerland*
SHIN-ICHI FUJITA, MD, PhD • *Department of Laboratory Medicine, Graduate School of Medical Science, Kanazawa University, Kanazawa, Japan*
KEIICHI HIRAMATSU, MD, PhD • *Department of Infection Control Science, Juntendo University, Tokyo, Japan*
DENIS HOCHSTRASSER, MD • *Biomedical Proteomics Research Group, Central Clinical Chemistry Laboratory, Geneva, Switzerland*
ANNE HOLMES, PhD • *Laboratory of HealthCare Associated Infections, Centre for Infections, London, United Kingdom*
TERUYO ITO, PhD • *Department of Infection Control Science, Graduate School of Medicine, Juntendo University, Tokyo, Japan*
YINDUO JI, PhD • *Department of Veterinary and Biomedical Sciences, University of Minnesota, St. Paul, MN*
KYOKO KUWAHARA, PhD • *Department of Infection Control Science, Graduate School of Medicine, Juntendo University, Tokyo, Japan*
CHIA Y. LEE, PhD • *Department of Microbiology and Immunology, University of Arkansas for Medical Sciences, Little Rock, AR*
JAMES S. LEWIS II, PharmD • *Pharmacy Service, University Hospital, San Antonio, TX*
XUDONG LIANG, MD • *Department of Veterinary and Biomedical Sciences, University of Minnesota, St. Paul, MN*
ELIZABETH PALAVECINO, MD • *Department of Pathology, Wake Forest University School of Medicine, Winston-Salem, NC*
GABRIEL QUINONES-NAZARIO, MD • *Division of Infectious Diseases, Department of Medicine, University of Texas Health Science Center at San Antonio, San Antonio, TX*

KURT D. REED, MD • *Emerging Infectious Disease Laboratory, Marshfield Clinic Research Foundation, Marshfield, WI*

NICHOLAS A. SAUNDERS, PhD • *Communicable Disease Microbiology Services Support Division, Centre for Infections, London, United Kingdom*

ALEXANDER SCHERL, PhD • *Biomedical Proteomics Research Group, Central Clinical Chemistry Laboratory, Geneva, Switzerland*

PATRICK M. SCHLIEVERT, PhD • *Department of Microbiology, University of Minnesota Medical School, Minneapolis, MN*

JACQUES SCHRENZEL, MD • *Department of Internal Medicine, Service of Infectious Diseases-Genomic Research Laboratory and Clinical Microbiology Laboratory, Geneva, Switzerland*

SANJAY K. SHUKLA, PhD • *Emerging Infectious Disease Laboratory, Marshfield Clinic Research Foundation, Marshfield, WI*

MARK S. SMELTZER, PhD • *Department of Microbiology and Immunology, University of Arkansas for Medical Sciences, Little Rock, AR*

MARY E. STEMPER, MS, MT (ASCP) • *Emerging Infectious Disease Laboratory, Marshfield Clinic Research Foundation, Marshfield, WI*

JUNSONG SUN, PhD • *Department of Veterinary and Biomedical Sciences, University of Minnesota, St. Paul, MN*

FUMIHIKO TAKEUCHI, PhD • *Department of Infection Control Science, Juntendo University, Tokyo, Japan*

LIANGSU WANG, PhD • *Infectious Disease Research, Merck & Co., Inc., Rahway, NJ*

J. SCOTT WEESE, DVM, DVSC, DIPACVIM • *Department of Clinical Studies, Ontario Veterinary College, University of Guelph, Guelph, Ontario, Canada*

CHUANXIN YU, PhD • *Department of Veterinary and Biomedical Sciences, University of Minnesota, St. Paul, MN*

LI ZHENG, MD • *Department of Veterinary and Biomedical Sciences, University of Minnesota, St. Paul, MN*

1

Clinical, Epidemiological, and Laboratory Aspects of Methicillin-Resistant *Staphylococcus aureus* (MRSA) Infections

Elizabeth Palavecino

Summary

Methicillin-resistant *Staphylococcus aureus* (MRSA) is a major pathogen responsible for both hospital- and community-onset disease. Resistance to methicillin in *S. aureus* is mediated by PBP2a, a penicillin-binding protein with low affinity to β-lactams, encoded by the *mecA* gene. Accurate susceptibility testing of *S. aureus* isolates and screening of patients for colonization with MRSA are important tools to limit the spread of this organism. This review focuses on the clinical significance of MRSA infections and new approaches for the laboratory diagnosis and epidemiological typing of MRSA strains.

Key Words: *Staphylococcus aureus*; antimicrobial resistance; methicillin-resistant *Staphylococcus aureus*; CA-MRSA; staphylococcal infections; susceptibility testing; molecular typing; virulence.

1. Introduction

Historically, *Staphylococcus aureus* has been recognized as an important cause of disease around the world and it has become a major pathogen associated with both hospital- and community-acquired infections. Before the availability of antibiotics, invasive infections caused by *S. aureus* were often fatal. The introduction of penicillin greatly improved the prognosis for patients with severe staphylococcal infections, but after a few years of clinical use, resistance appeared owing to production of β-lactamases. Methicillin was designed to resist β-lactamase degradation, but MRSA strains that were resistant to all β-lactam antibiotics were identified soon after methicillin was introduced into clinical practice. Until recently, MRSA was predominantly a nosocomial pathogen causing hospital-acquired infections, but MRSA strains are now being increasingly isolated from community-acquired infections as well.

From: *Methods in Molecular Biology: MRSA Protocols*
Edited by: Y. Ji © Humana Press Inc., Totowa, NJ

Vancomycin has been the antibiotic of choice to treat MRSA infections, and the emergence of vancomycin-nonsusceptible *S. aureus* reported in recent years is a cause of great public health concern and has made therapy of MRSA infections even more challenging for clinicians.

The purpose of this review is to discuss the clinical significance of MRSA infections, to present the mechanisms of antimicrobial resistance, and to comment on the current recommendations for susceptibility testing of MRSA strains.

2. Clinical Significance of MRSA Infections
2.1. Hospital-Associated MRSA Strains

Since the time that methicillin resistance emerged, MRSA has become widespread in hospitals worldwide, causing bacteremia, pneumonia, surgical site infections, and other nosocomial infections *(1–5)*. Nosocomial MRSA infections represent a burden for both patients and health care systems, because of their association with high morbidity and mortality and increased hospitalization costs *(6,7)*. Recent data from the Centers for Disease Control and Prevention showed that 59.5% of all health care–associated *S. aureus* infections in the United States are caused by MRSA *(8)*. In a nationwide surveillance study of nosocomial bloodstream infections, investigators reported that *S. aureus* was the second most common organism causing bloodstream infections and that the proportion of MRSA isolates increased from 22% in 1995 to 57% in 2001 *(9)*. Data from the SENTRY Antimicrobial Surveillance Program also demonstrated increasing rates of MRSA among *S. aureus* isolated from intensive care unit patients throughout the world *(10)*.

2.2. Community-Acquired MRSA Strains

Since the mid 1990s, MRSA strains have emerged in the community setting, causing infections in patients who do not have the risk factors usually associated with hospital associated (HA-MRSA), such as recent hospitalization, chronic diseases, kidney dialysis, human immunodeficiency virus infection, and iv drug use *(11,12)* Although community-acquired (CA-MRSA) strains cause mostly skin abscesses and furunculosis, severe necrotizing pneumonia and shock resulting in death has also been associated with CA-MRSA *(13,14)*. These new CA-MRSA strains are usually resistant to β-lactams but susceptible to other antimicrobial classes and carry mostly staphylococcal cassette chromosome *mec* (SCC*mec*) type IV. CA-MRSA strains are also more likely to possess unique combinations of virulence factors and seem to be genetically different from HA-MRSA *(11,13–18)*. Investigators have suggested that CA-MRSA strains have arisen from different genetic backgrounds rather than from the worldwide spread of a single clone *(19)*.

3. Virulence Factors

The pathogenicity and virulence of *S. aureus* is associated with the capacity of this organism to produce several virulence factors including enterotoxins serotypes A through Q (SEA-SEQ), toxic shock syndrome toxin-1 (TSST-1), cytolytic toxins (α and β hemolysins), exfoliative toxins, Panton-Valentine leukocidin (PVL), protein A, and several enzymes *(18,20)*.

The enterotoxins and the TSST-1 cause toxic shock and related illnesses through induction of massive cytokine release, from both macrophages and T-cells *(20)*. Recent CA-MRSA isolates have shown evidence of increased virulence resulting in increased prevalence of toxic shock cases and more severe soft-tissue infections and in many cases increased mortality. However, TSST-1 can be produced by HA-MRSA as well as methicillin-susceptible *S. aureus* (MSSA) strains and, therefore, TSST-1 production should not be considered a hallmark of CA-MRSA strains *(16)*.

Another important virulence factor in *S. aureus* is PVL, a member of the recently described family of synergohymenotropic toxins. PVL damages the membranes of host defense cells through the synergistic activity of two separately secreted but nonassociated proteins, LukS and LukF, causing tissue necrosis *(21)*. Although some investigators have suggested that PVL expression does not correlate directly with polymorphonuclear leukocyte lysis *(22)*, PVL producing CA-MRSA isolates has been increasingly associated with necrotizing pneumonia and necrotizing cutaneous infections *(13,14,21)*.

4. Mechanisms of Antibiotic Resistance

4.1. Mechanisms of β-lactam Resistance

S. aureus became resistant to penicillin owing to the production of β-lactamases that hydrolyze the penicillins. For that reason, penicillins that were resistant to the action of β-lactamases, such as methicillin, were developed to treat staphylococcal infection caused by β-lactamase-producing strains. However, *S. aureus* strains resistant to these agents soon appeared *(23)*.

Although there are three known mechanisms for which *S. aureus* becomes resistant to methicillin—hyperproduction of β-lactamases *(24)*; modification of normal penicillin-binding proteins (PBPs) *(25)*; and the presence of an acquired penicillin-binding protein, PBP2a *(26)*—most clinical isolates present the latter mechanism; therefore, my discussion focuses on this mechanism.

S. aureus strains have four normal PBPs anchored on the cytoplasmic membrane that participate in the crosslinking of the peptidoglycan of the bacterial cell wall. These normal PBPs have activity similar to that of serine proteases and have high affinity for β-lactam agents. When this binding occurs, the PBPs are not able to function in the assembly of the cell wall, causing bacterial death. PBP2a, on the other hand, is not part of the intrinsic set of PBPs of *S. aureus* but is a unique, inducible, acquired protein that has a molecular weight of approx 76 kDa, and it is produced only by methicillin-resistant staphylococci *(27)*. PBP2a has low affinity for β-lactam antibiotics and, therefore, is capable of substituting the biosynthetic functions of the normal PBPs even in the presence of the β-lactams, thereby preventing cell lysis. Isolates containing the PBP2a-mediated resistance mechanism are clinically resistant to all available β-lactams, including penicillins, cephalosporins, β-lactam/β-lactamase inhibitor combinations, monobactams, and carbapenems *(27,28)*.

PBP2a is encoded by the *mecA* gene, which is not present in methicillin-susceptible strains, and it is believed to have been acquired from a distantly related species, although the exact origin has not been found yet *(27,29)*. Song et al. *(30)* first sequenced *mecA* in 1987, and now it is known that this gene is carried on a mobile genetic element, the SCC*mec (31)*. In addition to carrying the *mecA* gene, the SCC contains regulatory genes; the IS431mec insertion sequence; and the recombinase genes *ccr*, which are responsible for the integration and excision of SCC*mec (32)*. The *mec* gene complex has been classified into four classes, and the *ccr* gene complex into three allotypes *(29)*. Different combinations of *mec* gene complex classes and *ccr* gene complex types have so far defined four types of SCC*mec* elements, labeled SCC*mec* types I–IV. Recently a novel type V SCC*mec* has been described in a CA-MRSA strain isolated in Australia *(33)*.

4.2. Mechanisms of Vancomycin Resistance

Until recently, vancomycin was the only antimicrobial agent that was active against all staphylococci; therefore, vancomycin has been the drug of choice to treat infection caused by MRSA. However, clinical strains of *S. aureus* with intermediate susceptibility to vancomycin (minimum inhibitory concentration [MIC] between 8 and 16 µg/mL) were first reported in Japan in 1997 *(34)*. Since then, four vancomycin-resistant *S. aureus* (VRSA) (MIC ≥ 32 µg/mL) strains have been documented in the United States in patients with clinical infections: two unrelated patients from Michigan, one from New York, and one from Pennsylvania *(35,36)*.

Vancomycin acts in the early stage of cell wall synthesis by binding to the C-terminal of the cell wall precursor pentapeptide complex and preventing it from being used for cell wall synthesis. Vancomycin-intermediate *S. aureus* strains have abnormal, thickened cell walls in the presence of vancomycin, and researchers have described two possible mechanisms of resistance in these strains: affinity trapping of vancomycin molecules by cell wall monomers and clogging of the outer layer of peptidoglycan by bound vancomycin molecules *(37)*.

VRSA strains carry the *van*A resistance determinant. The Michigan isolate (MI-VRSA) harbored a 57.9-kb multiresistant conjugative plasmid within which the van A transposon, Tn1546, was integrated. The structure of the Tn1546-like plasmid containing the *van*A resistance gene in the Pennsylvania isolate (PA-VRSA) showed several major differences from the prototypic Tn1546 seen in the MI-VRSA, including a deletion at the 5N end of the transposon. The differences observed in the plasmids of the MI-VRSA compared with the PA-VRSA may indicate two independent events of interspecies transfer, most likely from enterococci *(36,37)*.

To date, all VRSA isolates have also been resistant to oxacillin. Severin et al. *(38)* investigated the mechanism of expression of high-level vancomycin resistance using an oxacillin-resistant *S. aureus* carrying the *van*A gene complex and with inactivated *mecA*. They reported that the key penicillin-binding protein essential for vancomycin resistance and for the altered cell wall composition characteristic of VRSA is PBP2. They also concluded that although *mecA* is essential for oxacillin resistance, it is not involved in the expression of vancomycin resistance *(38)*.

Investigators have used a new technique to microscopically examine the cell wall and extracellular structures of the bacterial cell without the artifact produced by the fixation step needed for electron microscopy *(39)*. This new technique, atomic force microscopy (AFM), has the ability to measure surface topographic features and has proven highly useful for detection and characterization of extracellular matrices, as well as for understanding the mechanical and/or adhesive properties of the bacterial cell *(39,40)*.

AFM creates images by mechanically scanning a very sharp probe mounted on a flexible cantilever over a sample surface. The interaction forces between the scanning probe and the sample surface produce signals that are transformed into an image of the surface features *(41)*. The overall shape and general topography is shown by noncontact topography mode, in which the scanner adjusts the distance between the cell and cantilever such that the cantilever oscillation amplitude is constant during imaging, and an image is created from adjustment of the sample height (**Fig. 1**). To resolve fine surface features, the sample height is not adjusted and the cantilever oscillation amplitude is used to create an image (**Fig. 2**).

Using AFM to investigate the structural and topological characteristics of a glycopeptide-intermediate clinical strain, researchers found that this strain and its revertant had two parallel circumferential surface rings, whereas control strains had only one equatorial ring. Furthermore, in vancomycin-susceptible strains, additional rings were formed in the presence of vancomycin *(42)*. Further studies are needed to assess whether these observations are associated with the decreased susceptibility to vancomycin in these strains.

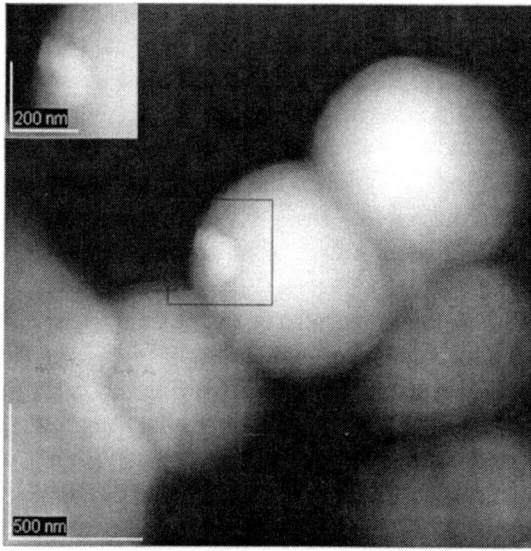

Fig. 1. AFM tapping mode topographical image of cells from a clinical MRSA isolate showing cocci arranged in clusters. AFM tapping mode provides very high-resolution topographical information without any fixation.

5. Emergence and Evolution of MRSA

Many studies have tried to elucidate the origin of MRSA strains and significant advances have been made in recent years. Most researchers seem to agree that MRSA emerged in the early 1960s, when it acquired the methicillin resistance gene *mecA*, which is carried by the genetic element now known as SCC*mec* *(29,43–48)*.

As already described, the origins of SCC*mec* are unknown, and although investigators have found that these elements are widely distributed in staphylococci, including *S. aureus*, they have not been found in any other genera of bacteria *(29,43)*. SCC*mec* is integrated near the *S. aureus* origin of replication, and this location might have been critical for providing MRSA with the ability to acquire other antibiotic-resistant genes *(29)*.

Crisostomo et al. *(43)* used multilocus sequence typing (MLST), spa typing, and pulsed-field gel electrophoresis (PFGE) to study the similarity of genetic backgrounds in historically early and contemporary European MSSA and MRSA epidemic clones *(43)*. They found that early MRSA isolates resembled early MSSA isolates in phenotypic and genetic characteristics, suggesting that these tested early MSSA isolates probably represent the progeny of a strain that served as one of the first *S. aureus* recipients of methicillin resistance in Europe. Enright et al. *(44)* used MLST data and a complex algorithm, denominated

Aspects of MRSA Infections 7

Fig. 2. AFM tapping mode amplitude image of MRSA isolate shown in **Fig. 1**. The amplitude image of tapping mode permits elucidation of fine surface structure not apparent from the topography image.

BURST analysis, to identify the ancestral MRSA clone and its MSSA ancestor using an international collection of MRSA and MSSA isolates. Based on their analysis, these investigators reported that methicillin resistance emerged in five phylogenetically distinct lineages and on multiple occasions within a given phylogenetic lineage *(44)*.

Although the frequency with which SCC*mec* is acquired is not completely known, most investigators agree that MRSA isolates are not all descendants of a single original clone and that horizontal transfer of SCC into epidemic MSSA isolates of different lineages may have played a significant role in the evolution of MRSA.

Robinson and Enright *(45)* proposed evolutionary models of the emergence of MRSA based on the application of MLST and SCC*mec* typing to an international collection of MRSA and MSSA isolates. On the basis of these models, they proposed that MRSA has emerged at least 20 times on acquisition of the *mecA* gene, and that SCC*mec* IV is the most frequently acquired element by methicillin-susceptible isolates *(45)*. This is very interesting because SCC*mec* IV has been found in most CA-MRSA isolates. However, pandemic clones associated with nosocomial infections have SCC*mec* I, II, or III, and the selection and spread of MRSA strains harboring these elements in hospitals probably occurs by exposure to antibiotics over time.

6. Susceptibility Testing

6.1. Detection of Oxacillin (Methicillin) Resistance

A distinctive characteristic of methicillin resistance is its heterogeneous expression, with the majority of cells susceptible to low concentrations of oxacillin, and only a small proportion of cells growing at an oxacillin concentration >50 µg/mL. Consequently, in vitro testing has been modified to enhance the expression of oxacillin resistance for detection of resistant strains *(28,49)*.

Most of the health care–associated MRSA strains are resistant to multiple classes of antimicrobial agents, including aminoglycosides, clindamycin, macrolides, quinolones, sulfonamides, and tetracycline *(27,28)*. However, as already discussed, most CA-MRSA isolates harboring SCC*mec* type IV are usually resistant to β-lactam and macrolide antibiotics, but susceptible to other classes *(11)*.

Currently, the Clinical and Laboratory Standard Institute (CLSI), formerly National Committee for Clinical Laboratory Standards (NCCLS) *(50)*, has recommendations for several standardized methods for detection of oxacillin resistance in *S. aureus*, including broth and agar dilution, disk diffusion, and agar screen methods. All these tests require incubation at temperatures no greater than 35°C and obtain final readings after a full 24 h of incubation. Supplementation of Mueller-Hinton broth or agar with 2% NaCl should be done for dilution tests *(50)*.

The agar screen test has been evaluated in numerous studies and found to be very good for detection of resistant strains. Infection control laboratories have used it for screening of colonized patients, and it is also the recommended method to use in addition to the dilution methods to confirm methicillin resistance in *S. aureus (50)*. For disk diffusion (Kirby-Bauer) testing, researchers have demonstrated that the performance of cefoxitin disk diffusion is equivalent to that of oxacillin broth microdilution and that it is easier to read than oxacillin disk diffusion *(51)*. Based on these findings, CLSI adopted the use of cefoxitin disk diffusion for predicting *mecA*-mediated oxacillin resistance in staphylococci, and the 2005 document M100-S15 states that cefoxitin disk diffusion is preferred over oxacillin disk diffusion for detecting oxacillin resistance in *mecA*-positive *S. aureus* and coagulase-negative staphylococci. The results should be reported for oxacillin and not for cefoxitin *(50)*.

Automated systems have achieved sensitivity and specificity for detecting staphylococcal oxacillin resistance at a level that is acceptable for clinical laboratory use *(52,53)*. However, none of the phenotypic methods are completely reliable, and clinical laboratories require several methods for confirmation of results.

6.2. Detection of Clindamycin Resistance

Macrolide resistance mechanisms in staphylococci are ribosomal methylase encoded by *erm* genes (MLS$_B$ phenotype), which could be constitutive or inducible, and efflux pump encoded by *msr* genes (M phenotype). When MLS$_B$ resistance is constitutive, staphylococci are resistant to erythromycin and clindamycin. When the resistance is inducible, the strains are resistant to erythromycin and induced resistant to clindamycin. Strains presenting the efflux pump mechanism are resistant to erythromycin and susceptible to clindamycin. Standard susceptibility broth methods cannot separate inducible resistance from susceptibility to clindamycin. Induction can be demonstrated using a disk approximation test by placing a 2-μg clindamycin disk 15 mm away from the edge of a 15-μg erythromycin disk on a standard blood agar or Mueller Hinton plate using a standard inoculum. Following incubation, organisms that do not show flattening of the clindamycin zone should be reported as clindamycin susceptible. Organisms that show flattening of the clindamycin zone adjacent to the erythromycin disk, usually described as "D-shaped," have inducible resistance and should be considered clindamycin resistant *(11)*.

Detection of inducible clindamycin resistance is very important in CA-MRSA because clindamycin is one of the antibiotics recommended to treat CA-MRSA infections and clinical laboratories are advised to perform the D test in macrolide-resistant MRSA isolates *(50)*.

6.3. Detection of Vancomycin Resistance

The current interpretative criteria for vancomycin susceptibility testing in *S. aureus* by CLSI classify as susceptible those isolates for which the vancomycin MICs are ≤4 μg/mL, as intermediate those with MICs between 8 and 16 μg/mL, and as resistant those with MICs ≥32 μg/mL *(50)*. These break points have been challenged based on reports of patients with MRSA infections who failed vancomycin therapy. The MICs of these isolates were ≤4 μg/mL and, therefore, they were considered susceptible by current interpretative criteria but found to have a subpopulation of cells with intermediate resistance to vancomycin. These isolates are now designated heterogeneously vancomycin-intermediate *S. aureus*, or hVISA, isolates *(54,55)*.

The existence of clinical hVISA isolates plus the fact that none of the VRSA isolates were detected by automated susceptibility instruments make detection of *S. aureus* nonsusceptible to vancomycin by clinical laboratories extremely challenging *(35,38)*.

Currently, the CLSI document states that VRSA strains with MICs ≥32 μg/mL are reliably detected by broth microdilution reference method using Mueller-Hinton broth and 24 h of incubation at 35°C. When using other MIC methods that have not been validated to detect VRSA, Brain Heart Infusion vancomycin agar screen plates containing 6 μg/mL of vancomycin should be inoculated to enhance the sensitivity of detecting vancomycin-resistant strains *(50)*. Any staphylococci determined to have an elevated MIC for vancomycin (MIC ≥ 4 μg/mL) should be sent to a reference laboratory for confirmation.

6.4. Rapid Methods for Detection of MRSA Strains

Antimicrobial susceptibility test methods such as disk diffusion, broth microdilution, and oxacillin screen plate require 24 h of incubation after having the organism grow in pure culture. Rapid and accurate identification of MRSA isolates is essential not only for patient care, but also for effective infection control programs to limit the spread of MRSA. In the last few years, several commercial tests for detection of *mecA* gene and PBP2a have been developed for use in clinical laboratories *(52,53)*. Real-time polymerase chain reaction (PCR) and other molecular tests are gaining popularity as MRSA screening tests to identify patients who are candidates for contact precaution at the time of admission, decreasing the risk of nosocomial transmission *(56,57)*.

7. Molecular Procedures for Epidemiological Studies

Study of the genetic relatedness of isolates obtained from an epidemiological cluster or during the course of an infection in a single patient is becoming a useful practice in many clinical and infection control laboratories today *(58)*. The goal of these techniques is to determine whether isolates recovered from different patients or sources represent a single strain or multiple different strains. Infection control practitioners use the information provided by molecular procedures to complement their epidemiological investigations, as well as to determine whether to initiate such investigations while clinicians may use the information in an individual patient to discriminate between relapse and reinfection.

Clinical microbiologists have used phenotypic methods to distinguish isolates of the same species. These phenotypic methods include serotyping, biotyping, bacteriophage typing, antimicrobial susceptibility profile, and MLST. Among these, in the past, reference laboratories used bacteriophage typing for differentiating unrelated *S. aureus* isolates, but because of the technical demands and poor reproducibility, this method is now rarely used. The antimicrobial susceptibility profile has been the phenotyping technique most frequently used by clinical microbiology laboratories because the data are readily available. However, the antimicrobial susceptibility typing method has not been very discriminatory for the analysis of nosocomial MRSA because most are resistant to many antibiotic classes; therefore, this method does not allow differentiation between related and unrelated isolated.

Owing to the poor discriminatory power of the phenotypic techniques, DNA-based, or genotypic techniques are now the strain-typing methods of choice for MRSA. Currently, the most commonly used genotypic techniques for epidemiological investigation of MRSA are PGFE and MLST, but more recently, investigators have evaluated other techniques, including arbitrarily primed PCR (AP-PCR), protein A gene (*spa*) typing, and gene chip–based techniques. A brief description of these techniques and their usefulness for the discrimination of MRSA strains is provided next.

7.1. PFGE of Chromosomal DNA

PFGE of chromosomal DNA is based on the digestion of bacterial DNA with restriction endonucleases with relatively few restriction sites generating fewer but much larger fragments than those generated by conventional, constant-field agarose gel electrophoresis. In PFGE, the orientation of electric field is changed periodically ("pulsed"), allowing the DNA fragments, embedded in agarose plugs, to be separated by size. PFGE analysis provides a restriction pattern of chromosomal DNA composed of well-defined fragments, facilitating the analysis and comparison of multiple isolates. This technique has been widely used for the epidemiological study of nosocomial and CA-MRSA isolates *(13–18)*. In addition, the interpretative scheme of PFGE pattern reported by Tenover et al. *(59)* has been very useful for determining the genetic relatedness of MRSA strains isolated during a relatively short period of time (1–3 mo), during which, presumably, the genetic variability is limited. PFGE has been tested and compared to several other typing methods and has been reported to be the most discriminatory method available for the epidemiological study of outbreaks in hospitals and communities. Furthermore, a national database of MRSA PFGE profiles has been assembled to facilitate the identification of major lineages of MRSA present in the United States *(60)*.

7.2. Multilocus Sequencing Typing

MLST is gaining popularity among researchers, particularly for studying long-term population relatedness and for understanding the emergence and evolution of MRSA clones *(43,44,46,61)*. In MLST, seven loci representing housekeeping genes for *S. aureus* are amplified by PCR. The PCR product is then sequenced and compared to known alleles, held at the MLST Web site (www.mlst.net), to obtain an allelic profile. This allelic profile consists of a string of seven numbers, which can be easily consulted over the Internet, unifying and standardizing epidemiological data collected all over the world. Although MLST provides information on strain lineage that is very important for understanding the overall epidemiology of MRSA infections, this technique may not be suitable for outbreak investigation in the clinical setting. It also requires performing PCR and sequencing of the PCR product using an automated sequencer, which is not readily available in most clinical laboratories.

7.3. Arbitrarily Primed PCR

The main feature of PCR is the ability to replicate a particular DNA sequence to obtain multiple copies of the target sequence. Among the typing techniques involving PCR, AP-PCR, or random amplified polymorphic DNA, has been used for the genetic analysis of *S. aureus (62)*. This technique involves the amplification of random chromosomal DNA sequences using a small primer (typically 10 bp) with an arbitrary sequence not directed to a specific region of the DNA target, but capable of hybridization at random chromosomal sites. The number and locations of these random sites will vary among different strains, generating a different AP-PCR profile based on the number and sizes of the fragments detected by electrophoresis. This technique has a lower discriminatory power than PFGE for typing of MRSA strains, but, owing to its simplicity, it could be useful for rapid differentiation of related from unrelated isolates during an outbreak.

7.4. spa Typing

The technique of *spa* typing involves sequencing of the polymorphic X region, or short sequence repeat region, of the protein A gene. These regions have a high degree of polymorphism and, therefore, are potentially suitable for discrimination in outbreak investigation. This typing method requires the ability to perform PCR and access to an automated sequencer for sequence typing of PCR products, as is also required for MLST. The information used for *spa* typing is obtained from a single locus, in contrast to MLST, which combines information from seven loci for typing of *S. aureus*. *spa* typing has been evaluated for typing well-characterized *S. aureus* strains and compared to PFGE *(16,43, 48,63)*. It was found that *spa* typing is rapid and apparently easier to perform and interpret than other available molecular techniques. However, its utility in the clinical setting for rapid screening of epidemiologically related strains during an outbreak needs to be evaluated.

7.5. Gene Chip–Based Techniques

More recently, investigators have reported the use of gene chips for studying the relatedness of MRSA strains. In this case, an Affymetric GeneChip that represented predicted open reading frames from six genetically divergent *S. aureus* strains and novel GenBank entries to analyze the relatedness of MRSA isolates were used. This new methodology has potential for evaluating MRSA lineages, but its complexity and cost make it unsuitable for clinical purposes at this time *(64)*.

8. Investigation of New Treatment Options

As discussed, the usefulness of the β-lactams has been compromised by the increasing isolation of MRSA strains not only in hospital settings but now also in the community. Clindamycin and trimethoprim-sulfamethoxazole are good options for treatment of skin infections caused by CA-MRSA strains susceptible to these agents, and, therefore, it is very important that laboratories test for inducible resistance to clindamycin to prevent treatment failures *(65)*. Overall, vancomycin is still the most widely used antibiotic for treating infections caused by MRSA. However, the isolation of clinical VRSA strains has led to an effort to limit the use of vancomycin to prevent further development and spread of vancomycin resistance in staphylococci. A description of some of the new agents being developed or evaluated for treatment of MRSA infections follows.

8.1. Non-β-lactams

Among the newer non-β-lactam agents either available or in advanced stages of development to treat MRSA infections are linezolid (an oxazolidinone); daptomycin (a cyclic lipopeptide); tygecycline (a tetracycline); and several glycopeptide agents, including oritavancin and dalvabancin *(66)*. However, the development of resistance in some of these newer agents magnifies the importance of the additional need for effective antimicrobial agents *(67,68)*.

8.2. β-lactams

As discussed, β-lactam antibiotics have too low affinity for PBP2a to be efficacious at clinically achievable concentrations. However, PBP2a-binding affinities vary among the various β-lactam agents, and in theory these agents might be clinically effective in high doses if hydrolysis of the agents is prevented by the addition of a β-lactamase inhibitor. Some studies have demonstrated that amoxicillin plus clavulanic acid, ampicillin plus sulbactam, and cefoperazone plus sulbactam have been efficacious in vitro and in animal models of MRSA infections *(69–71)*.

Several investigational β-lactam antibiotics with high-affinity binding by PBP2a are under development *(72–75)*. It is possible that these agents will become available for clinical use in the next few years, although PBP2a mutations producing very high-level resistance to PBP2a-binding β-lactams have already been reported in in vitro experiments *(76)*. BAL9141 (ceftobiprole) is one of these agents that it is in the advanced phases of clinical development *(77)*.

9. Conclusion

It is clear that the emergence of CA-MRSA and VRSA isolates is changing the management of clinical infections potentially caused by *S. aureus*. Rapid methods for accurate detection of MRSA are necessary to promptly identify patients and implement contact precautions as well as appropriate treatment. Molecular genotyping techniques have an important role in evaluating possible outbreaks and for understanding the emergence and evolution of MRSA strains.

Acknowledgments

I thank my research collaborators Faith Coldren and David Carroll for providing the AFM images, and Caryn Good and Michael Jacobs for helpful suggestions about the manuscript.

References

1. Panlilio, A. L., Culver, D. H., Gaynes, R. P., Banerjee, S., Henderson, S., Tolson, J. S., and Martone, W. J. (1992) Methicillin-resistant *Staphylococcus aureus* in U.S. hospitals, 1975–1991. *Infect. Control Hosp. Epidemiol.* **13**, 582–586.
2. Stefani, S. and Varaldo, P. E. (2003) Epidemiology of methicillin-resistant staphylococci in Europe. *Clin. Microbiol. Infect.* **9**, 1179–1186.
3. Kuehnert, M. J., Hill, H. A., Kupronis, B. A., Tokars, J. I., Solomon, S. L., and Jernigan, D. B. (2005) Methicillin-resistant-*Staphylococcus aureus* hospitalizations, United States. *Emerg. Infect. Dis.* **11**, 868–872.
4. Cuevas, O., Cercenado, E., Vindel, A., Guinea, J., Sanchez-Conde, M., Sanchez-Somolinos, M., and Bouza, E. (2004) Evolution of the antimicrobial resistance of *Staphylococus* spp. in Spain: five nationwide prevalence studies. *Antimicrob. Agents Chemother.* **48**, 4240–4245.
5. Simor, A. E., Ofner-Agostini, M., Bryce, E., et al. (2001) The evolution of methicillin-resistant *Staphylococcus aureus* in Canadian hospitals: 5 years of national surveillance. *CMAJ* **165**, 21–26.
6. Cosgrove, S. E., Qi, Y., Kaye, K. S., Harbarth, S., Karchmer, A. W., and Carmeli, Y. (2005) The impact of methicillin resistance in *Staphylococcus aureus* bacteremia on patient outcomes: mortality, length of stay, and hospital charges. *Infect. Control Hosp. Epidemiol.* **26**, 166–174.
7. Abramson, M. A. and Sexton, D. J. (1999) Nosocomial methicillin-resistant and methicillin-susceptible *Staphylococus aureus* primary bacteremia: at what costs? *Infect. Control Hosp. Epidemiol.* **20**, 408–411.
8. NNIS system. (2004) National Nosocomial Infections Surveillance (NNIS) System report, data summary from January 1992 through June 2004, issued October 2004. *Am. J. Infect. Control* **32**, 470–485.
9. Wisplinghoff, H., Bischoff, T., Tallent, S. M., Seifert, H., Wenzel, R. P., and Edmond, M. B. (2004) Nosocomial bloodstream infections in US hospitals: analysis of 24,179 cases from a prospective nationwide surveillance study. *Clin. Infect. Dis.* **39**, 309–317.
10. Jones, M. E., Draghi, D. C., Thornsberry, C., Karlowsky, J. A., Sahm, D. F., and Wenzel R. P. (2004) Emerging resistance among bacterial pathogens in the intensive care unit—a European and North American Surveillance study (2000–2002). *Ann. Clin. Microbiol. Antimicrob.* **29**, 3–14.
11. Palavecino, E. (2004) Community-acquired methicillin-resistant *Staphylococcus aureus* infections. *Clin. Lab. Med.* **24**, 403–418.
12. Herold, B. C., Immergluck, L. C., Maranan, M. C., et al. (1998) Community-acquired methicillin-resistant *Staphylococcus aureus* in children with no identified predisposing risk. *JAMA* **279**, 593–598.

13. Francis, J. S., Doherty, M. C., Lopatin, U., et al. (2005) Severe community-onset pneumonia in healthy adults caused by methicillin-resistant *Staphylococcus aureus* carrying the Panton-Valentine leukocidin genes. *Clin. Infect. Dis.* **40**, 100–107.
14. Miller, L. G., Perdreau-Remington, F., Rieg, G., et al. (2005) Necrotizing fasciitis caused by community-associated methicillin-resistant Staphylococcus aureus in Los Angeles. *N. Engl. J. Med.* **352**, 1445–1453.
15. Naimi, T. S., LeDell, K. H., Como-Sabetti, K., et al. (2003) Comparison of community- and health care-associated methicillin-resistant *Staphylococcus aureus* infection. *JAMA* **290**, 2976–2984.
16. Fey, P. D., Said-Salim, B., Rupp, M. E., et al. (2003) Comparative molecular analysis of community- or hospital-acquired methicillin-resistant *Staphylococcus aureus*. *Antimicrob. Agents Chemother.* **47**, 196–203.
17. Mulvey, M. R., MacDougall, L., Cholin, B., Horsman, G., Fidyk, M., Woods, S., and Saskatchewan CA-MRSA Study Group. (2005) Community-associated methicillin-resistant *Staphylococcus aureus*, Canada. *Emerg. Infect. Dis.* **11**, 844–850.
18. Vandenesch, F., Naimi, T., Enright, M. C., et al. (2003) Community-acquired methicillin-resistant *Staphylococcus aureus* carrying Panton-Valentine leukocidin genes: worldwide emergence. *Emerg. Infect. Dis.* **9**, 978–984.
19. Okuma, K., Iwakawa, K., Turnidge, J. D., et al. (2002) Dissemination of new methicillin-resistant *Staphylococcus aureus* clones in the community. *J. Clin. Microbiol.* **40**, 4289–4294.
20. McCormick, J. K., Yarwood, J. M., and Schlievert, P. M. (2001) Toxic shock syndrome and bacterial superantigens: an update. *Annu. Rev. Microbiol.* **55**, 77–104.
21. Lina, G., Piemont, Y., Godail-Gamot, F., et al. (1999) Involvement of Panton-Valentine leukocidin-producing *Staphylococcus aureus* in primary skin infections and pneumonia. *Clin. Infect. Dis.* **29**, 1128–1132.
22. Said-Salim, B., Mathena, B., Braughton, K., et al. (2005) Differential distribution and expression of Panton-Valentine leucocidin among community-acquired methicillin-resistant *Staphylococcus aureus* strains. *J. Clin. Microbiol.* **43**, 3373–3379.
23. Barber, M. (1961) Methicillin-resistant staphylococci. *J. Clin. Pathol.* **14**, 385–393.
24. McDougal, L. K. and Thornsberry, C. (1986) The role of beta-lactamase in staphylococcal resistance to penicillinase-resistant penicillins and cephalosporins. *J. Clin. Microbiol.* **23**, 832–839.
25. Tomasz, A., Drugeon, H. B., de Lencastre, H. M., Jabes, D., McDougal, L., and Bille, J. (1989) New mechanism for methicillin resistance in *Staphylococcus aureus*: clinical isolates that lack the PBP 2a gene and contain normal penicillin-binding proteins with modified penicillin-binding capacity. *Antimicrob. Agents Chemother.* **33**, 1869–1874.
26. Ubukata, K., Yamashita, N., and Konno, M. (1985) Occurrence of a beta-lactam-inducible penicillin-binding protein in methicillin-resistant staphylococci. *Antimicrob. Agents Chemother.* **27**, 851–857.
27. Chambers, H. F. (1997) Methicillin resistance in staphylococci: molecular and biochemical basis and clinical implications. *Clin. Microbiol. Rev.* **10**, 781–791.

28. Fasola, E. L. and Peterson, L. R. (1992) Laboratory detection and evaluation of antibiotic-resistant *Staphylococcus aureus* nosocomial infections. In: Weinstein RS, Gram AR (eds). Advances in Pathology, Volume V. Chicago, IL. Mosby-Year Book, Inc. 285–306.
29. Hiramatsu, K., Cui, L., Kuroda, M., and Ito, T. (2001) The emergence and evolution of methicillin-resistant *Staphylococcus aureus*. *Trends Microbiol.* **9**, 486–493.
30. Song, M. D., Wachi, M., Doi, M., Ischino, F., and Matsuhashi, M. (1987) Evolution of an inducible penicillin-target protein in methicillin-resistant *Staphylococcus aureus* by gene fusion. *FEBS Lett.* **221**, 167–171.
31. Katayama, Y., Ito, T., and Hiramatsu, K. (2000) A new class of genetic element, staphylococcus cassette chromosome mec, encodes methicillin resistance in *Staphylococcus aureus*. *Antimicrob. Agents Chemother.* **44**, 1549–1555.
32. Ito, T., Katayama, Y., Asada, K., Mori, N., Tsutsumimoto, K., Tiensasitorn, C., and Hiramatsu, K. (2001) Structural comparison of three types of staphylococcal cassette chromosome mec integrated in the chromosome in methicillin-resistant *Staphylococcus aureus*. *Antimicrob. Agents Chemother.* **45**, 1323–1336.
33. Ito, T., Ma, X. X., Takeuchi, F., Okuma, K., Yuzawa, H., and Hiramatsu, K. (2004) Novel type V staphylococcal cassette chromosome *mec* driven by a novel cassette chromosome recombinase, ccrC. *Antimicrob. Agents Chemother.* **48**, 2637–2651.
34. Hiramatsu, K., Hanaki, H., Ino, T., Yabuka, K., Oguri, T., and Tenover, F. C. (1997) Methicillin-resistant *Staphylococcus aureus* clinical strain with reduced vancomycin susceptibility. *J. Antimicrob. Chemother.* **40**, 135, 136.
35. Centers for Disease Control and Prevention. (2004) Vancomycin-resistant *Staphylococcus aureu*s: New York 2004. *MMWR Morb. Mortal. Wkly. Rep.* **53**, 322, 323.
36. Tenover, F. C. and McDonald, L. C. (2005) Vancomycin-resistant staphylococci and enterococci: epidemiology and control. *Curr. Opin. Infect. Dis.* **18**, 300–305.
37. Appelbaum, P. C. and Bozdogan, B. (2004) Vancomycin resistance in *Staphylococcus aureus*. *Clin. Lab. Med.* **24**, 381–402.
38. Severin, A., Wu, S. W., Tabei, K., and Tomasz, A. (2004) Penicillin-binding protein 2 is essential for expression of high-level vancomycin resistance and cell wall synthesis in vancomycin-resistant *Staphylococcus aureus* carrying the enterococcal *van*A gene complex. *Antimicrob. Agents Chemother.* **48**, 4566–4573.
39. Coldren, F. M., Palavecino, E., and Carroll, D. L. (2005) Atomic force microscopy as a potential diagnostic technique in staphylococcal infections. *Microsc. Microanal.* **11(Suppl. 2)**, 980, 981.
40. Touhami, A., Jericho, M. H., and Beveridge, T. J. (2004) Atomic force microscopy of cell growth and division in *Staphylococcus aureus*. *J. Bacteriol.* **186**, 3286–3295.
41. Tollersrud, T., Berge, T., Andersen, S. R., and Lund, A. (2001) Imaging the surface of *Staphylococcus aureus* by atomic force microscopy. *APMIS* **109**, 541–545.
42. Boyle-Vavra, S., Hahm, J., Sibener, S. J., and Daum, R. S. (2000) Structural and topological differences between a glycopeptide-intermediate clinical strain and glycopeptide-susceptible strains of *Staphylococcus aureus* revealed by atomic force microscopy. *Antimicrob. Agents Chemother.* **44**, 3456–3460.

43. Crisostomo, M. I., Westh, H., Tomasz, A., Chung, M., Oliveira, D. C., and de Lencastre, H. (2001) The evolution of methicillin resistance in *Staphylococcus aureus*: similarity of genetic backgrounds in historically early methicillin-susceptible and -resistant isolates and contemporary epidemic clones. *Proc. Natl. Acad. Sci. USA* **98,** 9865–9870.
44. Enright, M. C., Robinson, D. A., Randle, G., Feil, E. J., Grundmann, H., and Spratt, B. G. (2002) The evolutionary history of methicillin-resistant *Staphylococcus aureus* (MRSA). *Proc. Natl. Acad. Sci. USA* **99,** 7687–7692.
45. Robinson, D. A. and Enright, M. C. (2003) Evolutionary models of the emergence of methicillin-resistant *Staphylococcus aureus*. *Antimicrob. Agents Chemother.* **47,** 3926–3934.
46. Feil, E. J., Cooper, J. E., Grundmann, H., et al. (2003) How clonal is *Staphylococcus aureus*? *J. Bacteriol.* **11,** 3307–3316.
47. Oliveira, D. C., Tomasz, A., and de Lencastre, H. (2002) Secrets of success of a human pathogen: molecular evolution of pandemic clones of methicillin-resistant *Staphylococcus aureus*. *Lancet Infect. Dis.* **2,** 180–189.
48. Oliveira, D. C., Tomasz, A., and de Lencastre, H. (2001) The evolution of pandemic clones of methicillin-resistant *Staphylococcus aureus*: identification of two ancestral genetic backgrounds and the associated *mec* elements. *Microb. Drug Resist.* **7,** 349–361.
49. Chambers, H. F. and Hackbarth, C. J. (1987) Effect of NaCl and nafcillin on penicillin-binding protein 2a and heterogeneous expression of methicillin resistance in *Staphylococcus aureus*. *Antimicrob. Agents Chemother.* **31,** 1982–1988.
50. Clinical and Laboratory Standards Institute/NCCLS. (2005) Performance Standards for Antimicrobial Susceptibility Testing; Fifteenth Informational Supplement. CLSI/NCCLS document M100-S15. CLSI, Wayne, PA.
51. Swenson, J. M., Tenover, F. C., and Cefoxitin Disk Study Group. (2005) Results of disk diffusion testing with cefoxitin correlate with presence of *mec*A in *Staphylococcus* spp. *J. Clin. Microbiol.* **43,** 3818–3823.
52. Yamazumi, T., Furuta, I., Diekema, D. J., Pfaller, M. A., and Jones, R. N. (2001) Comparison of the Vitek gram-positive susceptibility 106 card, the MRSA-Screen latex agglutination test, and *mec*A analysis for detecting oxacillin resistance in a geographically diverse collection of clinical isolates of coagulase-negative staphylococci. *J. Clin. Microbiol.* **39,** 3633–3636.
53. Swenson, J. M., Williams, P. P., Killgore, G., O'Hara, C. M., and Tenover, F. C. (2001) Performance of eight methods, including two new rapid methods, for detection of oxacillin resistance in a challenge set of *Staphylococcus aureus* organisms. *J. Clin. Microbiol.* **39,** 3785–3788.
54. Charles, P. G., Ward, P. B., Johnson, P. D., Howden, B. P., and Grayson, M. L. (2004) Clinical features associated with bacteremia due to heterogeneous vancomycin-intermediate *Staphylococcus aureus*. *Clin. Infect. Dis.* **38,** 448–451.
55. Howden, B. P., Ward, P. B., Charles, P. G., et al. (2004) Treatment outcomes for serious infections caused by methicillin-resistant *Staphylococcus aureus* with reduced vancomycin susceptibility. *Clin. Infect. Dis.* **38,** 521–528.

56. Paule, S. M., Pasquariello, A. C., Hacek, D. M., Fisher, A. G., Thomson, R. B. Jr., Kaul, K. L., and Peterson, L. R. (2004) Direct detection of *Staphylococcus aureus* from adult and neonate nasal swab specimens using real-time polymerase chain reaction. *J. Mol. Diagn.* **6**, 191–196.
57. Warren, D. K., Liao, R. S., Merz, L. R., Eveland, M., and Dunne, W. M. Jr. (2004) Detection of methicillin-resistant *Staphylococcus aureus* directly from nasal swab specimens by a real-time PCR assay. *J. Clin. Microbiol.* **42**, 5578–5581.
58. Peterson, L. R., Petzel, R. A., Clabots, C. R., Fasching, C. E., and Gerding, D. N. (1993) Medical technologists using molecular epidemiology as part of the infection control team. *Diagn. Microbiol. Infect. Dis.* **16**, 303–311.
59. Tenover, F., Arbeit, R., Goering, R. V., Mickelsen, P. A., Murray, B. E., Persing, D. H., and Swaminathan, B. (1995) Interpreting chromosomal DNA restriction patterns produced by pulsed-field gel electrophoresis: criteria for bacterial strain typing. *J. Clin. Microbiol.* **33**, 2233–2239.
60. McDougal, L. K., Steward, C. D., Killgore, G. E., Chairtram, S. K., McAllister, S. K., and Tenover, F. C. (2003) Pulsed-field gel electrophoresis typing of oxacillin-resistant *Staphylococcus aureus* isolates from the United States: establishing a national database. *J. Clin. Microbiol.* **41**, 5113–5120.
61. Enright, M. C., Day, N. P., Davies, C. E., Peacock, S. J., and Spratt, B. G. (2000) Multilocus sequence typing for characterization of methicillin-resistant and methicillin-susceptible clones of *Staphylococcus aureus*. *J. Clin. Microbiol.* **38**, 1008–1015.
62. van Belkum, A., Kluytmans, J., van Leeuwen, W., et al. (1995) Multicenter evaluation of arbitrarily primed PCR for typing of *Staphylococcus aureus* strains. *J. Clin. Microbiol.* **33**, 1537–1547.
63. Shopsin, B., Gomez, M., Montgomery, S. O., et al. (1999) Evaluation of protein A gene polymorphic region DNA sequencing for typing of *Staphylococcus aureus* strains. *J. Clin. Microbiol.* **37**, 3556–3563.
64. Dunman, P. M., Mounts, W., McAleese, F., et al. (2004) Uses of *Staphylococcus aureus* GeneChips in genotyping and genetic composition analysis. *J. Clin. Microbiol.* **42**, 4275–4283.
65. Siberry, G. K., Tekle, T., Carroll, K., and Dick, J. (2003) Failure of clindamycin treatment of methicillin-resistant *Staphylococcus aureus* expressing inducible clindamycin resistance in vitro. *Clin. Infect. Dis.* **37**, 1257–1260.
66. Appelbaum, P. C. and Jacobs, M. R. (2005) Recently approved and investigational antibiotics for treatment of severe infections caused by Gram-positive bacteria. *Curr. Opin. Microbiol.* **8**, 510–517.
67. Wilson, P., Andrew, J. A., Charlesworth, R., Walesby, R., Singer, M., Farrell, D. J., and Robbins, M. (2003) Linezolid resistance in clinical isolates of *Staphylococcus aureus*. *J. Antimicrob. Chemother.* **51**, 186–188.
68. Tsiodras, S., Gold, H. S., Sakoulas, G., et al. (2001) Linezolid resistance in a clinical isolate of *Staphylococcus aureus*. *Lancet* **358**, 207, 208.
69. Hirano, L. and Bayer, A. S. (1991) Beta-lactam-beta-lactamase-inhibitor combinations are active in experimental endocarditis caused by b-lactamase-producing oxacillin-resistant staphylococci. *Antimicrob. Agents Chemother.* **35**, 685–690.

70. Cantoni, L., Wenger, A., Glauser, M., and Billie, J. (1989) Comparative efficacy of amoxicillin-clavulanate, cloxacillin, and vancomycin against methicillin-sensitive and methicillin-resistant *Staphylococcus aureus* in rats. *J. Infect. Dis.* **159,** 989–993.
71. Fasola, E. L., Fasching, C. E., and Peterson, L. R. (1995) Molecular correlation between in vitro and in vivo activity of beta-lactam and beta-lactamase inhibitor combinations against methicillin-resistant *Staphylococcus aureus*. *J. Lab. Clin. Med.* **125,** 200–211.
72. Miller, K., Storey, C., Stubbings, W. J., Hoyle, A. M., Hobbs, J. K., and Chopra, I. (2005) Antistaphylococcal activity of a novel cephalosporin CB-181963 (CAB-175). *J. Antimicrob. Chemother.* **55,** 579–582.
73. Entenza, J. M., Hohl, P., Heinze-Krauss, I., Glauser, M. P., and Moreillon, P. (2002) BAL9141, a novel extended-spectrum cephalosporin active against methicillin-resistant *Staphylococcus aureus* in treatment of experimental endocarditis. *Antimicrob. Agents Chemother.* **46,** 171–177.
74. Fung-Tomc, J. C., Clark, J., Minassian, B., et al. (2002) In vitro and in vivo activities of a novel cephalosporin, BMS-247243, against methicillin-resistant and -susceptible staphylococci. *Antimicrob. Agents Chemother.* **46,** 971–976.
75. Chambers, H. F. (2003) Solving staphylococcal resistance to beta-lactams. *Trends Microbiol.* **11,** 145–148.
76. Katayama, Y., Zhang, H. Z., and Chambers, H. F. (2004) PBP 2a mutations producing very-high-level resistance to beta-lactams. *Antimicrob. Agents Chemother.* **48,** 453–459.
77. Chambers, H. F. (2005) Evaluation of ceftobiprole in a rabbit model of aortic valve endocarditis due to methicillin-resistant and vancomycin-intermediate *Staphylococcus aureus*. *Antimicrob. Agents Chemother.* **49,** 884–888.

2

MRSA Case Studies

Kurt D. Reed, Mary E. Stemper, and Sanjay K. Shukla

Summary

Staphylococcus aureus is a versatile pathogen associated with diverse clinical presentations. Only recently have the genetic factors underlying the virulence of this bacterial species become understood in a significant way. Methicillin-resistant *S. aureus* (MRSA) strains have been extremely important as nosocomial pathogens in health care facilities for more than three decades. Additionally, infections resulting from community-associated MRSA strains have emerged in the last decade and become a public health problem of global proportions. This changing epidemiology has spurred renewed interest in translating knowledge of the molecular determinants of virulence into rational prevention and control strategies. Four case histories are provided (three involving MRSA and one involving a methicillin-sensitive strain of *S. aureus*) that highlight the diversity of clinical presentations and relative virulence of *S. aureus* infections in humans. The molecular characterization of clonality and virulence gene profile is compared among the four cases. Significant genetic diversity exists among MRSA and sensitive strains of *S. aureus*. It is obvious that various combinations of virulence factors contribute to disease manifestations of infected patients.

Key Words: Methicillin-resistant *Staphylococcus aureus*; community-associated MRSA; hospital-associated MRSA; virulence factors.

1. Introduction

For many years, methicillin-resistant *Staphylococcus aureus* (MRSA) was considered a multidrug-resistant pathogen that was strongly associated with infections in individuals with established risk factors, the most important of which was recent hospitalization *(1,2)*. Surveillance in U.S. hospitals indicated that staphylococcal infections resulting from MRSA increased from 2.4% in 1975 to 59.5% in 2004 *(3,4)*. More recently, there have been reports of community-associated MRSA (CA-MRSA) strains affecting individuals who have few risk factors for infection. As CA-MRSA strains have become more thoroughly characterized, it has been found that CA-MRSA strains are less likely to be resistant to multiple antibiotic classes and are more often associated with superficial skin and soft-tissue infections than hospital-associated MRSA (HA-MRSA) strains *(5,6)*.

Furthermore, strains of MRSA that originate in the community appear to have molecular features that are quite distinct from those of HA-MRSA. These include unique pulsed-field gel electrophoresis (PFGE) clonal types, the presence of type IV or type V staphylococcal cassette chromosome *mec* (SCC*mec*), and the presence of Panton-Valentine leukocidin (PVL) genes *(7–10)*. Additionally, much recent work has been directed toward understanding how various virulence factors carried in genomic islands, such as enterotoxins and exotoxins, differ between community-associated and hospital-associated strains. These molecular studies also provide valuable insight into the evolutionary and phylogenetic relationships between methicillin-sensitive *S. aureus* (MSSA) strains and MRSA.

The first two case studies described herein provide examples of important clinical and molecular features that differentiate HA-MRSA and CA-MRSA strains. Case three demonstrates a "transitional" form of MRSA that has clinical and laboratory features that seem to be intermediate between HA-MRSA and CA-MRSA. Finally, case four is provided as an example of a hypervirulent strain of MSSA associated with severe illness. This illustrates the important fact that the genomic background of *S. aureus* is a more important determinant of virulence than the presence or absence of *mec* elements.

2. Case Studies

2.1. Transcontinental Transmission of an HA-MRSA Strain

In 1994, a 36-yr-old white male was admitted to a hospital in eastern Wisconsin for treatment of a chronic hip wound infection. The patient had been a soldier in Bosnia and received multiple shrapnel injuries. He was transferred to the United States for care after having spent several months hospitalized in Europe. Prior treatment had included multiple surgical debridements, packing of the wound with gentamicin-impregnated beads, and multiple courses of various oral and iv β-lactam antibiotics.

Shortly after transfer to the United States, MRSA was isolated from the patient's hip wound. Surveillance cultures of nares, groin, and axilla indicated that the patient had widespread colonization with MRSA. In addition to being resistant to β-lactam antibiotics, the isolates were resistant to fluoroquinolones, clindamycin, erythromycin, tetracycline, gentamicin, and rifampin; they were susceptible to trimethoprin-sulfamethoxazole and vancomycin. Resistance to rifampin was a phenotypic marker rarely encountered among MRSA isolates in Wisconsin and was useful in helping laboratories identify this strain. Taken in aggregate the clinical history of prolonged hospitalization in Germany and the unique phenotype of the MRSA isolate suggested that the patient had acquired MRSA while in Europe.

Molecular characterization of the isolate indicated that it was sequence type 247 and had type 1a SCC*mec*. It had a unique plasmid profile never before seen among MRSA isolates in Wisconsin. The isolate was negative for PVL genes. The staphylococcal enterotoxin gene *sea* was present but all others (*seb* to *seo*) were absent. It harbored *lpl*10, a virulence gene initially thought to be unique to the hypervirulent MW2 strain as well as fibronectin-binding factors A and B. Epidemiological investigations performed over the next several months provided evidence that transmission of this highly resistant MRSA strain occurred among three patients within the same medical facility, four patients at another hospital within the same city, and two patients at a central Wisconsin hospital more than 100 miles away. In each instance of interinstitutional spread, there was documented evidence of patient transfers between facilities that enhanced the spread of the outbreak strain.

2.2. Index Case of CA-MRSA in a Native American Community Outbreak

In 1992, a 7-yr-old Native American girl presented to an outpatient clinic with a 2-d history of cellulitis involving the forearm. Her mother reported that several of the other children attending the same day care center were being treated for impetigo. The patient was treated with oral cephalexin, with only a partial clinical response. After returning to the clinic, a culture was taken that grew a mixture of *Streptococcus pyogenes* (group A streptococcus) and MRSA. In addition to β-lactam antibiotics, the MRSA isolate was resistant to erythromycin but sensitive to all other antimicrobial classes tested. Molecular characterization revealed that the isolate had a type IVa SCC*mec* and was positive for PVL genes. The patient had a complete recovery after receiving oral clindamycin.

Over the next 1-yr period, the outpatient clinic, which served a predominantly Native American community, documented more than 115 cases of superficial skin and soft-tissue infections owing to either MRSA alone or coinfections with group A streptococcus. Based on molecular characterization by plasmid analysis and PFGE, it was determined that two closely related clones of CA-MRSA were circulating in the community. These clones resembled the USA400 type strain (ST1) and were closely related to the MW2 strain. In addition to being positive for PVL genes, the strain was positive for the staphylococcal enterotoxin genes *sea, sec, seh, sek,* and *sel,* along with the virulence genes *seg2, ear, sel2, sec4, set16, lpl10,* and *bsa*—genes predominantly found in the MW2 lineage. Since the patients were being treated in outpatient clinics, the standard of care was initially empirical treatment with either an oral cephalosporin or erythromycin; cultures were generally performed only after treatment failure. When it became apparent that CA-MRSA was the cause of many of these infections, treatment with other antibiotic classes became first-line therapy, and the number of MRSA cases declined dramatically. Unfortunately, over the next decade, the same strains of CA-MRSA became widespread throughout Wisconsin, particularly among other Native American communities (11).

2.3. "Transitional" MRSA with Features of Both CA-MRSA and HA-MRSA

In 1992, an 11-mo-old Caucasian male with I-cell disease (mucolipidosis type II) was admitted for treatment of respiratory insufficiency. Over the prior 6 mo, the infant had been admitted to the hospital multiple times for failure to thrive and recurrent pulmonary infections. During his final admission, multiple bacterial pathogens were isolated from endotracheal aspirates and blood cultures, including *Stenotrophomonas maltophilia*, *Klebsiella pneumoniae*, and MRSA.

Clinically this patient had numerous risk factors for acquiring HA-MRSA. He had numerous hospital admissions, was chronically debilitated from his inherited genetic disorder, and had received multiple courses of oral and iv antibiotics. However, molecular characterization of his MRSA isolates revealed genetic features that were shared by both HA-MRSA and CA-MRSA. The hospital-associated characteristics included risk factors for acquiring MRSA, the nosocomial origin of infection, and a plasmid profile unique to hospitals in our surveillance area. The community-associated features included susceptibility to multiple classes of antimicrobial agents and genotypic markers such as PFGE type USA400 (ST1) and the presence of type IVa SCC*mec*. Interestingly, the strain lacked the PVL genes. The strain was positive for the staphylococcal enterotoxin genes *sec*, *sei*, and *sel*, and several of the MW2 genes associated predominantly with strain MW2 (*sel2*, *sec4*, *set16*, *lpl10*, and *bsa*).

2.4. Hypervirulent MSSA and Its Relationship to MRSA

In 2005, a 65-yr-old male with type 2 diabetes mellitus presented to his local physician with a 1-d history of cellulitis of the left forearm. He denied any trauma to the area. His admission temperature was 101.4°F and he had a leucocytosis of 14,400/mm^3. Two blood cultures drawn at the time of presentation grew MSSA that was susceptible to penicillin and all other antibiotic classes tested. Because the patient was allergic to penicillin, he initially received iv ciprofloxacin and clindamycin. However, there was no improvement, and over the next several days, he experienced increasing erythema and edema of his arm. The arm eventually became purpuric and there was necrosis along the medial aspect of the upper arm. He underwent multiple surgical debridements, which revealed necrotizing fasciitis and myonecrosis. Microbiological cultures obtained at the time of debridement showed pure growth of the same strain of MSSA even after the addition of vancomycin and rifampin to his treatment regimen. The patient eventually required amputation to control the infection.

The strain belonged to sequence type 45, and it lacked *mec* genetic elements, as expected. However, the strain was positive for uncommon staphylococcal enterotoxin genes such as *seg* and *sei* in addition to a group of genes that constitute the enterotoxin gene cluster (*sem*, *sen*, and *seo*). The strain was also positive for *set16*, *lpl10*, and *bsa*, but it lacked PVL genes.

3. Summary Comparison of Clinical and Molecular Features of MRSA

The diverse clinical presentations that we have described underscore the amazing versatility that *S. aureus* has acquired as a human pathogen. Although factors affecting the sequence of disease progression from colonization and invasion to disseminated infection have been studied for many years, only recently has the genetic basis underlying the virulence of *S. aureus* become understood in any detail. The changing epidemiology of MRSA, evidenced by the emergence of CA-MRSA as a significant global public health problem, has spurred renewed interest in determining more precisely how knowledge of molecular determinants of virulence can translate into better treatment and prevention.

Table 1 summarizes the clinical and molecular features associated with the case histories. Based on PFGE, multilocus sequence typing (MLST), and *spa* typing, each strain had a unique genetic background, with the CA-MRSA and transitional MRSA cases sharing some of those features. The MRSA strains in two clinical cases (cases 2 and 3) were from a related MRSA clone that was predominant in Native American communities in Wisconsin in the 1990s. The HA-MRSA case was owing to a strain that was probably introduced from Europe by the Bosnian soldier since the PFGE and MLST profile did not match with any of the common clones reported in the United States. It is important to remember that although significant genetic diversity exists among MRSA strains the breadth of diversity observed is less than that present for other species of bacteria. Further research will be needed to determine why some clones of MRSA are more successful as colonizers and pathogens in humans than others.

In this case series, only one of the two SCC*mec* IV–positive MRSA strains harbored the *lukSF*-PV genes. The "transitional" CA-MRSA had the SCC*mec* IV but not the *lukSF*-PV gene. Each strain had two or more staphylococcal enterotoxin genes but not the same set of genes. All four strains harbored *lpl10*, *fnaA*, *bsa*, *hlaA*, *hlaB*, and *hlaD*. The difference in the combination of virulence genes that were present or absent between the strains was substantial. Several of

Table 1
Comparison of Clinical and Molecular Features of Four Strains of *S. aureus*

Isolate	Infection type	Antimicrobial resistance	PFGE clonality	Sequence type	SCC*mec*	*spa* type	PVL genes	Select virulence genes
Case #1: HA-MRSA	Invasive wound infection	β-Lactams, fluoroquinolones, clindamycin, erythromycin, tetracycline, gentamicin, and rifampin	USA500	247	Ia	t051	Neg	*sea, fnbA, fnbB*
Case #2: CA-MRSA	Skin and soft-tissue infection	β-Lactams, erythromycin	USA400	1	IVa	t128	Pos	*sea, sec, seh, sek, and sel, seg2, ear, sel2, sec4, set16, lpll0,* and *bsa*
Case #3: transitional MRSA	Invasive respiratory infection	β-Lactams	USA400	12	IV	t213	Neg	*sec, sei,* and *sel, sel2, sec4, set16, lpll0,* and *bsa*
Case #4: hypervirulent MSSA	Skin and soft-tissue infection	Susceptible to all	USA600	45	None	t917	Neg	*seg,* and *sei,* and enterotoxin gene cluster—*sem, sen,* and *seo, set16, lpll0,* and *bsa*

the staphylococcal enterotoxins are known superantigens that generate excessive immunostimulatory responses in hosts, including a massive release of cytokines such as interleukin-2 (IL-2), interferon-γ, tumor necrosis factor-β (TNF-β) from T–cells, and IL-1β and TNF-α from macrophages, manifesting a toxic shock–like syndrome. However, it is obvious that *S. aureus* continues to acquire diverse arrays of virulence factors, and that many of them are capable of contributing to the disease manifestation in infected patients.

Acknowledgment

We gratefully acknowledge Jennifer Brady for excellent technical assistance. This work was funded in part by grant AI061385 from the National Institutes of Health.

References

1. Boyce, J. M. (1998) Are the epidemiology and microbiology of methicillin-resistant *Staphylococcus aureus* changing? *JAMA* **279,** 623, 624.
2. Chambers, H. M. (2001) The changing epidemiology of *Staphylococcus aureus*? *Emerg. Infect. Dis.* **7,** 178–182.
3. Panlilio, A. L., Culver, D. H., Gaynes, R. P., Banerjee, S., Henderson, T. S., Tolson, J. S., and Martone, W. J. (1992) Methicillin-resistant *Staphylococcus aureus* in U.S. hospitals, 1975–1991. *Infect. Control Hosp. Epidemiol.* **13,** 582–586.
4. National Nosocomial Infections Surveillance System. (2004) National Nosocomial Infections Surveillance (NNIS) System Report, data summary from January 1992 through June 2004, issued October 2004. *Am. J. Infect. Control* **32,** 470–485.
5. Gorak, E. J., Yamada, S. M., and Brown, J. D. (1999) Community-associated methicillin-resistant *Staphylococcus aureus* in hospitalized adults and children without known risk factors. *Clin. Infect. Dis.* **29,** 797–800.
6. Herold, B. C., Immergluck, L. C., Maranan, M. C., et al. (1998) Community-acquired methicillin-resistant *Staphylococcus aureus* in children with no identified predisposing risk. *JAMA* **179,** 593–598.
7. Shukla, S. K., Stemper, M. E., Ramaswamy, S. V., et al. (2004) Molecular characteristics of nosocomial and Native American community-associated methicillin-resistant *Staphylococcus aureus* clones from rural Wisconsin. *J. Clin. Microbiol.* **42,** 3752–3757.
8. McDougal, L. K., Steward, C. D., Killgore, G. E., Chaitram, J. M., McAllister, S. K., and Tenover, F. C. (2003) Pulsed-field gel electrophoresis typing of oxacillin-resistant *Staphylococcus aureus* isolates from the United States: establishing a national database. *J. Clin. Microbiol.* **41,** 5113–5120.
9. Vandenesch, F. T., Naimi, T., Enright, M. C., et al. (2003) Community-acquired methicillin-resistant *Staphylococcus aureus* carrying Panton-Valentine leukocidin genes: worldwide emergence. *Emerg. Infect. Dis.* **9,** 978–984.

10. Oliveira, D. C. and de Lencastre, H. (2002) Multiplex PCR strategy for rapid identification of structural types and variants of the mec element in methicillin-resistant *Staphylococcus aureus*. *Antimicrob. Agents Chemother.* **46,** 2155–2161.
11. Stemper, M. E., Shukla, S. K., and Reed, K. D. (2004) Emergence and spread of community-associated methicillin-resistant *Staphylococcus aureus* in rural Wisconsin, 1989 to 1999. *J. Clin. Microbiol.* **42,** 5673–5680.

3

Minimum Inhibitory Concentration (MIC) Analysis and Susceptibility Testing of MRSA

German Bou

Summary

The widespread occurrence of methicillin-resistant *Staphylococcus aureus* (MRSA) or oxacillin-resistant MRSA is a major cause of concern worldwide. Although mainly located in hospital environments, these microorganisms have been reported to have the capacity to cause infections in the community. Resistance to methicillin implies resistance to all β-lactam antibiotics and, furthermore, MRSA isolates normally harbor resistance to other families of antibiotics such as co-trimoxazole and aminoglycosides. Prompt and accurate detection of MRSA isolates is therefore extremely important in clinical microbiology laboratories. In this chapter, we review the most common methods of susceptibility testing for MRSA.

Key Words: *Staphylococcus aureus*; methicillin resistance; MIC analysis; *mecA*; cefoxitin; oxacillin.

1. Introduction

Staphylococcus aureus is one of the most common causes of nosocomial or community-based infections, leading to serious illnesses with high rates of morbidity and mortality. During recent years, the increase in the number of bacterial strains showing resistance to methicillin or oxacillin has become a serious clinical and epidemiological problem for several reasons. First, the latter antibiotic is considered to be the first option in the treatment of staphylococcal infections. Second, resistance to this antibiotic implies resistance to all β-lactam antibiotics. Third, methicillin-resistant *S. aureus* (MRSA) has many virulence factors that enable it to cause disease in normal hosts. For example, MRSA is a frequent cause of health care–associated infections of the blood as well as catheter-related infections. MRSA is also an emerging cause of community-associated infections, especially skin and soft-tissue

From: *Methods in Molecular Biology: MRSA Protocols*
Edited by: Y. Ji © Humana Press Inc., Totowa, NJ

infections and necrotizing pneumonia. For these reasons, accuracy and promptness in the detection of resistance of *S. aureus* to methicillin (MRSA) are of key importance in guaranteeing correct antibiotic treatment in patients infected with these strains and in checking the spread to MRSA isolates in hospital environments.

Strains that are oxacillin and methicillin-resistant, historically termed MRSA, although methicillin is no longer the agent of choice for testing or treatment, are resistant to all β-lactam agents, including penicillins, cephalosporins, carbapenems, and beta lactamase inhibitor combinations. Hospital-associated MRSA isolates often show multiple resistance to other commonly used antimicrobial agents, including aminoglycosides, erythromycin, clindamycin, co-trimoxazole and tetracycline, while community-associated MRSA isolates are often resistant only to β-lactam agents and erythromycin.

MRSA strains harbor the *mecA* gene, which encodes a modified PBP2 protein (PBP2N or PBP2a) with low affinity for methicillin and all β-lactam antibiotics. One important aspect is that phenotypic expression of methicillin resistance may vary depending on the growth conditions for *S. aureus*, such as the osmolarity or temperature of the medium, and this may affect the accuracy of detecting methicillin resistance *(1)*. Heteroresistant strains of bacteria may evolve into fully resistant strains and then multiply in patients receiving β-lactam antibiotics, causing therapeutic failure.

There are several methods of detecting methicillin resistance *(1–10)*, including classic methods for determining minimum inhibitory concentrations (MICs), i.e., the lowest concentration of an antimicrobial agent that prevents visible growth of a microorganism in a susceptibility test (disk diffusion, Etest®, broth dilution, or agar dilution); screening techniques with solid culture medium containing oxacillin or cefoxitin, and methods that detect the *mecA* gene or its protein product (PBP 2 protein) *(3,4)*.

2. Materials
2.1. Disk Diffusion

1. Mueller-Hinton agar plates (90 mm).
2. Oxacillin disks (1 μg).
3. Cefoxitin disks (30 μg).
4. Sterile loops, swabs (sterile, nontoxic, and not too tightly spun), sterile tubes, pipets, sterile forceps, and scissors.
5. Saline serum: 0.85 g of NaCl in 100 mL of distilled water.
6. Incubator (35–37°C).
7. Storage containers or tubes containing desiccant.
8. Quality control organisms: *S. aureus* ATCC 29213 (oxacillin susceptible) and 43300 (oxacillin resistant), and the latest NCCLS guidelines.

2.2. Etest

1. Mueller-Hinton agar plates (90 mm) + 2% Nacl.
2. Oxacillin Etest strips.
3. Sterile loops, swabs (sterile, nontoxic, and not too tightly spun), sterile tubes, pipets, sterile forceps, and scissors.
4. Etest manual applicator.
5. Saline serum: 0.85 g of NaCl in 100 mL of distilled water.
6. Incubator (35–37°C).
7. Storage containers or tubes containing desiccant.
8. Quality control organisms: *S. aureus* ATCC 29213 and 43300, and the latest NCCLS guidelines.

2.3. Broth Dilution Procedures (Macrodilution or Microdilution)

1. Mueller-Hinton broth supplemented with 2% NaCl.
2. Oxacillin powder of known potency.
3. Sterile loops, swabs (sterile, nontoxic, and not too tightly spun), sterile tubes, pipets, microtiter plates.
4. Saline serum: 0.85 g of NaCl in 100 mL of distilled water.
5. Incubator (35–37°C).
6. Storage containers or tubes containing desiccant.
7. Quality control organisms: *S. aureus* ATCC 29213 and 43300, and the latest NCCLS guidelines.
8. Microdilution trays or microtiter plates.

2.4. Agar Dilution

1. Mueller-Hinton agar supplemented with 2% NaCl.
2. Oxacillin powder of known potency.
3. Inoculum replicator available for transfer of 32–36 inocula to each plate; replicators with 3-mm-diameter pins, which will deliver approx 2 µL onto the agar surface. One-millimeter pins would deliver approx 0.1–0.2 µL of the inoculum.
4. Sterile loops, swabs (sterile, nontoxic, and not too tightly spun), sterile tubes, pipets.
5. Saline serum (0.85 g of NaCl in 100 mL of distilled water).
6. Incubator (35–37°C).
7. Storage containers or tubes containing desiccant.
8. Quality control organisms: *S. aureus* ATCC 29213 and 43300, and the latest NCCLS guidelines.

2.5. Agar Screening

1. Plate containing 6 µg/mL of oxacillin in Mueller-Hinton agar supplemented with NaCl (4%).
2. Sterile loops, swabs (sterile, nontoxic, and not too tightly spun), sterile tubes, pipets.
3. Saline serum (0.85 g of NaCl in 100 mL of distilled water).
4. Incubator (35–37°C).

5. Quality control organisms: *S. aureus* ATCC 29213 and 43300, and the latest NCCLS guidelines.

3. Methods
3.1. Disk Diffusion
3.1.1. Preparation of Mueller-Hinton Agar

Mueller-Hinton agar should be prepared from a commercially available dehydrated base according to the manufacturer's instructions.

1. After autoclaving, allow the agar to cool in a 45–50°C water bath.
2. Pour the freshly prepared and cooled medium into glass or plastic flat-bottomed Petri dishes on a level, horizontal surface (between 25 and 30 mL of agar for 9- to 10-cm-diameter plates).
3. Allow the agar medium to cool to room temperature and store in a refrigerator (2–8°C) until use.

Agar plates can be kept for a maximum of 1 wk after preparation. The pH of each batch of Mueller-Hinton agar should be between 7.2 and 7.4. Prior to use, the plates should be checked for excess moisture. If required, the plates can be placed in an incubator (35°C) or a laminar flow hood at room temperature to remove the moisture by evaporation. No droplets of moisture should be present on the surface of the medium when the plates are inoculated.

3.1.2. Storage of Antimicrobial Disks

Packets of oxacillin or cefoxitin disks should be stored frozen except for small supplies of ready-to-use disks, which may be refrigerated for a maximum of 1 wk. The disks should be equilibrated to room temperature prior to use. Once a packet of disks has been opened and removed from its package, it should be placed in a desiccated container in order to avoid accumulation of excessive moisture.

3.1.3. Preparation of Inoculum

Two main methods are used to prepare the inoculum. In the first method, three to five colonies from an agar plate culture grown overnight are inoculated into 5 mL of suitable liquid broth (Tryptic soy broth, Mueller-Hinton, Brain-Heart), and the culture is incubated at 35°C until a turbidity equivalent to 0.5 McFarland (approx 10^8 colony-forming units [CFU]/mL) is reached, which can occur within a few hours. The turbidity of the culture is adjusted with sterile saline serum or broth to reach the 0.5-McFarland standard. This can be checked using an optical device or by visual inspection. In the second method, colonies isolated from an agar culture plate grown overnight are suspended directly into broth or physiological serum (0.85% NaCl), then vortexed to achieve a clear suspension of 0.5-McFarland turbidity (10^8 CFU/mL).

MIC Analysis and Susceptibility Testing of MRSA

The Clinical and Laboratory Standards Institute (CLSI) *(11)* recommends the direct colony suspension method for testing staphylococci for potential methicillin or oxacillin resistance (*see* **Note 1**).

3.1.4. Turbidity Standard for Preparation of Inoculum

To prepare a 0.5-McFarland standard suspension, 0.5 mL of 0.048 M $BaCl_2$ (1.175% $BaCl_2 \cdot 2H_2O$) is added to 99.5 mL of 0.18 M H_2SO_4 (1% [v/v]) with constant mixing to maintain the solution in suspension. The appropriate density of the suspension should be verified by determining the absorbance in a spectrophotometer with a 1-cm light path. The absorbance should be between 0.08 and 0.1, at 625 nm. The $BaSO_4$ suspension used as turbidity standard should be well mixed before use by a mechanical vortex.

3.1.5. Inoculation of Plates

1. Once the inoculum is ready—optimally within 15 min of adjusting the turbidity—dip a cotton swab inside the suspension, remove, and rotate several times toward the wall of the tube to eliminate excess inoculum.
2. Inoculate the surface of the Mueller-Hinton agar plate with bacterial suspension by streaking the swab over the agar plate surface; be sure that no zone of the surface is left free of inoculum. Repeat this procedure several times, rotating the agar plate 60° each time to ensure even distribution of the inoculum to the edge of the agar.
3. Leave the plates to dry for 3–5 min to allow absorption of any moisture prior to applying the antibiotic disks.

3.1.6. Application of Disks to Inoculated Agar Plates

1. Place the oxacillin and/or cefoxitin disks onto the surface of the inoculated agar plate with sterile forceps. Press down each disk to ensure full contact with the surface of the agar. Either side of the disks can be placed on the surface of the plate. Be sure to leave at least 24 mm between the centers of the disks, and also no less than 15 mm from the border of the plate.
2. Invert the plates and place in an incubator within 15 min of applying the disks.
3. Incubate the plates for 24 h in ambient air at 33–35°C (the temperature should not exceed 35°C [*see* **Note 1**]).

3.1.7. MIC Reading and Interpretation of Results

After 24 h of incubation the plates are examined. The diameters of the complete inhibition zones, including the diameter of the disk, are measured (with a ruler or sliding calipers) on the reverse of the plate. With oxacillin disks, the plates are held up to light to allow observation of small colonies that may grow inside the apparent inhibition zones. Any discernible growth within the zone of inhibition is indicative of methicillin resistance. For interpretation

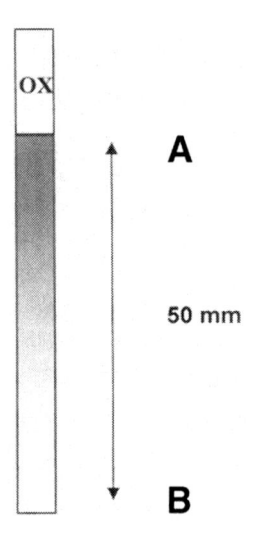

Fig. 1. Etest strip.

of the MICs of oxacillin and cefoxitin, the criteria provided by CLSI *(11)* are followed. With 1-µg oxacillin disks, diameters of inhibition zones of ≤10, 11–12, or ≥13 mm correspond to categorization of these isolates as resistant, intermediate, or susceptible to oxacillin, respectively. With the 30-µg cefoxitin disks, diameters of the inhibition zones of ≤19 and ≥20 mm correspond to classification of the staphylococci as resistant or susceptible to oxacillin, respectively. There is no intermediate category with the cefoxitin disk diffusion test (*see* **Notes 2–7**).

3.2. Etest

Etest (AB BIODISK, Solna, Sweden) is a predefined gradient technique based on a combination of the concepts of both dilution and diffusion tests and directly quantifies antimicrobial susceptibility in terms of discrete MIC values. Standard dilution methods for MIC determinations are based on discontinuous twofold serial dilutions. By contrast, Etest MICs are determined from a predefined and continuous concentration gradient of the antibiotic and are therefore more precise values.

Etest consists of a thin, inert, nonporous plastic strip 5 mm wide and 60 mm long (**Fig. 1**) (antibiotic gradient across 50 mm). One side of the strip carries the MIC reading scale in micrograms per milliliter and a two- or three-letter code to designate the identity of the antibiotic, e.g., "OX" for oxacillin. A predefined, exponential gradient of the antibiotic is immobilized on the other side or surface of the strip with the concentration maximum at a/ and the minimum at b/ (**Fig. 1**). The gradient covers a continuous concentration range across 15 twofold dilutions of a conventional MIC method (*see* **Notes 3** and **8**).

3.2.1. Test Medium for MRSA

The appropriate medium to use with Etest Oxacillin for phenotypic detection of MRSA strains is Mueller-Hinton agar supplemented with 2% NaCl. It is also important to use a reliable brand of Mueller-Hinton agar to ensure that the agar plates prepared or purchased commercially fulfill quality control specifications before use. *See* **Subheading 3.1.1**.

3.2.2. Storage of Etest Strips

Etest strips should be stored according to the product insert specifications. When stored cold (i.e., refrigerated or frozen), ensure that the package reaches room temperature before opening (approx 30 min if stored at −20°C). Etest strips also must be kept dry. Leftover Etest strips from an opened package should be stored in an airtight storage container with desiccant under the same conditions as the unopened package.

3.2.3. Preparation of Inoculum

Prepare the inoculum suspension using saline to achieve a turbidity equivalent to a 0.5- to 1-McFarland standard. A heavier inoculum may be preferable to improve detection of low-level heteroresistance. *See* **Subheading 3.1.3**.

3.2.4. Inoculation of Agar Plates

1. Streak the agar surface in three directions. Rotate the plate approx 60° each time to evenly distribute the inoculum. Alternatively, use the Retro C80™ (rota-plater; AB BIODISK) to streak the plate.
2. Allow the agar surface to dry completely prior to applying the Etest strips. *See* **Subheading 3.1.5**.

3.2.5. Application of Etest Strips

1. When using Etest strips supplied in a blister package, open a single blister by cutting along the dotted line according to the instructions in the product insert. Remove the Etest strips required using forceps and place them on a dry, clean surface, such as a sterile Petri dish or in the Etest applicator tray (**Fig. 2**). Alternatively, if the product is supplied in a foil pouch, cut across the end of the pouch and withdraw the foam cartridge containing the Etest strips.
2. Apply the Etest strips to the agar surface using sterile forceps, an Etest manual applicator, or Nema C88™ (vacuum pen; AB BIODISK) with the MIC scale facing upward and the top end of the strip (antibiotic code) positioned at the edge of the plate. Ensure that each strip is in complete contact with the agar surface and remove any air pockets underneath the strip by pressing on it gently, moving from

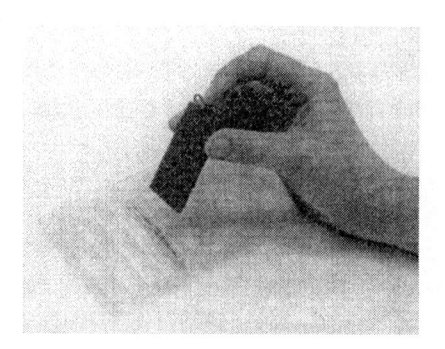

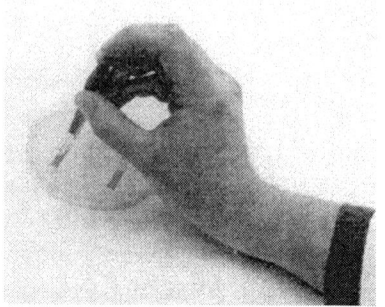

Fig. 2. Etest applicator tray.

the low concentration and upward. Once applied, the antibiotic gradient is instantaneously released into the agar and the strip cannot be moved. One to two strips can be placed on a 90-mm agar plate; for detection of MRSA phenotype one Etest Oxacillin strip is sufficient.
3. Invert the agar plates and incubate for a full 24 h in ambient air at 35°C (the temperature should not exceed 35°C).

3.2.6. Reading and Interpretation of MIC Results (see **Note 8**)

After the full period of incubation, the Etest MIC result is read where the edge of the inhibition ellipse intersects the MIC scale on the strip. If growth is observed along the entire edge of the Etest strip and no inhibition ellipse is seen, the MIC is reported as greater than the highest value on the scale. If the edge of the inhibition ellipse is below the lower end of the strip (i.e., it does not intersect the strip), the MIC is reported as less than the lowest value on the scale.

MIC break points for defining susceptibility categories provided by the CLSI (USA) or other national reference group can be used to interpret Etest MIC results. Etest half-dilution values should always be rounded up to the next upper twofold value before susceptibility categorization. For example, an Etest Oxacillin MIC of 3 µg/mL is rounded to 4 µg/mL and the *S. aureus* isolate is categorized as resistant (R ≥ 4 µg/mL) to oxacillin (MRSA). If the MIC of oxacillin is ≤ 2 µg/mL, the *S. aureus* isolate is categorized as susceptible (S ≤ 2 µg/mL) to oxacillin (methicillin-susceptible *S. aureus* [MSSA]).

3.3. Broth Dilution Procedures (Macrodilution and Microdilution) (see **Note 9**)

Broth dilution procedures are used for quantitative measurement of the in vitro activity of an antimicrobial agent against a given bacterial isolate. For this, a set of tubes is prepared with a broth medium to which various concentrations of the antimicrobial agent (oxacillin) are added. The tubes are inoculated with the target microorganisms to which antibiotic susceptibility has to be determined, and after overnight incubation at 35°C, the MIC is determined.

3.3.1. Antimicrobial Agent

Oxacillin can be obtained directly from the corresponding manufacturer or purchased from companies as antimicrobial standard or powder. Powder must be stored as recommended by the manufacturer or at ≤−20°C in a desiccator, to exclude moisture. It is very important to consider the potency supplied by the manufacturer, which may be expressed as a percentage, or in micrograms per milligram (w/w). The following equation can be used to assess the amount of powder or diluent needed for a standard solution:

$$\text{Weight (mg)} = \text{Volume (mL)} \times \text{Concentration (μg/mL)}/\text{Potency (μg/mg)}$$

Regarding the stock solutions to be assayed, oxacillin should be prepared at 10 times the highest concentration to be tested. Distilled water can be used to dissolve and/or dilute this antibiotic. The antibiotic solution should be sterilized by membrane filtration and maintained frozen in sterile vials, preferably at −60°C or below but never at a temperature higher than −20°C. Stock solutions can be stored at −60°C or below for 6 mo or more without significant loss of activity (these times are reduced to 3 mo for storage at −20°C, or to 1 wk at 4°C). Frozen vials can be thawed as needed and used the same day.

Concerning the number of oxacillin concentrations assayed, they should encompass the interpretative end points to the corresponding antibiotic, as reported by CLSI, as well as a range that allows at least one quality control organism to have on-scale values. For *S. aureus* ATCC 29213, which is included as a quality control strain, appropriate concentrations would be 16, 8, 4, 2, 1, 0.5, 0.25, 0.12, and 0.06 μg/mL.

3.3.2. Mueller-Hinton Broth Medium

CLSI (formerly National Committee for Clinical Laboratory Standards [NCCLS]) recommends the use of cation-adjusted Mueller-Hinton broth (CAMHB) plus 2% NaCl for determining oxacillin MICs. Briefly, the medium is made as follows:

1. Prepare magnesium stock solution by dissolving 8.36 g of $MgCl_2\text{-}6H_2O$ in 100 mL of deionized water. This solution contains 10 mg of Mg^{2+}/mL. Sterilize the solution by membrane filtration and store at between 2 and 8°C.
2. Prepare calcium stock solution by dissolving 3.68 g of $CaCl_2\text{-}2H_2O$ in 100 mL of deionized water. This solution contains 10 mg of Ca^{2+}/mL. Sterilize the solution by membrane filtration and store at between 2 and 8°C.

3. Prepare Mueller-Hinton broth following the manufacturer's instructions. Sterilize the broth by autoclaving, and chill overnight at 2–8°C prior to the addition of cations.
4. For each desired increase of 1 mg/L in the final concentration, add 0.1 mL of chilled Mg^{2+} and Ca^{2+} stock solution/L of broth. This medium is now designated CAMHB and is recommended for use by CLSI. The pH of the medium must be maintained at between 7.2 and 7.4.

3.3.3. Manipulation of Antimicrobial Agents (Oxacillin)

3.3.3.1. MACRODILUTION BROTH METHOD

Tests are performed in sterile 13 × 100 mm test tubes. Tubes must be closed during the assay using, e.g., cotton plugs or caps. Two control tubes should be used in the assay with broth without oxacillin, one to check the viability of the bacterial strain, and another one that is not inoculated to check the sterility of the medium.

The final twofold dilutions of oxacillin are prepared volumetrically in the broth. Dilutions are those indicated in **Subheading 3.3.1**. A minimum volume of 1 mL of each dilution is required for the test. The ratio of volume of the microorganism:dilution of the drug will be 1:1. Thus, the antimicrobial dilutions are often prepared at double the desired final concentration; that is, 1 mL of 320 µg/mL oxacillin is mixed with 9 mL of CAMHB, yielding a final concentration of 32 µg/mL, and then will yield 16 µg/mL (assay concentration) after adding 1 mL of bacterial suspension. Different pipettes can be used for adding the stock antimicrobial solution to the first tube and then for remaining dilutions. The original stock is then appropriately diluted and mixed with CAMHB to reach oxacillin concentrations of 32, 16, 8, 4, 2, 1, 0.5, 0.25, and 0.12 µg/mL prior to adding the bacterial inoculum (final concentration tested as described in **Subheading 3.3.1.**).

3.3.3.2. MICRODILUTION BROTH METHOD

The term *microdilution* is used because of the small volumes of broth dispensed in sterile, plastic microtiter plates or microdilution trays with round- or conical-bottomed wells. Each well should contain 0.1 mL of broth. Oxacillin dilutions are performed as described in **Subheading 3.3.3.1**. Antimicrobial dilutions (0.1 mL) are dispensed by a dispensing device using a starting volume that depends on the number of samples to be analyzed. Once prepared, the filled trays should be sealed in plastic bags and maintained frozen, preferably at ≤−60°C, until use. The trays will remain stable for several months. Trays must not be stored in a self-defrosting freezer, and thawed microtiter plates must not be refrozen, because cycles of severe temperature changes may accelerate the degradation of oxacillin.

3.3.4. Broth Dilution Assay (Macrodilution and Microdilution)

3.3.4.1. Preparation of Inoculum

The standarized inoculum for either macrodilution or microdilution broth methods can be prepared by either method described in **Subheading 3.1.3.**, although CLSI recommends direct colony suspension. Within 15 min of preparation, the adjusted inoculum suspension should be diluted in broth (macrodilution method), water or saline (microdilution method) so that after inoculation each tube or well contains approx 5×10^5 CFU/mL. To achieve this concentration, the dilution procedure will vary according to the method used to deliver the inoculum to the wells or tubes. With microdilution and considering a volume of medium in the well of 0.1 mL and an inoculum volume of 0.005 mL, the 0.5-McFarland suspensions (10^8 CFU/mL) should be diluted 1:10. The addition of 0.005 mL of the suspension to the broth will result in a final concentration of bacteria of 5×10^5 CFU/mL or 5×10^4 CFU/well.

3.3.4.2. Macrodilution

Fifteen minutes after the inoculum has been standardized, 1 mL of bacterial inoculum is added and mixed with 1 mL of each antimicrobial solution at different dilutions. This results in a 1:2 dilution of each antibiotic concentration and the same ratio for the inoculum. A tube without oxacillin should also be inoculated, as well as a tube containing only broth medium. At the same time, it is important to check the purity of the bacterial suspension by subculturing an aliquot of the suspension in, e.g., sheep blood agar, and incubating this with the macrodilution tubes. All tubes are incubated at 35°C for 24 h in an ambient air incubator. The MIC value is the lowest concentration of oxacillin that completely inhibits growth of the microorganism in the tubes, as detected by the unaided eye (**Fig. 3**). The growth in tubes used to assess the growth end points is determined by comparing with the control tubes that were inoculated without oxacillin or not inoculated. For the results to be considered valid, acceptable growth (≥2-mm button or clear turbidity) must be visualized in the positive control wells. End points are as defined by CLSI, and *S. aureus* isolates for which the MICs of oxacillin are ≤2 or ≥4 µg/mL are considered susceptible or resistant to this antibiotic, respectively.

3.3.4.3. Microdilution

Fifteen minutes after the inoculum has been standardized as described, each well of the microtiter plate or microdilution tray can be inoculated using an inoculator device that applies a volume not exceeding 10% of the volume in the well (≤10 µL of bacterial suspension in 0.1 mL). Alternatively, 0.05 mL of inoculum with the same volume of antimicrobial agent can be delivered, thus

MIC= 1 mg/L

0   0.06 0.12 0.25 0.5   1   2   4   8   16
Oxacillin conc (mg/L)

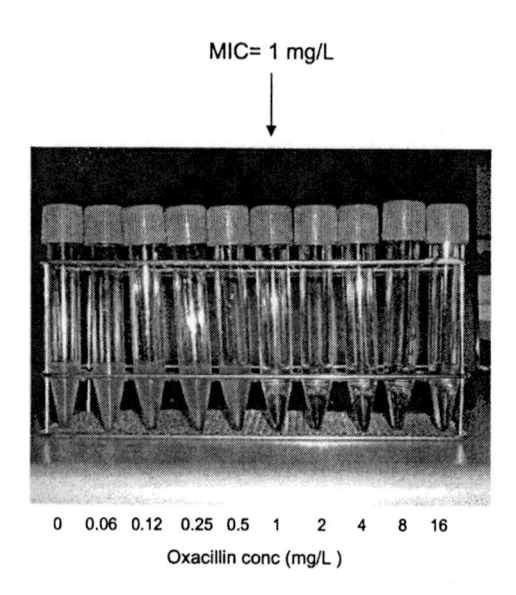

Fig. 3. Macrodilution scheme.

resulting in a 1:2 dilution. At the same time, it is important to check the purity of the bacterial suspension by subculturing an aliquot in, e.g., sheep blood agar, and incubating this along with the microdilution tray. To prevent desiccation, microtiter plates or microdilution trays should be sealed with plastic tape before incubation. All tubes are incubated at 35°C for 24 h in an ambient air incubator. The MIC value is the lowest concentration of oxacillin that completely inhibits growth of the microorganisms in the tubes, as detected by the unaided eye. Special devices programmed to read the microdilution test can be used to facilitate the reading. As with the macrodilution method, growth in the wells with the oxacillin dilutions should be compared with the growth in the control wells (no oxacillin or no microorganism) for determining the end points. For assay results to be valid, growth in the positive control well must occur (≥2-mm button or clear turbidity). End points are identical to those cited for macrodilution.

3.4. Agar Dilution (see Note 10)

Agar dilution is another method used to determine the antibiotic susceptibility of a specific microorganism. With this method, the antibiotic (oxacillin) is incorporated into the agar medium. Each agar plate contains a certain dilution of the antimicrobial agent.

3.4.1. Mueller-Hinton Agar Medium

CLSI recommends the use of Mueller-Hinton agar supplemented with 2% NaCl (without supplemented cations) as the best method for routine testing of susceptibility of *S. aureus* to oxacillin.

3.4.2. Preparation of Agar Dilution Plates

Oxacillin dilutions are added to previously sterilized molten Mueller-Hinton agar, which has been allowed to equilibrate in a water bath at 45–50°C. The agar and the oxacillin are mixed thoroughly and then poured into Petri dishes on a level surface to obtain a depth of at least 3–4 mm of agar. This process should be done as quickly as possible to avoid partial solidification of the agar in the mixing flask and to avoid formation of bubbles. The agar should be allowed to solidify at room temperature. If plates are not used immediately, they should be stored at 2–8°C in sealed plastic bags for up to 5 d; β-lactams are generally fairly susceptible to degradation. The stored agar plates should be equilibrated at room temperature prior to use. Is important to eliminate any moisture from the surface of the agar plate before bacterial inoculation.

For dilution of oxacillin, one method is to add 2 mL of oxacillin stock solution (or dilutions of the solution) to each 18 mL of Muller-Hinton molten agar (i.e., to achieve a 1:10 dilution). The oxacillin stock solution should contain 160 µg/mL and be diluted twofold until reaching a concentration of 0.6 µg/mL so that after following the aforementioned procedure, the final concentrations of oxacillin on the agar plates will be 16, 8, 4, 2, 1, 0.5, 0.25, 0.12, and 0.06 µg/mL. Sterile distilled water can be used to prepare the oxacillin stock solutions and dilutions.

3.4.3. Preparation of Inoculum

As previously mentioned in **Subheading 3.1.3.**, a culture adjusted to 0.5 McFarland should contain approx 10^8 CFU/mL. The final concentration on the agar surface should be approx 10^4 CFU/spot. If 3-mm pin replicators are used (delivering 2 µL), a previous 1:10 dilution of the inoculum should be made in order to obtain 10^4 CFU/spot. If 1-mm pin replicators are used (delivering ca. of 0.2 µL), no dilution of the initial suspension is required. Inoculation must be carried out no later than 15 min after preparation of the suspension.

3.4.4. Inoculation and Incubation of Agar Dilution Plates

All bacterial suspensions (10^7 CFU/mL) to be inoculated should be kept in order in a rack. An aliquot will be placed into the corresponding well of the replicator inoculum block.

1. Mark the Mueller-Hinton agar plates containing different concentrations of oxacillin for orientation of the inoculum spots.
2. Apply a small aliquot onto the agar surface as described above (**Subheading 3.4.3.**), or with standarized loops or pipets.
3. First inoculate a growth control plate (no antimicrobial agent) and then inoculate the plates containing the different concentrations of oxacillin starting with the lowest concentration. Finally, inoculate a second growth control plate, to ensure no contamination or antimicrobial carryover during inoculation.

4. Test the inoculum by streaking it on appropriate agar plates and incubating overnight to detect bacterial contamination.
5. Leave the agar plates at room temperature until the inoculum has been absorbed onto the surface of the agar, but for no longer than 30 min. Then invert the plates and incubate at 35°C for 24 h.

3.4.5. Determination of End Points

After incubation, the plates should be placed on a dark, nonreflecting surface for MIC determination. MIC is defined as the lowest concentration of oxacillin that completely inhibits growth.

3.5. Agar Screening (see Notes 11 and 12)

The oxacillin-salt agar screening-plate procedure may be used in addition to the procedures described above (**Subheadings 3.1., 3.2., 3.3.,** and **3.4.**) to detect and confirm the presence of MRSA.

1. Inoculate the *S. aureus* isolate onto plates of Muller-Hinton agar supplemented with NaCl (4% [w/v]) containing 6 mg/L of oxacillin.
2. Inoculate the bacterial strain from a direct colony suspension equivalent to a 0.5-McFarland standard by using either a swab or a 1-µL loop. Inoculate the isolate in an area of 10–15 mm in diameter or streak it within a quadrant of the plate.
3. Incubate the plate in ambient air at 35°C (no higher) for 24 h, and then visualize carefully using transmitted light to observe small colonies or light films of growth, which would indicate the presence of oxacillin resistance (MRSA).

4. Notes

1. As stated above, cells expressing heteroresistance grow more slowly than oxacillin-susceptible cells, and they can be missed at temperatures higher than 35°C. This is why CLSI *(11)* recommends incubating the isolates being tested against oxacillin, methicillin, or nafcillin at 33–35°C (maximum 35°C) for 24 h before reading. For this reason, it is also important that the inoculum of staphylococci for determination of oxacillin resistance be prepared by direct suspension of colonies rather than the inoculum growth method.
2. The use of other penicillinase-stable penicillins such as cloxacillin, dicloxacillin, methicillin, or nafcillin, rather than oxacillin, may yield false positive results, because these penicillins are more susceptible to degradation during storage. Oxacillin testing (which can be extrapolated to other penicillinase-stable penicillins) is therefore preferred.
3. The use of oxacillin (disks or Etest) can also fail to detect oxacillin-resistant *S. aureus* strains when the expression of PBP2a, the protein encoded by the *mecA* gene and associated with resistance to oxacillin, is very low. **Figure 4** shows an MRSA strain expressing the PBP2a protein at a low level. After 24 h of incubation, measurement of the diameter of inhibition zones or MIC (Etest) of oxacillin allows categorization as susceptible to oxacillin. However, applying the CLSI criteria, the

diameter of inhibition zones measured in the cefoxitin disk clearly indicates resistance to oxacillin. This is owing to induction of PBP2a in the presence of cefoxitin *(7,12–14)*. Cefoxitin disk diffusion is already preferred over oxacillin disk diffusion for predicting *mecA*-mediated oxacillin resistance in *S. aureus*, including those with low-level methicillin resistance or class 1 MRSA *(6–8,12–14)*. Selective media based on cefoxitin (such as mannitol salt agar-cefoxitin) are therefore superior to those based on oxacillin for the detection of MRSA *(15)*.

4. Resistance to oxacillin and susceptibility to cefoxitin may be related to other mechanisms not associated with the presence of the *mecA* gene, such as β-lactamases.
5. Although detection of oxacillin resistance can be reported after a minimum of 16 h of incubation, most methods require an incubation period of 24 h.
6. Although MRSA isolates should exhibit an antibiotic multiresistant pattern, the presence of a multiresistant antibiotic profile does not ensure the presence of oxacillin resistance, and this must always be confirmed.
7. For performing an oxacillin or cefoxitin disk diffusion test, the following points must be taken into consideration:

   a. Because antibiotic drugs diffuse almost instantaneously, a disk must not be repositioned at the same site once it makes contact with the agar surface. If required, the disk can be moved to another part of the plate.
   b. Regarding the inoculum, poor turbidity may alter the diameter of the inhibition zones. It is important to maintain 0.5-McFarland turbidity. After 24 h of incubation, if isolated colonies are observed on the surface of the Mueller-Hinton agar plates, this indicates that the inoculum was too light, and the zone of inhibition of growth will be large, resulting in quality control values that are out of range, and inappropriate susceptibility test results (false susceptibility results).
   c. Regarding the number of disks per 90- to 100-mm-diameter plate, generally up to 5 disks are permitted for *S. aureus*. If larger, 150-mm-diameter plates are used, 10–12 disks may be tested on one plate.
   d. As stated above (**Subheading 3.1.7.**), transmitted light (plate held up to light) is used to examine the oxacillin inhibition zones. Careful inspection for light growth or for inner colonies is especially required for susceptibility testing of staphylococci against oxacillin (**Fig. 4B**).
   e. The thickness of the medium should be checked for acceptability. Inadequately filled plates (depth of <3.5 mm) are associated with a tendency for a larger diameter of the zone of inhibition.
   f. Note that with NCCLS disk diffusion susceptibility testing of oxacillin against staphylococci, the addition of salt to the Mueller-Hinton broth is not recommended. Confusion on this point is common because the addition of 2% NaCl to Mueller-Hinton broth is recommended for NCCLS broth and agar dilution oxacillin susceptibility testing.
   g. If oxacillin or cefoxitin disks are not correctly stored or are used after the expiry date, they can yield false positive results—assigning susceptible isolates as MRSA because of the antibiotic inactivation. It is important to include control bacterial strains as indicated.

A 24 hours

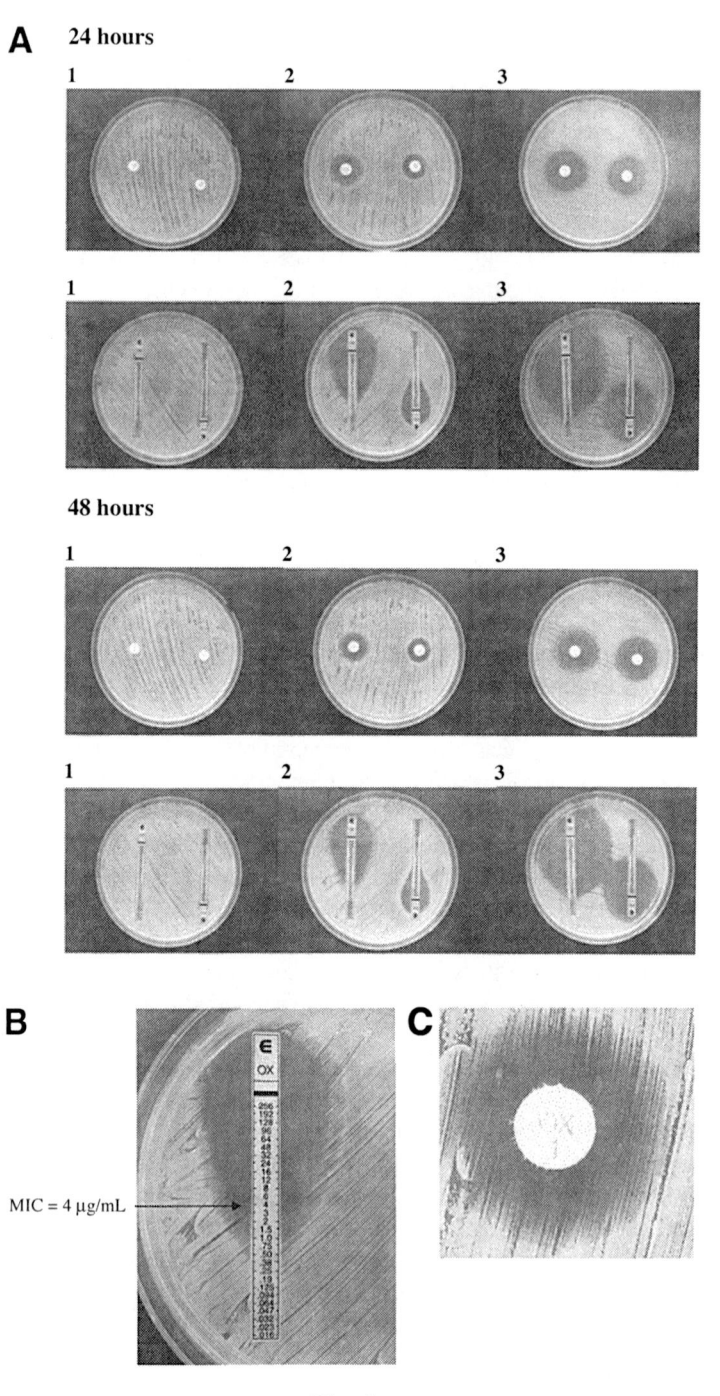

48 hours

B

MIC = 4 µg/mL

C

Fig. 4.

h. It is extremely important to confirm that the antimicrobial content of the disks used meets that specified in the NCCLS tables.
i. If the 1-µg oxacillin disk yields doubtful or intermediate results (diameters of inhibition zones of 11 to 12 mm or borderline results), the results must be confirmed by cefoxitin disk or an alternative method, such as oxacillin screening test (outlined in NCCLS document M7), PBP2a protein detection, or *mecA* gene detection *(13)*.

8. When using Etest Oxacillin to detect MRSA, observe the following:
   a. Isolated colonies within the inhibition ellipse may represent subpopulations of MRSA (**Fig. 5**). Read the MIC at the point of complete inhibition of all growth, including hazes and micro- or macrocolonies. Refer to the Etest Technical Guide (ETG 9, downloadable from www.abbiodisk.com) for different MRSA growth patterns that may be observed.
   b. When the inhibition ellipse intersects the Etest strip in between the gradings on the MIC scale, the next upper value should be reported (**Fig. 6**). MIC of bacterial isolate 0.19 µg/mL.
   c. The result should not be read if the culture appears mixed or the lawn of growth is too light or too heavy; repeat the test.

9. When carrying out the dilution test (macrodilution or microdilution), the following considerations should be taken into account:
   a. The chemical characteristics of Mueller-Hinton broth should be routinely monitored. The pH of each batch of Mueller-Hinton broth should be checked with a pH meter after the medium is prepared; the pH should be between 7.2 and 7.4 at room temperature (25°C).
   b. The viability of a batch of Mueller-Hinton broth can be evaluated by using a standard set of control organisms, such as *S. aureus* ATCC 25923 and ATCC 29213 (the latter is a β-lactamase-producing strain). If the new batch of Mueller-Hinton broth does not yield the expected MICs, the test components as well as cation component should be analyzed.

Fig. 4. (**A**) Susceptibility to oxacillin and cefoxitin determined by disk diffusion and Etest in three staphylococci strains: 1, MRSA; 2, MRSA with low-level expression of PBP2a; 3, MSSA. On the left side of the plates is oxacillin, either disk or Etest, and on the right side of the plates is cefoxitin, either disk or Etest. After 24 h of incubation, strain 2 yielded an inhibition zone of 15 mm for oxacillin (susceptible following CLSI criteria), whereas the zone was 14 mm for cefoxitin (resistance to oxacillin following CLSI criteria). With Etest, MICs of 2 and 16 µg/mL for oxacillin and cefoxitin, respectively, were obtained. Strains 1 and 3 were perfectly classified as MRSA and MSSA, respectively, following CLSI criteria by either method. After 48 h of incubation, strain 2 was perfectly classified as oxacillin resistant by disk diffusion and Etest (view magnification in [B] and [C]). (**B**) Oxacillin Etest (µg/mL) and (**C**) oxacillin disk diffusion of strain 2 after 48 h of incubation. Colonies growing as diffuse zone edge are visualized.

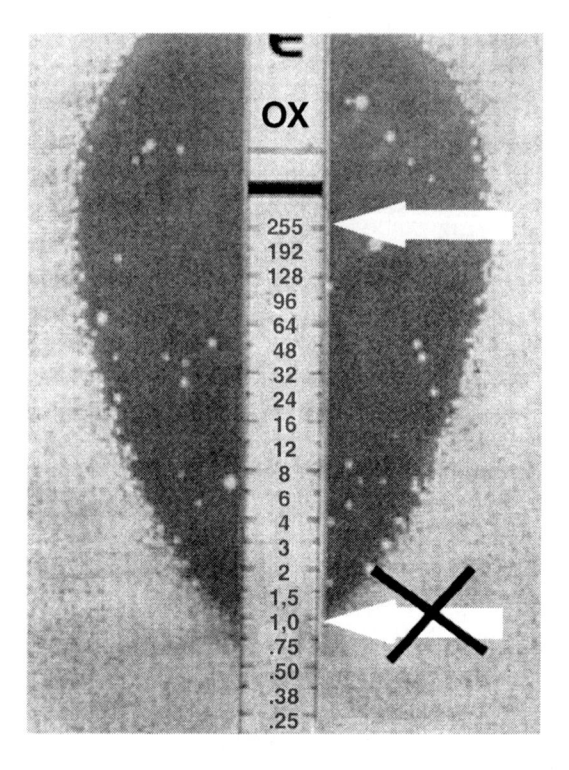

Fig. 5. MRSA subpopulations.

    c. Laboratory investigators are encouraged to perform colony counts on inoculum suspensions to ensure that the final inoculum concentration routinely obtained is close to 5×10^5 CFU/mL. This can be easily done by removing a 0.01-mL aliquot from a growth control well or any tube just after inoculation and diluting it in 10 mL of 0.9% saline. After mixing, an aliquot of 0.1 mL is spread over the surface of a suitable agar medium. After 16–18 h of incubation, the presence of approx 50 colonies would indicate the presence of 5×10^5 CFU/mL.

    d. When there is a single skipped well in a microdilution test, the highest MIC should be read. Results for drugs for which there is more than one skipped well should not be reported.

10. For performing the agar dilution test, the following considerations should be addressed:

    a. Antibiotic (weighed on an analytical balance) should be added to molten Mueller-Hinton agar cooled to 45–50°C. The addition of antibiotic to agar at higher temperatures may cause inactivation of the antibiotic.

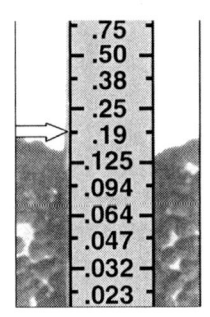

Fig. 6. Interpretative criteria for Etest reading.

    b. In this step, the antibiotic and the agar must be very well mixed prior to being poured into Petri dishes. Otherwise, an antibiotic gradient can be formed in the Petri dishes that may alter the end points and determination of MIC.
    c. Storing agar plates with the antibiotics for a long period of time (even refrigerated) may lead to antibiotic inactivation, and then inaccurate MIC data. Therefore, with this method it is very important to evaluate the stability of the plates prior to use by using control strains or applying the user's criteria.
    d. After application, the final inoculum on the agar should be approx 10^4 CFU/spot. If the initial inoculum is not correctly prepared, the final amount of inoculated cells will be out of the range that may be interpreted as MIC variations and lack of reproducibility.
    e. On reading MICs, single colonies or a faint haze caused by the inoculum should not be considered. In addition, if ≥2 colonies are kept in concentrations of the oxacillin beyond an obvious end point, or if there is any growth at lower concentrations but growth at higher concentrations, the test should be repeated, with all parameters tested (bacterial purity, quality of plates) and reanalyzed.

11. For performing the oxacillin-salt agar screening test, the following considerations should be addressed:
    a. The expiry date of agar plates should be checked before use. In addition, incorrect storage of the plates (i.e., not between 2 and 8°C) can cause the same type of false positive results as oxacillin and can become degraded. Be sure to follow the manufacturer's instructions.
    b. The concentration of the inoculum should be checked before inoculation. At concentrations higher than 0.5 McFarland, excess bacterial growth on the agar plate may yield false positive results. Any such growth in an agar plate should be confirmed by another method, such as oxacillin Etest or disk diffusion test (1 μg of oxacillin or 30 μg of cefoxitin).

12. If the MIC test result is in doubt with an MRSA, it is important to perform confirmatory tests, such as oxacillin-salt agar screening test as described above.

13. The aforementioned control strains of *S. aureus* should be used in all methods described.

Acknowledgments

I wish to thank David Velasco for assisting in the preparation of some of the figures and for helpful discussion. This work was financially supported by grants FIS PI040514, PI061368, and REIPI-RD06/0008.

References

1. Chambers, H. F. (1997) Methicillin resistance in staphylococci: molecular and biochemical basis and clinical implications. *Clin. Microbiol. Rev.* **10**, 781–791.
2. Swenson, J. M., Williams, P., Killgore, G., et al. (2001) Performance of eight methods, including two new methods for detection of oxacillin resistance in a challenge set of *Staphylococcus aureus* organisms. *J. Clin. Microbiol.* **39**, 3785–3788.
3. Van Leeuwen, W. B., Van Pelt, C., Luijendijk, A., et al. (1999) Rapid detection of methicillin resistance in *Staphylococcus aureus* isolates by the MRSA-screen latex agglutination test. *J. Clin. Microbiol.* **37**, 3029, 3030.
4. Louie, L., Majury, A., Goodfellow, J., et al. (2001) Evaluation of a latex agglutination test (MRSA-Screen) for detection of oxacillin resistance in coagulase-negative Staphylococci. *J. Clin. Microbiol.* **39**, 4149–4151.
5. National Committee for Clinical Laboratory Standards. (2003) Methods for dilution Antimicrobial Susceptibility tests for bacteria that grow aerobically. Document M7-A6 and Document M100-S13 Wayne, PA.
6. Skov, R., Smyth, R., Clausen, M., et al. (2003) Evaluation of cefoxitin 30 µg disc on Iso-Sensitest agar for detection of methicillin-resistant *Staphylococcus aureus*. *J. Antimicrob. Chemother.* **52**, 204–207.
7. Felten, A., Grandry, B., Lagrange, P. H., et al. (2002) Evaluation of three techniques for detection of low-level methicillin-resistant *S. aureus* (MRSA): a disc diffusion method with cefoxitin and moxalactam, the Vitek 2 system, and the MRSA-screen latex agglutination test. *J. Clin. Microbiol.* **40**, 2766–2771.
8. Cauwelier, B., Gordts, B., Descheemaecker, P., et al. (2004) Evaluation of a disk diffusion method with cefoxitin (30 µg) for detection of methicillin-resistant *Staphylococcus aureus*. *Eur. J. Clin. Microbiol. Infect. Dis.* **23**, 389–392.
9. Kluytmans, J., Van Griethuysen, A., Willemse, P., et al. (2002) Performance of CHROM agar selective medium and oxacillin resistance screening agar base for identifying *Staphylococcus aureus* and detecting methicillin resistance. *J. Clin. Microbiol.* **40**, 2480–2482.
10. Louie, L., Matsumura, S. O., Choi, E., et al. (2000) Evaluation of three rapid methods for detection of methicillin resistance in *S. aureus*. *J. Clin. Microbiol.* **38**, 2170–2173.
11. CLSI. (2005) Performance standards for antimicrobial susceptibility testing. CLSI approved standard M100-S15. Clinical and Laboratory Standards Institute, Wayne, PA.

12. Pottumarthy, S., Fritsche, T. R., and Jones R. N. (2005) Evaluation of alternative disk diffusion methods for detecting mecA-mediated oxacillin resistance in an international collection of staphylococci: validation report from the SENTRY Antimicrobial Surveillance Program. *Diag. Microbiol. Infect. Dis.* **51,** 57–62.
13. Velasco, D., Tomas, M., Cartelle, M., et al. (2005) Evaluation of different methods for detecting methicillin (oxacillin) resistance in *Staphylococcus aureus. J. Antimicrob. Chemother.* **55,** 379–382.
14. Swenson, J. M., Tenover, F. C., and the cefoxitin disk study group. (2005) Results of disk diffusion testing with cefoxitin correlate with presence of *mecA* in *Staphylococcus* spp. *J. Clin. Microbiol.* **43,** 3818–3823.
15. Smyth, R. W. and Kahlmeter G. (2005) Mannitol salt agar-cefoxitin combination as a screening medium for methicillin-resistant *Staphylococcus aureus. J. Clin. Microbiol.* **43,** 3797–3799.

4

Internal Transcribed Spacer (ITS)-PCR Identification of MRSA

Shin-ichi Fujita

Summary

Polymerase chain reaction (PCR) analysis of the 16S-23S rRNA gene internal transcribed spacer (ITS) followed by microchip gel electrophoresis was useful for identification of staphylococci and for strain delineation of *Staphylococcus aureus*. In the study presented in this chapter, 40 ITS patterns were demonstrated among 228 isolated colonies of *S. aureus*: 26 patterns for methicillin-susceptible *S. aureus* (MSSA); 11 patterns for methicillin-resistant *S. aureus* (MRSA); and 3 patterns for both MSSA and MRSA, highlighting the inability of ITS pattern analysis to differentiate the MSSA and MRSA strains. To overcome this problem, simultaneous PCR amplification of the ITS region and the *mecA* gene was applied to isolated colonies of staphylococcus species and positive-testing blood culture bottles.

Key Words: Methicillin-resistant *Staphylococcus aureus*; 16S-23S rRNA gene; internal transcribed spacer; microchip gel electrophoresis; polymerase chain reaction; *mecA* gene.

1. Introduction

Infections produced by *Staphylococcus aureus* are often acute and pyogenic and, if left untreated, can spread to surrounding tissue or other organs. Therefore, rapid identification of *S. aureus*, especially methicillin-resistant *S. aureus* (MRSA), from blood culture bottles is important for the establishment of effective antibiotic therapy. Although several molecular techniques have been reported for the identification of MRSA, it takes 4–8 h to obtain the results *(1–3)*.

The internal transcribed spacer (ITS) separating the 16S rRNA and 23S rRNA genes is characterized by a high degree of sequence and length variation at both the genus and species levels *(4–7)*. Regarding *S. aureus*, however, the ITS- polymerase chain reaction (PCR) patterns do not allow discrimination of

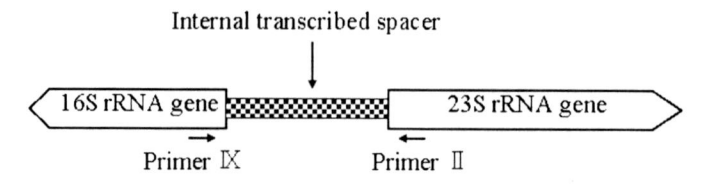

Fig. 1. Schematic representation of bacterial ribosomal genes containing primer target areas.

the methicillin-susceptible *S. aureus* and MRSA strains *(8)*. However, Fujita et al. have shown that PCR detection of the *mecA* gene, a methicillin-resistant gene, and an rRNA gene spacer length polymorphism from positive-testing blood culture bottles followed by microchip gel electrophoresis (MGE) is useful for rapid identification of MRSA *(7)*. The time course of the PCR-MGE assay for identification and delineation of MRSA was about 1 h (*see* **Note 1**).

2. Materials
2.1. Preparation of DNA

1. 0.5-McFarland standard (Eiken Chemical, Tokyo, Japan).
2. Achromopeptidase (Sigma, St. Louis, MO): Dissolve at 10 mg/mL in 10 mM NaCl, and store in 50-μL aliquots at −20°C (*see* **Note 2**).
3. TE buffer: 10 mM Tris-HCl (pH 8.0), 1 mM EDTA (pH 8.0). Autoclave before storage at room temperature.
4. Sodium dodecyl sulfate (SDS) solution: 0.1% (w/v) SDS. Store at room temperature.
5. Proteinase K solution (20 mg/mL; Takara Shuzo, Otsu, Japan). Store 50-μL aliquots at −20°C (*see* **Note 3**).
6. *Takara Z-Taq*™ polymerase (5 U/μL; Takara Shuzo).
7. 10X *Z-Taq*™ PCR buffer containing 30 mM Mg^{2+} (supplied with *Z-Taq* DNA polymerase kit).
8. dNTP mixture (2.5 mM of each dNTP) (Takara Shuzo).
9. DNA size markers: 100 bp (100 ng/mL) and 1000 bp (100 ng/mL) (GenSura, San Diego, CA). Dilute each DNA size marker with TE buffer at 1:5, and mix the diluted markers at an equal volume (each 10 ng/mL concentration). Store at −20°C.
10. DNA extraction kit (MagExtractor-Genome; Toyobo, Tokyo, Japan) containing lysis buffer, washing buffer, and magnetic silica beads (*see* **Note 4**).
11. Primer design for ITS-PCR *(9)*: primer IX (100 μM; GGTGAAGTCGTAACAAG), primer II (100 μM; TGCCAAGGCATCCACC). Store at −20°C. The primer target areas are shown in **Fig. 1**.
12. Primer design for amplification of the *mecA* gene *(7)*: forward primer (100 μM; 5′-AGAAATGACTGAACGTCCG); reverse primer (100 μM; 5′-GCGATCAAT-GTTACCGTAG). Store at −20°C.
13. PCR master mix per reaction: 5 μL of 10X *Z-Taq* PCR buffer, 4 μL of dNTP mix, 0.5 μL of *Z-Taq* polymerase. Store at −20°C.

2.2. Special Equipment

1. Thermal cycler (TGradient 96; Biometra, Rudolf-Wissell-Str., Göttingen, Germany). This thermocycler has a fast ramp speed.
2. Microchip Electrophoresis Analysis System (model SV1210; Hitachi Electronics Engineering, Tokyo, Japan).
3. Magnetic particle separator (Magical Trapper™; Toyobo).
4. Dry bath incubater (FastGene™; Nippon Genetics, Tokyo, Japan).

3. Methods

3.1. Cell Lysis

1. Suspend a colony grown on blood agar plates in 0.2 mL of TE buffer at a density of 0.5-McFarland standard (about 1 to 2×10^8 CFU/mL).
2. Add a 10-µL aliquot of achromopeptidase solution to the resuspended colony and incubate at 60°C for 5 min.
3. Lyse the cells by adding 5 µL of proteinase K solution and incubating at 60°C for 5 min.
4. Incubate the solution for 7 min in a boiling water bath.
5. Pellet cell debris by centrifuging for 5 min at 10,000g.
6. Use the supernatant as template DNA for PCR. It can be stored at –20°C. The expected DNA yield is 30–50 ng/µL.

3.2. DNA Extraction from Positive-Testing Blood Culture Bottle

1. Add 10–15 mL of 0.1% SDS to 0.1–0.2 mL of blood culture fluids.
2. Centrifuge the mixture at 4000g for 5 min.
3. Suspend the pellets in 1.5 mL of deionized H_2O, and transfer the solution into a 1.5-mL microfuge tube.
4. Centrifuge at 9000g for 5 min and discard the supernatant.
5. Suspend the pellet in 100 µL of deionized H_2O.
6. Add 10 µL of achromopeptidase solution to the sample.
7. Incubate at 60°C for 5 min in a dry bath incubator.
8. Spin the specimen down by brief centrifugation.
9. Add 750 µL of lysis buffer and 40 µL of magnetic silica beads to the sample (supplied with the DNA extraction kit).
10. Vortex vigorously for 1 min.
11. Spin the specimen down by brief centrifugation.
12. Set the sample tube onto the magnetic stand and leave the magnetic silica beads collected.
13. Remove the supernatant completely.
14. Add 900 µL of washing buffer to the tube.
15. Vortex well until the magnetic beads pellet are resuspended completely.
16. Spin the specimen down by brief centrifugation.
17. Set the sample tube onto the magnetic stand to allow collection of the magnetic silica beads.

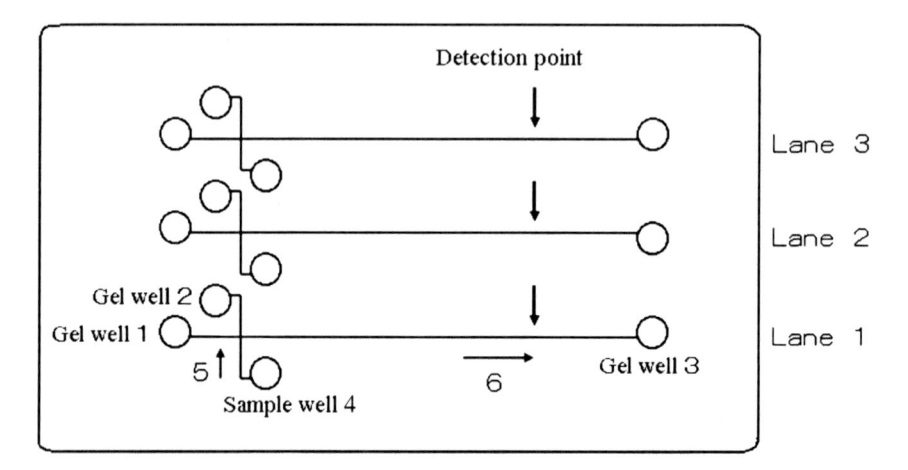

Fig. 2. Microchip (i-chip 3DNA^R) viewed from the top. One i-chip has three lanes. Electrophoresis can be performed three times by switching the lanes. 1, gel well 1; 2, gel well 2; 3, gel well 3; 4, sample well; 5, sample injection channel; 6, separation phoresis channel.

18. Remove the supernatant completely.
19. Add 900 µL of 70% ethanol to the tube.
20. Vortex well until the magnetic beads pellet are resuspended completely.
21. Spin the specimen down by brief centrifugation.
22. Remove the supernatant completely (see **Note 5**).
23. Add 100 µL of deionized H$_2$O.
24. Vortex vigorously for 1 min.
25. Spin the specimen down by brief centrifugation.
26. Collect the magnetic beads as before and then transfer about 100 µL of the supernatant containing DNA to a fresh microcentrifuge tube. The DNA yield at this stage is expected to be 20–30 ng/µL.

3.3. Multiplex PCR

1. Thaw the master mix and primer solutions on ice.
2. Add 1 µL each of four primers and 2 µL of template DNA to the PCR master mix, and make up to a 50-µL final reaction volume with deionized H$_2$O.
3. Program the thermal cycler to denature initially for 10 s at 95°C then for 25 cycles of 5 s at 99.9°C for denaturation, 4 s at 50°C for annealing, and 10 s at 72°C for primer extension. Set the final extension period for 2 min at 72°C.

3.4. Microchip Gel Electrophoresis

1. Take 10 µL of electrophoresis gel and add to gel well 3 of the i-chip. A scheme of the i-chip viewed from the top is shown in **Fig. 2** (see **Note 6**).

(ITS)-PCR Identification of MRSA

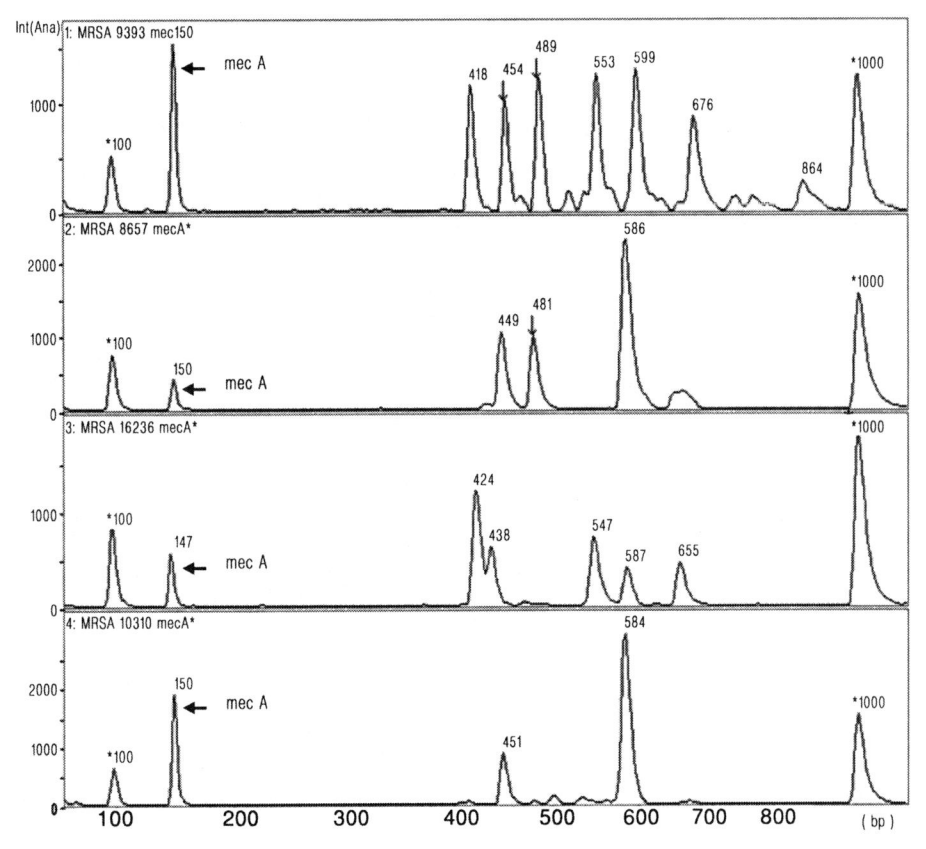

Fig. 3. Representative ITS PCR patterns of MRSA. Asterisks indicate DNA size markers of 100 (*left*) and 1000 (*right*) bp. Int., intensity.

2. Cover gel well 3 with air from the attached syringe.
3. Push the syringe slowly until the electrophoresis gel reaches each well from the separation phoresis channel.
4. Take 10 µL of electrophoresis gel and add to gel wells 1 and 2.
5. Mix 10 µL of TE buffer, 0.4 µL of PCR sample, and 0.4 µL of size marker in a 0.7-mL microfuge tube.
6. Take 10 µL of the mixture and inject it into the sample well.
7. Run the electrophoresis at 300 V for 1 min (injection time), then at 565 V for 4 min (separation time). The i-chip temperature is 30°C. The analyzing software starts at the completion of electrophoresis, and the measured waveform is automatically analyzed. The results displayed are base size, emission intensity, concentration, and tone (*see* **Notes 7–9**). Examples of results are shown in **Fig. 3**.

4. Notes

1. It takes 20–25 min to extract DNA from isolated colonies or positive-testing blood culture bottles. The PCR procedure requires 30 min, and MGE takes 5 min. Therefore, the overall turnaround time of the PCR-MGE assay is about 1 h.
2. Achromopeptidase solution can be kept for several months at –20°C.
3. Proteinase K solution can be kept for several months at –20°C.
4. Any commercially available DNA extraction kit is suitable for extraction of bacterial genomic DNA.
5. The ethanol wash removes excess salt; this salt may interfere with PCR.
6. Take out the electrophoresis gel from the refrigerator and leave it at room temperature for about 5 min. The electrophoresis gel includes a mutagen. Special care should be taken when handling it. If the electrophoresis gel adheres to your skin, carefully wash it off.
7. It is important to ensure that the PCR product (151 bp) of the *mecA* gene is well separated from both amplified products of ITS and the DNA size markers.
8. Electrophoretic analysis of the amplified products consistently shows two to seven intense, sharp fragments for each sample, ranging from 233 to 845 bp.
9. The fragments with intensities of <20% of the peak intensity should not be taken into account for analysis.

References

1. Levi, K. and Towner, K. J. (2003) Detection of methicillin-resistant *Staphylococcus aureus* (MRSA) in blood with the EVIGENE MRSA detection kit. *J. Clin. Microbiol.* **41,** 3890–3892.
2. Maes, N., Magdalena, J., Rottiers, S., Gheldre, Y. D., and Struelens, M. J. (2002) Evaluation of a triplex PCR assay to discriminate *Staphylococcus aureus* from coagulase-negative staphylococci and determine methicillin resistance from blood cultures. *J. Clin. Microbiol.* **40,** 1514–1517.
3. Mason, W. J., Blevins, J. S., Beenken, K., Wibowo, N., Ojha, N., and Smeltzer, M. (2001) Multiplex PCR protocol for the diagnosis of staphylococcal infection. *J. Clin. Microbiol.* **39,** 3332–3338.
4. Gurtler, V. and Barrie, H. D. (1995) Typing of *Staphylococcus aureus* by PCR-amplification of variable-length 16S-23S rDNA spacer regions: characterization of spacer sequences. *Microbiology* **141,** 1255–1265.
5. Gurtler, V. and Stanisich, V. A. (1996) New approaches to typing and identification of bacteria using the 16S-23S rDNA spacer region. *Microbiology* **142,** 3–16.
6. Mendoza, M., Meugnier, H., Bes, M., Etienne, J., and Freney, J. (1998) Identification of Staphylococcus species by 16S-23S rDNA intergenic spacer PCR analysis. *Int. J. Syst. Bacteriol.* **48,** 1049–1055.
7. Fujita, S., Senda, Y., Iwagami, T., and Hashimoto, T. (2005) Rapid identification of staphylococcal strains from positive-testing blood culture bottles by internal transcribed spacer PCR followed by microchip gel electrophoresis. *J. Clin. Microbiol.* **43,** 1149–1157.

8. Dolzani, L., Tonin, E., Lagatolla, C., and Monti-Bragadin, C. (1994) Typing of *Staphylococcus aureus* by amplification of the 16S-23S rRNA intergenic spacer sequences. *FEMS Microbiol. Lett.* **119,** 167–174.
9. Saruta, K., Matsunaga, T., Kono, M., et al. (1997) Rapid identification and typing of *Staphylococcus aureus* by nested PCR amplified ribosomal DNA spacer region. *FEMS Microbiol. Lett.* **146,** 271–278.

5

Pulsed-Field Gel Electrophoresis of MRSA

Kurt D. Reed, Mary E. Stemper, and Sanjay K. Shukla

Summary

Pulsed-field gel electrophoresis (PFGE) is a genetic typing method that is widely used as a molecular epidemiological tool for studying the genetic diversity of *Staphylococcus aureus* and numerous other bacterial pathogens. For PFGE, intact bacterial cells are embedded in soft agarose plugs followed by lysis of the cell wall *in situ* to minimize shearing of the chromosome. The genome, which for *S. aureus* is approx 2.8 Mb, is then digested with a rare cutting restriction endonuclease and separated by agarose gel electrophoresis. The restriction fragments generated are too large to be resolved by conventional electrophoresis. Therefore, resolution of the *bands* is achieved in a "contour-clamped homogeneous electrical field" where electrical current to the gel switches direction between multiple electrodes over a period of time. Initially, current switches are short (pulsed) but become longer (ramped) as electrophoresis continues. Banding patterns are captured by an imaging system and comparisons are made based on the Dice coefficient and the unweighted pair group method using arithmetic averages with BioNumerics software.

Key Words: Pulsed-field gel electrophoresis; methicillin-resistant *Staphylococcus aureus*; macrorestriction; chromosomal typing; molecular epidemiology; Tenover criteria; BioNumerics dendrogram.

1. Introduction

Since the 1970s, a number of techniques have been used to establish the relatedness of isolates of methicillin-resistant *Staphylococcus aureus* (MRSA) for hospital infection control and other epidemiological purposes. For many years phage typing was a method of choice. Unfortunately, the results of phage typing were not always reproducible, owing to phenotypic changes, and its use required reagents that were often available only to large public health laboratories *(1)*. Molecular techniques such as restriction fragment length polymorphism analysis of plasmid DNA became available in the 1980s and early

1990s and proved to be a useful genetic-typing methodology that was practical for a wider spectrum of laboratories. Limitations of plasmid analysis included the fact that a significant number of MRSA strains are devoid of plasmids and that bacterial strains have a tendency to lose their plasmids over time in the absence of selective pressure, making this type of analysis less useful for long-term epidemiological studies *(2–4)*.

Since the mid-1990s, pulsed-field gel electrophoresis (PFGE) has been shown to be a highly versatile genetic-typing method and particularly effective for MRSA. PFGE involves isolating intact bacterial chromosomes by lysing the cell walls of organisms embedded in soft agarose plugs *(5)*. The chromosome is then digested using a rare cutting restriction endonuclease selected to produce 12 or more high molecular weight DNA fragments on gel electrophoresis. Separating these large restriction fragments necessitates the use of pulsed fields of electrical current from 24 electrodes spaced in a hexagonal contour that alternate direction at a 120°C fixed angle over a prolonged electrophoresis time. The advantages of PFGE include simple equipment requirements and technical resources that are within reason for most moderate-sized microbiology laboratories. Furthermore, the parameters for performing and interpreting PFGE for MRSA have become highly standardized over the past few years *(6–8)*. In addition, the interpretive criteria of PFGE patterns published by Tenover et al. *(9)* are used widely in infection control and other outbreak investigations. The availability of digital image analysis software now allows results to be compared between runs in a laboratory and between laboratories *(10)*. Although newer sequence-based genetic typing methods such as multilocus sequence typing and surface protein A typing are becoming increasingly popular as highly discriminatory methods for typing MRSA, PFGE is often the benchmark to which these newer techniques are compared *(11,12)*.

2. Materials

2.1. Subculturing of Isolates

1. *S. aureus* control strain NCTC 8325 (American Type Culture Collection, Manassas, VA).
2. Blood agar plates with 5% sheep blood (Remel, Lenexa, KA).
3. Incubator set at 37°C and 5% CO_2 atmosphere.

2.2. Preparation of Agarose Plugs

1. SeaPlaque agarose (BMA/Cambrex, Walkersville, ME).
2. TE buffer: 10 m*M* Tris-HCl, 1 m*M* EDTA (pH 8.0). Prepare by mixing 10 mL of 1 *M* Tris-HCl (pH 8.0) (Amresco, Solon, OH) with 2 mL of 0.5 *M* EDTA (pH 8.0) (Amresco) and QS to 1000 mL with distilled water (*see* **Note 1**). Autoclave. Store at room temperature for up to 6 mo.
3. Heat block set at 60°C.

2.3. Cell Suspensions

1. Disposable 12 × 75 mm glass tubes (Cardinal Health, McGaw Park, IL).
2. TEN buffer: 0.1 M Tris-HCl, 0.15 M NaCl, 0.1 M EDTA (pH 7.5). Autoclave. Store at room temperature for up to 6 mo.
3. Cotton swabs.
4. Vitek colorimeter (bioMerieux, Durham, NC) or equivalent.
5. Microcentrifuge tubes (2 mL) (Sarstedt, Newton, NC).

2.4. Preparation of Plugs

1. Large plug mold (Bio-Rad, Hercules, CA).
2. Lysostaphin enzyme (Sigma, St. Louis, MO). Prepare a 1 mg/mL suspension in sterile distilled water, aliquot, and freeze at −20°C for up to 6 mo.
3. Plastic conical tubes (50 mL) (Sarstedt).
4. Stainless steel spatulas, 5–8 mm wide, depending on plug size.

2.5. Plug Lysis

1. EC buffer: 6 mM Tris-HCl, 1.0 M NaCl, 0.1 M EDTA, 0.5% Brij 58, 0.2% deoxycholate, 0.5% sarkosyl (pH 7.5). Autoclave. Store at room temperature for up to 6 mo.
2. Water bath set at 37°C.
3. ESP buffer: 10 mM Tris-HCl, 1.0 mM EDTA, 1% sodium dodecyl sulfate (SDS), 1 mg/mL of proteinase K (pH 8.0). Prepare by adding 0.788 g of Tris-HCL and 0.186 g of EDTA to approx 480 mL of distilled water in a 500-mL volumetric flask. Mix well to dissolve. Adjust the pH to 8.0 using concentrated NaOH and QS to 500 mL. Combine 100 mL of this buffer with 1 g of SDS and 1 vial (100 mg) of proteinase K (Roche, Indianapolis, IN). Mix well, aliquot into 10- to 20-mL volumes, and store at −20°C for up to 6 mo.
4. Water bath set at 55°C.

2.6. Washing of Plugs

1. TE buffer.
2. Variable speed rotator.

2.7. Restriction Digest

1. Restriction endonuclease *Sma*I (Promega, Madison, WI) with packaged 10X restriction buffer and bovine serum albumin (BSA).
2. Reference strain *S. aureus* NCTC 8325: Previously prepared plugs serve as the global reference standards in multiple lanes of the gel during each PFGE procedure.
3. Microcentrifuge tubes (2 mL) with screw cap (Sarstedt).
4. Sterile disposable Petri dish.
5. Scalpel or razor blade.

2.8. Casting of Gel

1. 10X TBE buffer: Dissolve 54 g of Tris base, 27.5 g of boric acid, and 4.65 g of EDTA in 500 mL of distilled water. Adjust the pH to 8.0. Autoclave and store at room temperature.

2. Seakem GTG agarose (BMA/Cambrex).
3. CHEF-DR III gel (21 H 41 cm)–casting accessories (Bio-Rad): casting stand, platform, leveling bubble, comb, and comb holder.

2.9. Loading of Samples

1. Sterile Petri dish.
2. Stainless steel, 3- to 5-mm spatula.
3. 1.8% SeaPlaque agarose (remaining from plug preparation) for sealing plugs in the wells.

2.10. Pulsed-Field Gel Electrophoresis

1. 10X TBE buffer.
2. CHEF-DRIII PFGE system (Bio-Rad) (or comparable PFGE system).
3. Leveling bubble.
4. Gel frame.

2.11. Gel Staining

1. Covered Pyrex 23 H 33 cm glass staining dish.
2. Ethidium bromide (Sigma): Prepare a 1 mg/mL stock solution. Store in a refrigerator and protect from light (*see* **Note 2**).

2.12. Image Capture

1. GelDoc 2000 imaging system (Bio-Rad) (or comparable imaging system).

2.13. Fingerprint and Cluster Analysis

1. BioNumerics software version 4.0 (Applied Maths, Belgium) or comparable imaging software.

3. Methods

3.1. Subculturing of Isolates

1. Inoculate a single colony of each MRSA strain to a blood agar plate and streak for isolation.
2. Incubate the plates at 37°C in 5% CO_2 for 18–24 h.

3.2. Preparation of Agarose Plugs

1. Prepare 1.8% SeaPlaque agarose by weighing out 0.18 g of SeaPlaque agarose and mixing the agarose in a small flask containing 10 mL of TE buffer. Bring the agarose to a boil.
2. Aliquot the molten agarose into 2-mL microcentrifuge tubes and cap. Allow the tubes of agarose to equilibrate in a 60°C heat block. This volume is sufficient for 20–25 plugs and sealing the plugs later during the gel-loading process.

3.3. Cell Suspensions

1. Label a 12 × 75 mm glass tube for each isolate to be typed.
2. Transfer 2 mL of TEN buffer to each of the glass 12 × 75 mm tubes.
3. With a sterile swab (**Fig. 1**), pick approx 10–12 isolated MRSA colonies and resuspend in the TEN buffer to a transmittance of 20% using a Vitek colorimeter (*see* **Note 3**).
4. Transfer 250 µL of this suspension to a labeled 2-mL microcentrifuge tube.

3.4. Preparation of Plugs

1. Assemble the plug mold and finger tighten the screws to prevent leaking (*see* **Note 4**). Place the mold upright on the benchtop with the screws to the back.
2. Prepare one plug at a time. Add 5 µL of lysostaphin enzyme to the cell suspension followed by 250 µL of 60°C 1.8% SeaPlaque agarose. Quickly mix well by pipetting up and down, and transfer 250 µL to the plug mold (*see* **Note 5**) (**Fig. 1**). Retain the labeled microcentrifuge tube in a rack to identify the order of the plugs in the mold.
3. After all of the plugs have been made, allow them to solidify for 10 min in a refrigerator.
4. Place the plug mold facedown on the benchtop. Carefully loosen the screws and take the plug mold apart.
5. Carefully transfer the plugs to their respectively labeled 50-mL conical tube using a spatula.

3.5. Plug Lysis

1. Submerge the plug in 2 mL of EC buffer (prewarmed in a 37°C water bath). Make sure the plug is covered by buffer and not sticking to the side of the tube.
2. Incubate the tubes in a 37°C water bath for 4 h.
3. Decant off the EC buffer and replace it with 2 mL of ESP buffer.
4. Incubate the plug in a 55°C water bath overnight.

3.6. Washing of Plugs

1. Wash the plugs by decanting off the ESP buffer and adding 10 mL of TE buffer.
2. Place the tubes in a tray on a variable speed rotor at 60 rpm for 30 min. Repeat this wash four times, decanting and adding fresh TE buffer each time. After the final wash, store the tubes at 4°C in a refrigerator or proceed with restriction digest.

3.7. Restriction Digest

1. Determine the total number of MRSA isolates and *S. aureus* NCTC 8325 global standards that will be digested and loaded on a gel (*see* **Note 6**). Prepare a sufficient amount of 1X concentration of *Sma*I restriction buffer to cover each plug in 125 µL of digest buffer. Calculate the amount based on a single reaction of 12.5 µL of 10X buffer, 1 µL of BSA, and 111.5 µL of distilled water per reaction. Mix well.

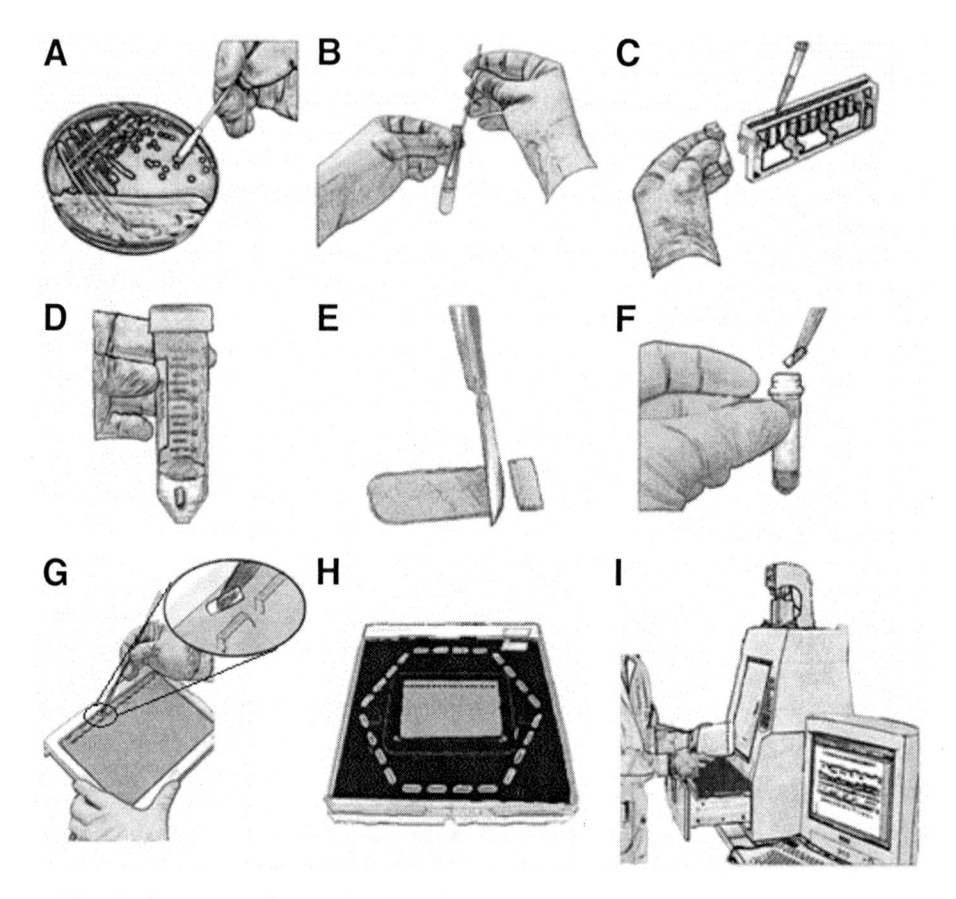

Fig. 1. Sequence of steps involved in PFGE: (**A**), Ten to 12 isolated colonies of MRSA are picked with a cotton-tipped swab. (**B**) The colonies are suspended in TEN buffer. (**C**) An aliquot of the bacterial suspension is added to molten agarose along with lysostaphin and pipetted into a plug mold. (**D**) The agarose plugs are lysed sequentially in a mixture of EC buffer detergent, proteinase K buffer, and then a TE buffer wash. (**E**) A portion of the plug is cut to correspond to the size of the electrophoresis well. (**F**) The cut portion is digested with a restriction endonuclease. (**G**) The digested plug is inserted into the well (inset shows a close-up view). (**H**) PFGE is performed in a CHEF DR III unit for 20 h. (**I**) The gel is stained with ethidium bromide and the image is recorded digitally with the GelDoc 2000.

2. Label a 2-mL microcentrifuge tube for each isolate and global standard. Add 125 μL of the 1X restriction buffer to each tube.
3. Carefully remove the washed plug from its storage tube and place it on a sterile Petri dish.
4. Cut off and discard the uppermost uneven portion of the plug using a scalpel (**Fig. 1**). Cut a 3 × 5 mm slice off the plug (*see* **Note 7**) and transfer the slice to its

appropriately labeled microcentrifuge tube (**Fig. 1**). Check to make certain that the plug is beneath the level of the buffer.
5. Add 30 U of *Sma*I enzyme to each tube. Cap the tubes and mix gently. Incubate the plugs for 2 to 3 h at room temperature (25°C). During the digestion time, prepare and prechill the electrophoresis running buffer described in **Subheading 3.10., step 1**.

3.8. Casting of Gel

1. Prepare a 1% gel by weighing out 1.5 g of Seakem GTG agarose and mixing it in 150 mL of 0.5X TBE buffer in a 250-mL flask. Bring the agarose to a boil on a stirring hot plate (or in a microwave oven). Place the flask in a 60°C water bath for approx 10 min to equilibrate the temperature.
2. Assemble the running platform inside the casting stand and tighten the screws. Set the assembly on a level surface and place the leveling bubble in the center of the tray to confirm that it is level. Adjust the sides of the stand if necessary. Attach the comb to the comb holder and insert it into the uppermost slots on the side of the casting tray with the comb side closest to the top of the stand. Ensure that the height of the comb is 2 mm above the surface of the platform.
3. Pour the molten agarose into the casting stand. Allow the gel to solidify at room temperature for at least 1 h. Carefully remove the comb just prior to loading.

3.9. Loading of Samples

1. Place the tubes containing the digested plugs in their loading order and create a gel record to document the samples and reference lanes on the gel.
2. Load one plug at a time by pouring it out on a disposable Petri dish. Transfer the plug into the well using a thin spatula (**Fig. 1**). Orient the plug in such a way that the 3 H 5 mm surface is inserted against the front wall of the gel (*see* **Note 8**).
3. Remelt the 1.8% SeaPlaque agarose left over from preparation of the plugs. Dilute the agarose to 1.0% with distilled water. Seal the plugs in the wells by pipetting agarose down the back corner of the well being careful not to dislodge the plug from the front wall or introduce air bubbles. Allow the agarose to harden for 5 min.
4. Remove one side of the casting stand. Lift the casting platform and gel out of the stand being careful not to separate the gel from the platform. Wipe off any excess gel on the bottom of the platform and around the edges of the gel.

3.10. Pulsed-Field Gel Electrophoresis

1. Prepare 2000 mL of 0.5X TBE by mixing 100 mL of 10X TBE with 1900 mL of ultrapure distilled water (*see* **Note 9**).
2. Before using the CHEF-DRIII electrophoresis chamber, remove the lid, place the leveling bubble in the center of the chamber, and adjust the corner leveling screws if necessary. Place the gel frame into the chamber. Turn on the main power of the CHEF module.
3. Pour the prechilled 0.5X running buffer into the electrophoresis chamber, turn on the pump, and then turn on the chiller unit (*see* **Note 10**).

4. Insert the loaded gel into the gel frame making certain that it is seated level inside the frame (**Fig. 1**).
5. Set the CHEF-DRIII run parameters as follows: 5-s initial switch time, 40-s final switch time, 20-h duration of run, 120°C included angle, 6-V/cm gradient, and chiller at 14°C. Begin the electrophoresis.

3.11. Gel Staining

1. After the electrophoresis is complete, turn off the equipment. Carefully remove the gel from the chamber and place it in a glass Pyrex dish.
2. Drain off 200 mL of the running buffer into a flask and add 175 µL of 1 mg/mL stock ethidium bromide solution. (Drain and discard the remaining running buffer in the electrophoresis chamber. Rinse the unit with distilled water and drain completely.)
3. Pour the ethidium bromide stain solution over the gel. Cover the Pyrex dish and place it on a rotator at 40 rpm for 30 min.
4. Destain the gel in distilled water for 15–20 min.

3.12. Image Capture

1. Place the gel on the transilluminator surface of the GelDoc 2000 (**Fig. 1**).
2. Using the Quantity One imaging software (version 4.3.0), adjust the gel in Live/Focus mode so that it is straight and has the best fit in the viewing window to capture all the lanes. Turn on the ultraviolet transilluminator and select Auto Expose to visualize the gel. Freeze the image and store it as an xxx.1sc file. Also export the image and save it as an xxx.tif file for the fingerprint analysis (**Fig. 2A**).

3.13. Fingerprint and Cluster Analysis

1. Open the BioNumerics Software (version 4.0).
2. Import the xxx.tif image of the gel to be analyzed.
3. Process the gel by going through the following steps according to the BioNumerics software (version 2.5) manual: Step 1. Strips—defining the lanes; Step 2. Curves—defining the densitometric curves; Step 3. Normalization; Step 4. Bands—defining bands and quantification.
4. To create a dendrogram, select the isolates to be entered into the comparison. Enter the Calculate Cluster Analysis. The primary comparison settings are selected as follows: Dice coefficient, unweighted pair group method using averages (UPGMA) with a position tolerance setting of 1.0% and an optimization of 0.5%. Click OK to create the cluster analysis. An example cluster analysis of some of the major MRSA clones in the United States is shown in **Fig. 2B**. Interpretation of PFGE results may vary based on the application; however, for clinical purposes the criteria by Tenover et al. *(9)* are commonly applied.

4. Notes

1. All buffers made throughout the procedure are prepared with ultrapurified distilled water.

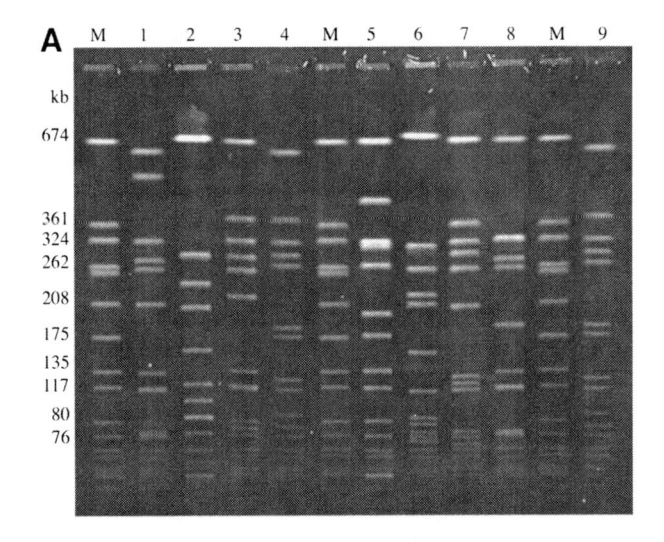

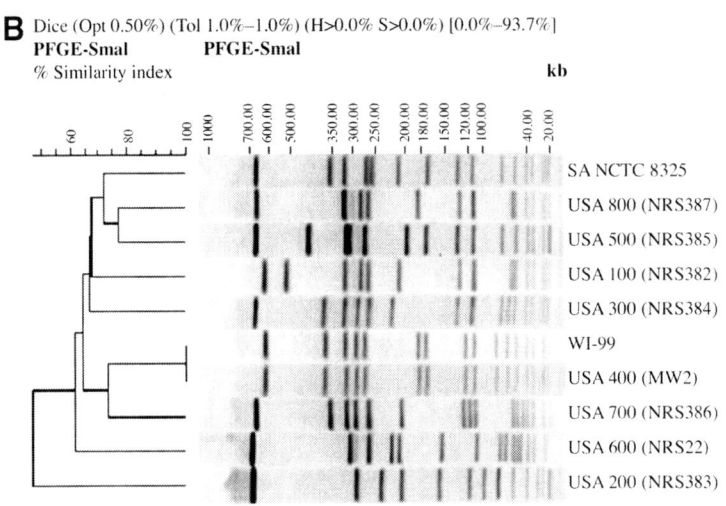

Fig. 2. (**A**) Gel image of *Sma*I restricted genomic DNA from representative strains of major MRSA clones (USA100 to USA800) from the United States. Lane M, *S. aureus* NCTC 8325 reference strain; lane 1, NRS382; lane 2, NRS383; lane 3, NRS384; lane 4, MW2; lane 5, NRS385; lane 6, NRS22; lane 7, NRS386; lane 8, NRS387; lane 9, WI-99. The image shown is an xxx.tif file obtained from the GelDoc 2000. (**B**) Example of dendrogram based on percentage of genetic relatedness of MRSA strains. The horizontal bar above the dendrogram shows the percentage of genetic relatedness among different strains. The numbers (1000 to 20 kb) above the PFGE bands denote the sizes of the DNA marker used to standardize the image. Strain designations are given in the far right column. Dendrograms were created by means of the Dice coefficient and UPGMA. The position tolerance identifying similar sized fragments between two strains was set at 1.0%.

2. Ethidium bromide is a carcinogen. Wear gloves while performing all steps involving staining and imaging of the gel.
3. Uniform cell suspensions are necessary to standardize the amount of DNA in the plugs. In the absence of a Vitek colorimeter, resuspend cells to approx 3.0 McFarland.
4. The plug mold should be washed with mild detergent, rinsed with distilled water followed by 70% alcohol, and allowed to dry before use.
5. Isolation of intact DNA is essential. Minimize shearing of the DNA by using 1000-μL pipet tips and avoid excessive pipet mixing.
6. For proper normalization of the gel, a global reference standard should be run in approximately every fifth lane of the gel. The molecular weights of the *Sma*I restriction fragments of *S. aureus* NCTC 8325 are shown in **Fig. 2A**. (The first 11 bands are 674, 361, 324, 262, 208, 175, 135, 117, 80, and 76 kb in size.)
7. It is important to consistently cut the same size plugs to standardize the amount of DNA loaded on the gel. Placing a thin plastic ruler under the Petri dish serves as a good gauge.
8. Placement of the plug on the front wall of the well ensures that DNA migration begins from a consistent starting position.
9. Prechill the buffer in the refrigerator for approx 2 h prior to electrophoresis. This will minimize the time it takes for the chiller to cool the buffer to the appropriate start temperature.
10. A common problem in the initial setup of the CHEF system occurs if the chiller is turned on and allowed to run a few minutes before the pump is turned on. This causes the buffer in the chiller to freeze enough that it prohibits any circulation once the pump is turned on. If this occurs, turn the chiller off and allow the buffer to thaw prior to restarting the run.

Acknowledgments

We gratefully acknowledge Sharon Brock and Michelle Wellsandt for help in preparing the manuscript and Tamara Kronenwetter-Koepel for preparation of **Fig. 1**.

References

1. Bannerman, T. L., Hancock, G. A., Tenover, F. C., and Miller, J. M. (1995) Pulsed-field gel electrophoresis as a replacement for bacteriophage typing of *Staphylococcus aureus*. *J. Clin. Microbiol.* **33**, 551–555.
2. Maslow, J. N., Mulligan, M. E., and Arbeit, R. D. (1993) Molecular epidemiology: application of contemporary techniques to the typing of microorganisms. *Clin. Infect. Dis.* **17**, 153–162.
3. Reed, K. D., Stemper, M. E., Vandermause, M. F., and Mitchell, P. D. (1993) Evaluation of a commercial DNA purification system for plasmid analysis of nosocomial bacterial pathogens. *Am. J. Clin. Pathol.* **100**, 304–307.
4. Tenover, F. C., Arbeit, R. D., Archer, G. (1994) Comparison of traditional and molecular methods of typing isolates of *Stalphylococcus aureus*. *J. Clin. Microbiol.* **32**, 407–415.

5. Matushek, M. G., Bonten, M. J., and Hayden, M. K. (1996) Rapid preparation of bacterial DNA for pulsed-field gel electrophoresis. *J. Clin. Microbiol.* **10,** 2598–2600.
6. Goering, R. V. and Winters, M. A. (1992) Rapid method for the epidemiological evaluation of gram-positive cocci by field inversion gel electrophoresis. *J. Clin. Microbiol.* **30,** 577–580.
7. McDougal, L. K., Steward, C. D., Killgore, G. E., Chaitram, J. M., McAllister, S. K., and Tenover, F. C. (2003) Pulsed-field gel electrophoresis typing of oxacillin-resistant *Staphylococcus aureus* isolates from the United States: establishing a national database. *J. Clin. Microbiol.* **41,** 5113–5120.
8. Stemper, M. E., Shukla, S. K., and Reed, K. D. (2004) Emergence and spread of community-associated methicillin-resistant *Staphylococcus aureus* in rural Wisconsin, 1989–1999. *J. Clin. Microbiol.* **42,** 5673–5680.
9. Tenover, F. C., Arbeit, R. D., Goering, R. V., et al. (1995) Interpreting chromosomal DNA restriction patterns produced by pulsed-field gel electrophoresis: criteria for bacterial strain typing. *J. Clin. Microbiol.* **33,** 2233–2239.
10. Applied Maths. (2000) BioNumerics software v 2.5 Manual: the integral study of biological relationships. Applied Maths, Belgium. www.applied-maths.com.
11. Enright, M. C., Day, N., Davies, C. E., Peacock, S. J., and Spratt, B. G. (2000) Multilocus sequence typing for characterization of methicillin-resistant and methicillin-susceptible clones of *Staphylococcus aureus. J. Clin. Microbiol.* **38,** 1008–1015.
12. Koreen, L., Ramaswamy, S. V., Graviss, E. A., Naidich, S., Musser, J. M., and Kreiswirth, B. N. (2004) *spa* typing method for discriminating among *Staphylococcus aureus* isolates: implications for use of a single marker to detect genetic micro- and macrovariation. *J. Clin. Microbiol.* **42,** 792–799.

6

Multilocus Sequence Typing (MLST) of *Staphylococcus aureus*

Nicholas A. Saunders and Anne Holmes

Summary

Multilocus sequence typing (MLST) is a widely accepted method of DNA sequence-based typing that relies on analysis of relatively conserved genes that encode essential proteins. For *Staphylococcus aureus*, the level of discrimination provided by MLST is sufficient to provide a relatively detailed picture of the global dissemination of the organism. The technique is not restrictive in the precise methodology used to acquire the sequences, but the method of assigning types requires that the data be of high quality. Excellent Web-based tools have been developed and are curated by the groups that launched MLST. These tools have allowed the scheme to be maintained as a coherent global asset and assist users in the analysis of their data.

Key Words: *Staphylococcus aureus*; multilocus sequence typing; polymerase chain reaction; sequencing; epidemiology; eBURST.

1. Introduction

Multilocus sequence typing (MLST) is a nucleotide sequence-based method for characterizing, subtyping, and classifying members of bacterial populations *(1,2)*. It is a modification of multilocus enzyme electrophoresis in which, instead of comparing the electrophoretic mobilities of housekeeping enzymes, allelic variation is determined by sequencing internal fragments of the encoding genes. Housekeeping genes are used because they are essential to cell function and thus are present in every organism, and sequence variations evolve slowly and are likely to be selectively neutral.

In 2000, an MLST scheme was developed and validated for *Staphylococcus aureus (1)*. It involves polymerase chain reaction (PCR) amplification and sequencing of internal fragments (~450 bp) of seven housekeeping genes (*arcC*, carbamate kinase; *aroE*, shikimate dehydrogenase; *glp*, glycerol kinase, *gmk*, guanylate kinase; *pta*, phosphate acetyltransferase; *tpi*, triosephosphate isomerase; and *yqiL*, acetyl coenzyme A acetyltransferase). These genes were selected because they provided the greatest number of alleles of the 14 genes investigated and enabled adequate resolution to characterize the genetic diversity of the *S. aureus* population. For each of the loci, different sequences are assigned arbitrary allele numbers, and the seven assigned numbers form the allelic profile, or the sequence type (ST). Sequences with a single nucleotide difference are considered distinct, and no weighting is applied to reflect the number of nucleotide differences between alleles.

A major strength of MLST is that unambiguous, portable data are generated that can easily be compared among laboratories. With the help of a Web-based database *(3)*, the technique has been extremely useful for global epidemiology, and in conjunction with SCC*mec* typing *(4,5)* it has provided a common international nomenclature for *S. aureus* strains. Furthermore, the method has been invaluable for providing insight into the origin and evolution of *S. aureus (6–10)*.

1.1. Analysis of MLST Data

On completion of MLST, the result is an ST (e.g., ST36) that is underlain by the allelic profile for the seven loci (i.e., 2, 2, 2, 2, 3, 3, 2 for ST36). Although this is clearly useful for typing, relationships between different STs are not immediately apparent from these notations. Useful information about the relationships between STs is nevertheless clearly available within the MLST data. Consequently, methods have been developed to process the raw data so that underlying relationships between strains are conveniently displayed.

Two methods have been exploited for the purpose of showing relationships between STs. Legitimate arguments in favor of either method are available. The first method uses all of the sequence data and gives weight to each nucleotide difference. The allele sequences for all seven loci of each strain are compiled in series, and the resulting concatamers are then compared pairwise using a suitable sequence analysis tool to estimate the number of nucleotide changes required for a common ancestor to have evolved to give both descendants. These values are then used to construct phylogenetic trees that show the relationships between strains. The main advantage of this approach is that it will still show a close relationship between two strains that differ by just three nucleotides even when each change occurs at a different locus (i.e., triple-locus variants, or TLVs). The main disadvantage of the sequence concatenation approach for analysis of MLST data is that it does not take into account recombination events. The homologous recombination events known to shuffle MLST loci between strains in *S. aureus* might be expected to cause the relationships estimated using concatenated sequences to be inaccurate *(11)*. For example, a single recombination that replaced an allele with another that had multiple nucleotide differences would change the topology of the phylogenetic tree and prevent detection of the close relationship between the two strains. To avoid this artefact the second method of analysis, which take into account only the number of alleles shared, has been developed *(11)*. One approach available through the MLST Web site *(3)* is to select STs that have a minimum degree of similarity in their allelic profiles to a query strain. A dendrogram (tree) based on the pairwise differences between profiles is then drawn. However, it can be argued that trees are a poor representation of the way in which bacterial lineages emerge and diversify. In the BURST method *(11)*, relationships are presented in a way that is arguably more appropriate. The method assumes that selection of strains proceeds to the emergence of a genotype that is present within the population at relative high frequency. This strain is termed a *founding type* and might achieve relative dominance owing to, e.g., the acquisition of improved colonization potential or antibiotic resistance. Once established, the founding type will diversify over time as a result of the accumulation of mutations and by the occurrence of recombination events. BURST diagrams attempt to represent founding types and their progeny, as shown in **Fig. 1**.

1.2. Need for Accuracy and Quality Control

Day et al. *(12)* have illustrated the need for very high standards of accuracy in sequencing for MLST. They studied the link between virulence and ecological abundance of *S. aureus* by genotyping (MLST) strains from nasal carriage and episodes of severe disease within a defined population. The data appeared to show that the most frequently carried genotypes were disproportionately common as causes of disease. In addition, recombination appeared to be a more frequent cause of the diversification of clonal complexes than mutation. However, in a later retraction that was possible as a result of careful work by the same investigators *(13)*, it was reported that the apparent virulent subgroup of the earlier study was the result of sequencing errors involving approx 0.1% of nucleotides. Unfortunately, errors had occurred disproportionately within the nasal carriage strains, leading to the unsupported conclusion. The reanalysis also showed that mutation was a more common cause of the observed variation within clonal complexes. Clearly, a relatively low rate of sequencing error can be amplified by the process of allele and ST assignment. For this reason, only sequencing chromatograms of good quality are acceptable for MLST.

1.3. Worldwide Coordination of S. aureus *MLST Scheme*

The value of the *S. aureus* MLST scheme is greatly enhanced by the creation and maintenance of a Web site (http://saureus.mlst.net/) that provides a central point for the rapid designation of new alleles and STs *(3)*. The Web site also collects MLST and additional relevant data (including clinical and drug resistance

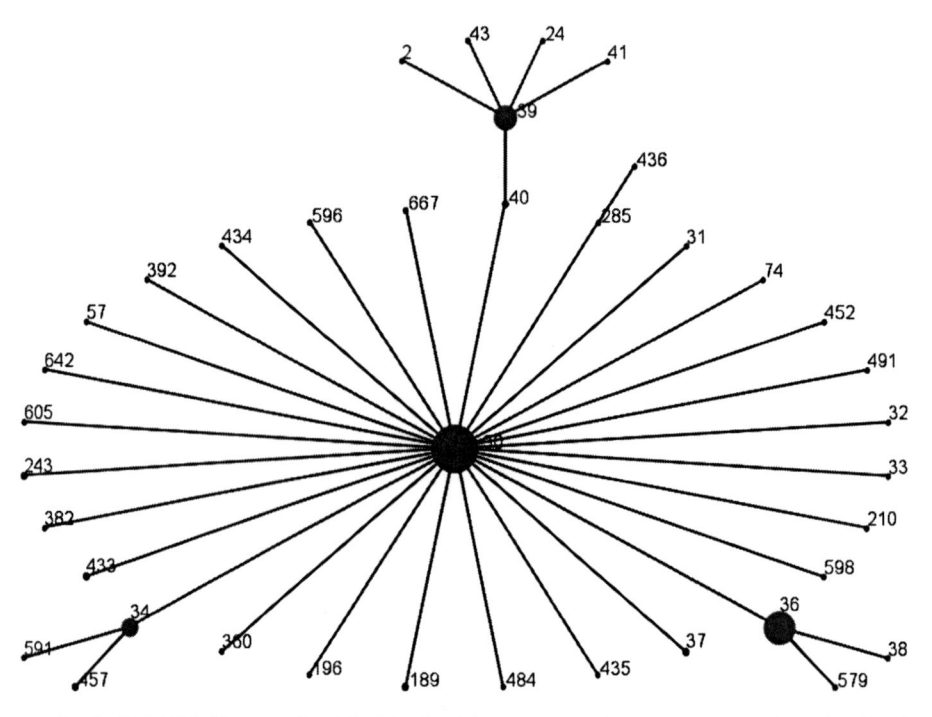

Fig. 1. BURST diagram for 39 STs. In this representation the primary founder ST30 (large black circle) has 28 SLVs, 5 double-locus variants (DLVs), and 4 TLVs. ST39, ST36, and ST34 are subgroup founders (gray circles). This group was found in an analysis of all STs. Six identical loci was the minimum number required to define a group.

data) on strains isolated worldwide. The site is hosted at Imperial College and development is funded by the Wellcome Trust. The *S. aureus* site is one of a growing number of species-specific sites accessed through the gateway www.mlst.net/ *(3)*. The species-specific sites use a common set of tools and procedures to facilitate access to and control of the databases. To ensure the quality of the data, the databases are curated manually. The curator assigns new allele numbers and maintains the allele database. When a potential new allele is identified, the user is prompted to check the sequencing plot at nucleotide sites that differ from the most similar alleles in the database. This is facilitated by use of the Jalview *(3)* alignment editor and tools provided on the Web site. The sequence traces are then submitted to the curator for a final quality check, before the new allele is numbered and included in the allele database. ST designation of a strain proceeds when all seven alleles have been identified. The allelic profile can be checked against the database using a query tool to determine whether it is identical or similar to that of any strains already in the database. The curator assigns new ST numbers.

1.4. MLST Direct from Specimens

A logical development of MLST is to modify the MLST PCRs by making them nested or hemi-nested. As expected, this increases the sensitivity of the PCRs so that sufficient sequencing template can be obtained by direct amplification of bacterial DNA within either clinical or environmental specimens. Thus, it becomes unnecessary to culture the bacterium and extract DNA. Modifications of the *Streptococcus pneumoniae (14)* and *Neisseria meningitidis (15)* schemes have been described that are applicable to direct typing of these organisms in cerebrospinal fluid. It is likely that equivalent modifications to the *S. aureus* MLST protocol would allow typing direct from swabs or clinical samples. However, the circumstances in which this would be required appear to be rather limited.

1.5. Resequencing Arrays for MLST

MLST has many advantages, but one of the main barriers to acceptance has been that it is necessary to obtain good quality sequence data on all seven alleles so that each ST equates to 14 sequence reads. In many laboratories, this is an expensive and time-consuming process. However, it can now be reasonably argued that sequencing is a largely automated process and that the materials costs have been greatly reduced by the introduction of a new generation of parsimonious capillary sequencers. Arrays for resequencing the seven alleles have nevertheless been developed *(16)*.

The array developed by van Leeuwen et al. *(16)* uses the Affymetrix platform. Each nucleotide position is interrogated using four oligonucleotide probes that are identical except for the central nucleotide. One probe in each set of four matches a sequence in at least one MLST allele, with the other three mismatching at the central base. For *S. aureus* MLST *(16)* only four probes are used for each position, rather than the five (includes a base deletion probe) usual for resequencing. This is reasonable because single-base deletions have not been reported in these essential genes. When the fluorescently labeled target sequence is bound to the four probes, the perfectly matched probe is expected to give the strongest hybridization signal. van Leeuwen et al. *(16)* reported encouraging levels of base call accuracy ranging from 98.7 to 99.6% for different centers and strain groups using the best available target-labeling protocol. Further improvement in accuracy will be required to meet the exacting standards needed for MLST. A clear advantage of using the array is that the seven genes may be amplified in a single multiplex PCR, rather than the seven reactions advised for the standard sequencing method.

Resequencing array devices are currently considerably more expensive than the capillary sequencing method but have been developed on the basis that they may be cost-effective as the price of array fabrication falls, as it is likely to do.

2. Materials

2.1. Culture and Sample Preparation

1. Nutrient agar plates.
2. Enzymatic lysis buffer: 20 mM Tris-HCl (pH 8.0), 2 mM EDTA, 1.2% Triton X-100, 30 µg/mL of lysostaphin (Sigma, Poole, Dorset, UK), 300 µg/mL of lysozyme (Sigma) (see **Note 1**).
3. DNeasy Tissue Kit (Qiagen, Crawley, UK).
4. 99.7–100% Ethanol (Sigma): Prepare 95% (v/v) and 70% (v/v) ethanol/water (see **Note 2**).

2.2. Polymerase Chain Reaction

1. Primers (see **Table 1**; MWG-Biotech, Milton Keynes, UK): Prepare 10 µM of each primer pair.
2. *Taq* DNA Polymerase (5 U/µL; Invitrogen, Paisley, UK) supplied with 10X PCR buffer and 50 mM MgCl$_2$.
3. 100 mM dNTP set (Invitrogen): Prepare 10 mM dNTP master mix.
4. Thin-walled 96-well plate (0.2 mL) (Abgene, Epsom, Surrey, UK).
5. Heat-sealing foil (Abgene).
6. Thermal cycler with heated lid (e.g., Eppendorf MasterCycler).
7. Molecular biology agarose.
8. 10X TBE buffer: Dilute to 1X TBE using distilled water.

2.3. PCR Cleanup

1. MultiScreen PCR$_{96}$ filter plates (Millipore [UK], Watford, UK).
2. MultiScreen™ Vacuum Manifold 96-well (Millipore).
3. Vacuum pump (220 V/50 Hz).
4. Plate-sealing tape.
5. Plate shaker.

2.4. Sequencing Reaction Primers and Materials

1. 3.2 µM of each primer (see **Table 1**).
2. CEQ Dye Terminator Cycle Sequencing (DTCS) Quick Start Kit (Beckman Coulter, High Wycombe, UK).
3. Half CEQ (Genetix, New Milton, Hampshire, UK).
4. Sterile, thin-walled thermal cycling plates (Beckman Coulter).
5. Thermal cycler with heated lid.

2.5. Sequencing Reaction Reagents and Equipment

1. 3 M sodium acetate (pH 5.2) (Sigma).
2. 0.5 M Na$_2$-EDTA (Sigma): Prepare 100 mM Na$_2$-EDTA (pH 8.0).
3. Glycogen (20 mg/mL; supplied with DTCS Quick Start Kit).
4. Refrigerated microplate centrifuge, e.g., Allegra™ X-22R (Beckman Coulter).
5. 99.7–100% Ethanol (Sigma): Prepare 95 and 70% (v/v) ethanol/water (see **Note 3**).

Multilocus Sequence Typing of S. aureus

Table 1
PCR and Sequencing Primers

Primer	Sequence 5'–3'	Length (bp)
arcC-Up	TTG ATT CAC CAG CGC GTA TTG TC	456
arcC-Dn	AGG TAT CTG CTT CAA TCA GCG	
aroE-Up	ATC GGA AAT CCT ATT TCA CAT TC	456
aroE-Dn	GGT GTT GTA TTA ATA ACG ATA TC	
glpF-Up	CTA GGA ACT GCA ATC TTA ATC	465
glpF-Dn	TGG TAA AAT CGC ATG TCC AAT TC	
gmk-Up	ATC GTT TTA TCG GGA CCA TC	429
gmk-Dn	TCA TTA ACT ACA ACG TAA TCG TA	
Pta-Up	GTT AAA ATC GTA TTA CCT GAA GG	474
Pta-Dn	GAC CCT TTT GTT GAA AAG CTT AA	
tpi-Up	TCG TTC ATT CTG AAC GTC GTG AA	402
tpi-Dn	TTT GCA CCT TCT AAC AAT TGT AC	
yqiL-Up	CAG CAT ACA GGA CAC CTA TTG GC	516
yqiL-Dn	CGT TGA GGA ATC GAT ACT GGA AC	

6. Sample Loading Solution (SLS) (supplied with DTCS Quick Start Kit) (*see* **Note 4**).
7. Mineral oil (supplied with DTCS Quick Start Kit).

2.6. Sequencing

1. CEQ™ 8000 Genetic Analysis System (Beckman Coulter).
2. CEQ Capillary Array (Beckman Coulter).
3. CEQ Separation Gel LPA-1 (Beckman Coulter).
4. CEQ Separation Buffer (Beckman Coulter).
5. Buffer Microtiter Plates (Beckman Coulter).

3. Methods

3.1. Culture and Isolation of Genomic DNA (DNeasy Tissue Kit Handbook)

For MLST, various methods may be used to extract *S. aureus* DNA. If relatively small sample numbers are being processed, a single-tube method such as that described next is suitable. However, a microtiter plate or automated approach (e.g., Magnapure; Roche) is more appropriate for large sample numbers. Regardless of which method is used, a prelysis step using lysostaphin is essential.

1. Subculture bacteria onto nutrient agar using a sterile disposable loop (*see* **Note 5**).
2. Incubate at 37°C overnight.
3. Using a sterile loop, pick approx 5–10 bacterial colonies and suspend in 180 μL of enzymatic lysis buffer.
4. Incubate at 37°C for at least 30 min to lyse the bacterial cells.
5. Add 25 μL of proteinase K and 200 μL of Buffer AL. Mix by vortexing.

Table 2
Preparation of PCR Master Mix

Reagent	For 1 test (μL)	For 48 tests (μL)	For 96 tests (μL)
Buffer (10X)	5	240	480
MgCl (50 mM)	1.5	72	144
dNTP (10 mM)	1	48	96
Primer mix (10 μM)	1	48	96
Taq polymerase (5 U/μL)	0.2	9.6	19.2
H$_2$O	39.3	1886.4	3772.8

6. Incubate at 70°C for 30 min.
7. Add 200 μL of 100% ethanol to the sample and mix thoroughly by vortexing.
8. Pipet the mixture into a DNeasy mini column placed in a 2-mL collection tube. Centrifuge at 6000g for 1 min. Discard the flow-through and the collection tube.
9. Place the DNeasy mini column into a new 2-mL collection tube, add 500 μL of buffer AW1, and centrifuge at 6000g for 1 min. Discard the flow-through and the collection tube.
10. Place the DNeasy mini column into a new 2-mL collection tube, add 500 μL of buffer AW2, and centrifuge at full speed (i.e., 14,000g) for 3 min. Discard the flow-through and the collection tube.
11. Place the DNeasy mini column into a clean 1.5-mL microfuge tube and add 200 μL of Buffer AE directly onto the membrane to elute the DNA. Incubate at room temperature for 1 min and then centrifuge at 6000g for 1 min.
12. Store the DNA at –20°C until ready for use.

3.2. PCR Amplification of MLST Genes

1. Thaw PCR reagents and prepare seven master mixes with the primer pairs by combining the reagents in **Table 2**. Include a negative PCR control, which consists of the reaction components and no added template DNA.
2. Aliquot 48 μL of master mix into each well of a microtiter plate.
3. Add 2 μL of 1:10 (~40 ng/μL) DNA to the wells.
4. Seal the plate with foil and load onto a PCR machine.
5. Perform PCR amplification by running the following cycling program: 94°C for 5 min; 94°C for 30 s, 55°C for 30 s, and 72°C for 30 s for 35 cycles; 72°C for 10 min and a 4°C hold.
6. Run the products on a 1.5% agarose gel to confirm the presence of the desired product (*see* **Note 6**).

3.3. Purification of PCR Products

Because seven PCR reactions are prepared for each test sample, a high-throughput method is most appropriate, especially for large sample numbers. The method described next uses filter plates, which are available in micro96-,

Table 3
Preparation of Sequencing Reaction

Master mix reagent	Full reaction (μL)	Half reaction (*see* **Note 9**) (μL)
Sequencing primer (3.2 μ*M*)	3	3
DTCS Quick Start Master Mix	8	8
Genetix halfCEQ (*see* **Note 9**)	—	4
Sterile water (adjust total volume to 20 μL depending on DNA volume)	8	8

96-, and 384- well formats. They are cheap, quick and easy to use, and compatible with liquid handling systems. Alternatively an enzymatic method may be used (e.g., ExoSap-IT; Amersham BioSciences, Chalfont, UK) or spin columns (e.g., QIAquick PCR purification kit; Qiagen).

1. Make PCR reactions up to 100 μL and load into the wells of a Multiscreen PCR$_{96}$ filter plate (*see* **Note 7**). Cover the wells that are not in use with plate-sealing tape.
2. Place the Multiscreen PCR$_{96}$ filter plate on top of the vacuum manifold.
3. Apply the vacuum at 20 in. of Hg for 7–12 min, or until the wells are empty. Allow 1 min extra under vacuum after the wells appear empty, to be sure all liquid has filtered. Filters appear shiny when they are dry.
4. Remove the plate and blot from underneath with an absorbent material.
5. Add 40 μL of water to the wells, cover the wells, and place on a plate shaker for 20 min at 1100 rpm.
6. Retrieve the purified PCR products from each well by pipetting and place in a microtube.
7. Store the purified DNA at –20°C until use.

3.4. Sequencing Reaction

Quantify purified PCR products on a 1.5% agarose gel using a DNA mass ladder (e.g., BioLine Hyperladder IV).

The following protocol is for use on a Beckman capillary sequencer. This machine gives excellent results in our hands. Other alternative sequencing platforms can be used (*see* **Note 8**).

1. Prepare a master mix using a single MLST primer as shown in **Table 3**, setting up both a forward primer and a reverse primer reaction for each PCR reaction.
2. Aliquot 19 μL of the master mix into each well of a 96-well plate and add 1 μL of ~30 ng/μL purified PCR products.
3. Cover the wells with plastic strip caps and amplify on an Eppendorf thermocycler using the following conditions: 96°C for 20 s, 50°C for 20 s, and 60°C for 4 min for 30 cycles; and a 4°C hold.

3.5. Electrophoresis (Beckman Platform)

1. Spin a plate at 4°C for 1 min at 1000 rpm in a centrifuge with a microtiter plate rotor.
2. Prepare the stop solution (2 μL of 3.0 M sodium acetate, pH 5.2; 2 μL of 0.1 M EDTA, pH 8.0; 1 μL of 20 mg/mL glycogen), and add 5 μL to each well containing sequencing reaction.
3. Add 60 μL of 95% (v/v) cold ethanol and vortex.
4. Centrifuge the plate in a refrigerated centrifuge with a microplate rotor (e.g., Beckman Coulter, Allegra X-22R Centrifuge) at 3000 rpm for 30 min at 4°C.
5. A pellet should now be seen in each of the wells. Carefully remove the supernatant and rinse the pellets two times with 200 μL of 70% (v/v) cold ethanol. For each rinse, centrifuge immediately at 3000 rpm for 5 min and carefully remove the supernatant by inverting and gently shaking three times. After the last spin, place the inverted plate onto a blue paper towel—*do not turn the plate upright*—and spin at 300 rpm for 15 s.
6. Leave the plate for 30 min to air-dry before resuspending the pellets in 40 μL of SLS (*see* **Note 10**).
7. Add 1 drop of mineral oil per well.
8. Make up a buffer plate in a flat-bottomed 96-well plate by filling the same number of wells as the sample plate with CEQ Separation Buffer.
9. Load both plates onto the CEQ 8000 Genetic Analysis System.
10. Run the sample plate using preprogrammed conditions designed for sequencing PCR products.

3.6. Assessment of Data Quality

The first step in the analysis is to assess the quality of the raw and analyzed data. The signal from all four dyes should be sufficient to give high signal-to-noise ratios. As a general guide, it is best to reject sequences that do not achieve an average Phred score (see "Phred—Quality Base Calling" at www.phrap.com/phred/) of at least 20 for every 20 contiguous bases throughout the sequence required for MLST. This means that the minimum base call accuracy of any subsequence will be >99%.

The next step is to align and compare the forward and reverse sequences. This can be done easily with a convenient sequence alignment editor such as BioEdit *(17)*. BioEdit can also be used to trim the sequences to the bases required for the *S. aureus* MLST scheme. It is useful to import a known allele to make trimming easier. Discrepancies between the forward- and reverse-strand sequence should be resolved by referring to the Phred scores and carefully analyzing the original electrophoretograms.

3.7. Assigning Allele Numbers and STs

Alleles and allelic profiles can be assigned using the tools provided on the MLST Web site. A single- or batch-locus query allows a single sequence or a batch of sequences for the same locus to be compared with all alleles in the database. A multiple-locus query identifies the allelic profile, and a batch-strain query returns the allelic profile for multiple strains. In most cases, the software is expected to return an allele number for each sequence submitted.

If the allele sequence is novel, the sequence should be checked against the closest alleles in the database that are returned by the Web site. This should involve a reexamination of the Phred scores, and if necessary the original data should be checked. If the new allele is confirmed, the forward and reverse trace files are submitted to the curator.

When all seven alleles for a strain or batch of strains are confirmed, the data can be entered into the allelic profile query tool. This query determines whether the strains are identical, or similar, in profile to any strains in the database. If a match is found, the ST number is returned. For strains with no exact match, data on the closest available matches can be displayed.

Local handling of this process can be streamlined using a BioNumerics database. This process is described in **Note 11**.

3.8. Comparison of Strains Using eBURST

The eBURST *(11)* application is integrated into the *S. aureus* MLST site. On starting the application, a number of options are presented. To enter profile data, the input is in the form of tab-delimited text with the data for each strain on a line. The data consist of the ST followed by the allele numbers for the seven MLST loci in the standard scheme order. An example of the correct format is shown here for *S. aureus*:

15	13	13	1	1	12	11	13
16	15	13	1	1	12	11	13
17	2	23	2	2	19	20	2
18	13	15	1	1	12	11	13
18	13	15	1	1	12	11	13
19	2	2	2	2	7	17	19
20	4	9	1	8	1	10	8

New MLST types and alleles that are not yet in the curated database can be given temporary numbers (*see* **Note 12**). Often the input file will consist of the data for the complete MLST database with new types appended.

When the profile data have been entered, it is possible to analyze the data separately or with STs in the database. When the dataset has been defined the eBURST applet is started. A window with profile, analysis, and diagram tabs is opened together with a profile, analysis, or diagram panel depending on which tab is selected. The various options are explained in the eBURSTv2 manual, which is available at http://eburst.mlst.net.

The analysis panel allows selection of the minimum number of identical loci to define a group, the minimum number of single-locus variants (SLVs) to define a group, and the number of bootstrap replicates to perform. Clicking on "compute" returns the groups. Groups can be drawn using the diagram panel. Tools are provided to manipulate eBURST diagrams including reassignment of the group founder. These tools are explained in detail in the eBURSTv2 manual. **Figure 1** presents an eBURST diagram showing a group found in an analysis of the complete *S. aureus* MLST database.

3.9. Comparison of Strains Using Concatenated Allele Sequences

The MLST Web site provides an option to output concatenated allele sequences in FASTA format. The FASTA file can then be input into a suitable tree-drawing program such as those provided in the Phylip suite of programs *(18)*. **Figure 2** presents a typical distance tree.

4. Notes

1. Lysostaphin (1 mg/mL) and lysozyme (10 mg/mL) working stock solutions should be prepared and stored in single-use aliquots at −20°C. Always prepare fresh lysis buffer.
2. Unless stated otherwise, all solutions should be prepared using sterile, molecular-grade water. This standard is referred to as "water" in the text.
3. When using ethanol precipitation for purification of sequencing reactions, always use molecular biology–grade ethanol and store working solutions at −20°C. Ultrapure ethanol that has been treated to remove traces of contaminating water should not be used, because some batches appear to contain quenchers that reduce the fluorescent signal. The quality of the ethanol is essential for obtaining clean sequencing reactions. Avoid the use of old bottles of ethanol, which are likely to be <99% owing to the absorption of atmospheric water. Alternative methods of removing the dye terminators from sequencing reactions prior to electrophoresis are acceptable and may save time. We have found CleanSEQ kits (Agencourt) to be particularly effective and convenient.
4. SLS should not be subjected to repeated freeze/thaw cycles, because this may cause the formamide to break down to ammonia and formic acid, which destroys the fluorescent dyes. SLS should therefore be aliquoted into single-use amounts (~350 µL).
5. It is essential that the initial culture be pure. This should be ensured by picking a single well-defined colony.
6. The same primers are used for amplification and sequencing, so it is important that only a single DNA fragment be amplified. If more than one fragment is observed, the PCR conditions may need to be reoptimized.
7. Replicate reactions may be added to each well to increase the yield, although this should not be necessary under normal circumstances for MLST.

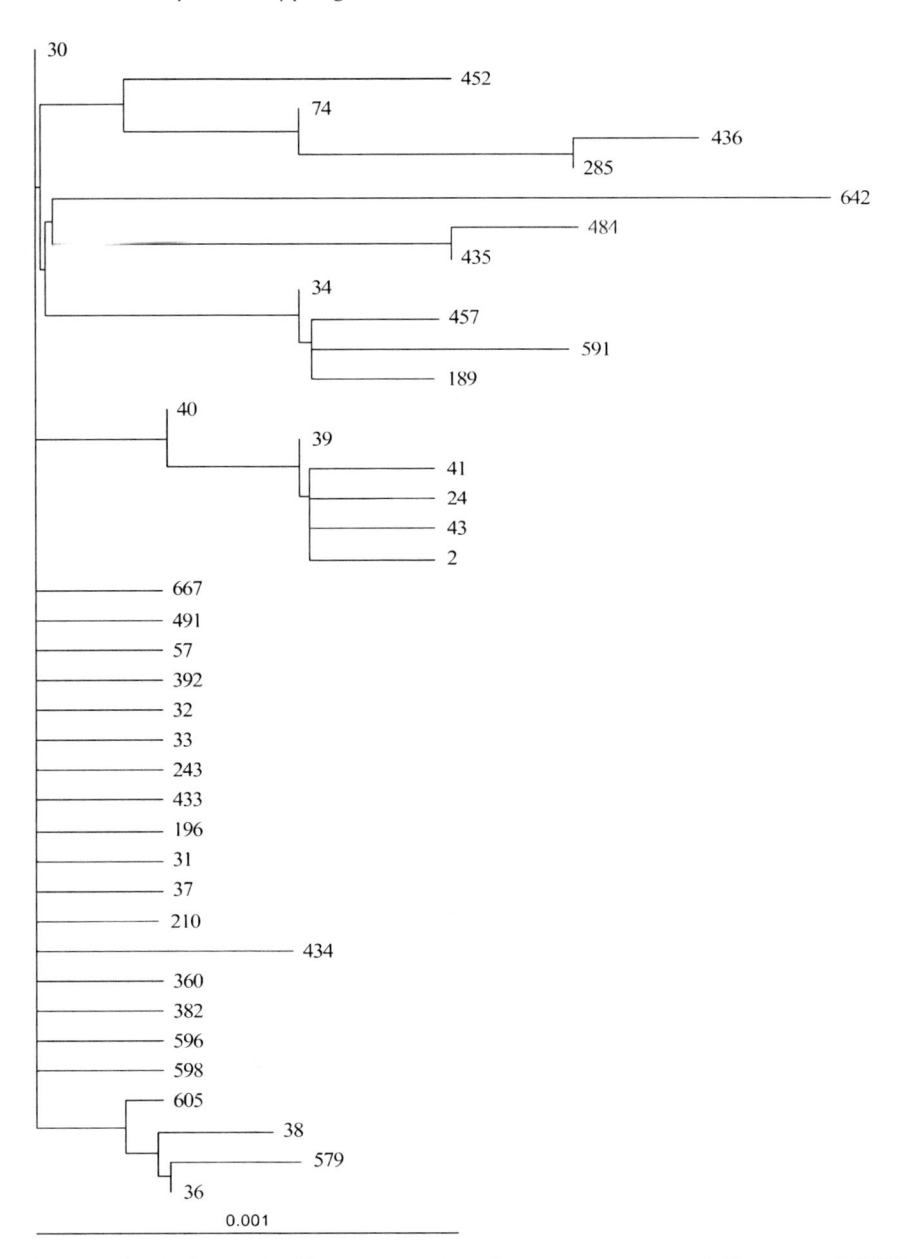

Fig. 2. A phylogenetic distance tree based on concatamers of *S. aureus* MLST sequences is shown. The STs included were those analyzed in the BURST diagram in **Fig. 1**. The tree was calculated using the Phylip package (Felsenstein). Distances were first computed using the DNADIST program (Kimura two-parameter model) and then used to draw a tree using the FITCH program. The treefile was drawn using TreeView v2. The scale bar shows Knuc distances.

8. Alternative sequencing platforms may be used including the Applied Biosystems capillary sequencer(s) or the MegaBACE™ 4000 DNA Analysis System (GE HealthCare).
9. HalfCEQ (Genetix) is a specially formulated sequencing reagent that when mixed in equal volume with DCTS Quick Start mix can reduce sequencing costs without loss of resolution, read length, and accuracy. Similar products are available for use with other commercial sequencing kits. Additionally, it is possible to halve the size of the reaction mixture to reduce costs further. Greater reductions in reaction mixture volume (and consequent increases in economy) are possible with newer, more sensitive, capillary sequencing systems such as the Applied Biosystems 3730.
10. Care should be taken to ensure that the wells are completely dry before adding the SLS. Ethanol may quench the fluorescent signal.
11. To streamline data analysis and reduce errors owing to copying and pasting sequences, a local database may be used to assign alleles and STs. For example, BioNumerics (Applied Maths) is a software platform that enables trace files to be analyzed using a sequence alignment editor (Genebuilder), and the resulting unknown sequences to be assigned allele numbers by using a script to compare them with known sequences stored as a simple format file in BioNumerics. Similarly, the ST of a given strain may be determined once the seven alleles are assigned by using a script to compare the unknown profile with known profiles stored as a space-delimited file. This is possible because the alleles and STs are available for download from the MLST Web site.
12. It is suggested that the user-defined temporary numbers be greater than 1000 for alleles and greater than 10,000 for STs in order to avoid confusion. These numbers should not be used in publications external to the user's laboratory, for obvious reasons.

Acknowledgment

We acknowledge the support of the Health Protection Agency.

References

1. Enright, M. C., Day, N. P., Davies, C. E., Peacock, S. J., and Spratt, B. G. (2000) Multilocus sequence typing for characterization of methicillin-resistant and methicillin-susceptible clones of *Staphylococcus aureus*. *J. Clin. Microbiol.* **38,** 1008–1015.
2. Enright, M. C. and Spratt, B. G. (1999) Multilocus sequence typing. *Trends Microbiol.* **7,** 482–487.
3. Aanensen, D. M. and Spratt, B. G. (2005) The multilocus sequence typing network: mlst.net. *Nucleic Acids Res.* **33,** W728–W733.
4. Ito, T., Katayama, Y., Asada, K., Mori, N., Tsutsumimoto, K., Tiensasitorn, C., and Hiramatsu, K. (2001) Structural comparison of three types of staphylococcal cassette chromosome mec integrated in the chromosome in methicillin-resistant *Staphylococcus aureus*. *Antimicrob. Agents Chemother.* **45,** 1323–1336.

5. Oliveira, D. C. and de Lencastre, H. (2002) Multiplex PCR strategy for rapid identification of structural types and variants of the mec element in methicillin-resistant *Staphylococcus aureus*. *Antimicrob. Agents Chemother.* **46**, 2155–2161.
6. Howe, R. A., Monk, A., Wootton, M., Walsh, T. R., and Enright, M. C. (2004) Vancomycin susceptibility within methicillin-resistant *Staphylococcus aureus* lineages. *Emerg. Infect. Dis.* **10**, 855–857.
7. Feil, E. J. and Enright, M. C. (2004) Analyses of clonality and the evolution of bacterial pathogens. *Curr. Opin. Microbiol.* **7**, 308–313.
8. Enright, M. C., Robinson, D. A., Randle, G., Feil, E. J., Grundmann, H., and Spratt, B. G. (2002) The evolutionary history of methicillin-resistant *Staphylococcus aureus* (MRSA). *Proc. Natl. Acad. Sci. USA* **99**, 7687–7692.
9. Robinson, D. A. and Enright, M. C. (2004) Evolution of *Staphylococcus aureus* by large chromosomal replacements. *J. Bacteriol.* **186**, 1060–1064.
10. Robinson, D. A. and Enright, M. C. (2004) Multilocus sequence typing and the evolution of methicillin-resistant *Staphylococcus aureus*. *Clin. Microbiol. Infect.* **10**, 92–97.
11. Feil, E. J., Li, B. C., Aanensen, D. M., Hanage, W. P., and Spratt, B. G. (2004) eBURST: inferring patterns of evolutionary descent among clusters of related bacterial genotypes from multilocus sequence typing data. *J. Bacteriol.* **186**, 1518–1530.
12. Day, N. P. J., Moore, C. E., Enright, M. C., et al. (2001) A link between virulence and ecological abundance in natural populations of *Staphylococcus aureus*. *Science* **292**, 114–116.
13. Day, N. P. J., Moore, C. E., Enright, M. C., et al. (2002) Retraction of Day et al., Science 292 (5514) 114–116. *Science* **295**, 971.
14. Enright, M. C., Knox, K., Griffiths, D., Crook, D. W., and Spratt, B. G. (2000) Molecular typing of bacteria directly from cerebrospinal fluid. *Eur. J. Clin. Microbiol. Infect. Dis.* **19**, 627–630.
15. Birtles, A., Hardy, K., Gray, S. J., Handford, S., Kaczmarski, E. B., Edwards-Jones, V., and Fox, A. J. (2005) Multilocus sequence typing of *Neisseria meningitidis* directly from clinical samples and application of the method to the investigation of meningococcal disease case clusters. *J. Clin. Microbiol.* **43**, 6007–6014.
16. van Leeuwen, W. B., Jay, C., Snijders, S., et al.(2003) Multilocus sequence typing of *Staphylococcus aureus* with DNA array technology. *J. Clin. Microbiol.* **41**, 3323–3326.
17. Hall, T. A. (1999) BioEdit: a user-friendly biological sequence alignment editor and analysis program for Windows 95/98/NT. *Nucleic Acids Symp. Ser.* **41**, 95–98.
18. Felsenstein, J. (1993) PHYLIP (Phylogeny Inference Package) version 3.5c. Distributed by the author. Department of Genetics, University of Washington, Seattle.

7

Staphylococcal Cassette Chromosome *mec* (SCC*mec*) Analysis of MRSA

Teruyo Ito, Kyoko Kuwahara, and Keiichi Hiramatsu

Summary

Methicillin-susceptible *Staphylococcus aureus* changes to methicillin-resistant *S. aureus* (MRSA) on acquisition of the staphylococcal cassette chromosome *mec* (SCC*mec*). At least five types of SCC*mec* elements have been reported. All the SCC*mec* elements share four common characteristics: (1) the elements carry the *mec* gene complex (*mec*); (2) they carry the *ccr* gene complex (*ccr*); (3) the elements are flanked by characteristic nucleotide sequences, both inverted repeats and direct repeats, at both ends; and (4) the SCC*mec* elements are integrated at the 3′-end of *orfX*. In the two essential components, *mec* and *ccr*, four classes of *mec* and five types of *ccr* have been identified. SCC*mec* elements can be defined by the different combinations of *mec* and *ccr* types. Regions other than *mec* and *ccr* within the element are designated junkyard regions (J regions). Even in the same SCC*mec* type, these regions are not always identical and have therefore been regarded as good targets for subtyping SCC*mec* elements in epidemiological studies. Nucleotide differences in the J1 region and/or the presence of inserted plasmids and transposons, most of which encode resistant determinants integrated in the J2 and J3 regions, can be used to further classify SCC*mec* types. In this chapter, we describe polymerase chain reaction methods to type SCC*mec* elements by first identifying the *mec* and *ccr* type, and subsequently identifying genes in the J regions.

Key Words: *mecA*; *mec* gene complex; *ccr* gene complex; cassette chromosome recombinase; staphylococcal cassette chromosome *mec*.

1. Introduction

The methicillin resistance gene, *mecA*, is located in the large chromosomal region called *mec*DNA, which is absent in the methicillin-susceptible *Staphylococcus aureus* chromosome. The *mec*DNA of preMRSA N315 has been sequenced, and the characteristic structure of *mec*DNA has been elucidated *(1)*. After the *mec* element was shown to be a mobile genetic element, it was designated the staphylococcal cassette chromosome *mec* (SCC*mec*), and the two site-specific recombinases that are responsible for the mobility of the element were designated cassette chromosome recombinase A and B (*ccrA* and *ccrB*, respectively) *(2)*. The SCC*mec* of N315 carries the *mec* gene complex, which is composed of *mecA*; its regulatory genes, *mecI* and *mecR1*; and insertion sequence IS*431*; and the *ccr* gene complex (*ccr*), which is composed of *ccrA* and *ccrB* genes, and flanking sequences. Subsequently, two structurally different SCC*mec* elements were identified in two methicillin-resistant *Staphylococcus aureus* (MRSA) strains: NCTC10442, isolated in England in 1960; and 85/2082, isolated in New Zealand in 1985 *(3)*. These elements carried *ccrA* and *ccrB* genes, which exhibited nucleotide identities of 72.5–86.5% to the respective genes in the SCC*mec* of N315. Therefore, *ccrA* and *ccrB* genes were designated *ccrA1ccrB1* for NCTC10442, *ccrA2ccrB2* for N315, *ccrA3ccrB3* for 85/2082, respectively. NCTC10442 carried class B *mec*, whereas 85/2082 carried class A *mec*. Based on the differences in the combinations of *mec* and *ccr*, we designated three SCC*mec* elements as type I, type II, and type III, respectively. Type I SCC*mec* (in NCTC10442) carried class B *mec* and type 1 *ccr*, type II SCC*mec* (in N315) carried class A *mec* and type 2 *ccr*, and type III SCC*mec* (in 85/2082) carried class A *mec* and type 3 *ccr*.

In 2001, Oliveira et al. *(4)* found *ccrA* and *ccrB* genes that showed rather weak similarity to three previously reported *ccr* genes in a pediatric strain, HDE288, and designated them *ccrA4* and *ccrB4* *(4)*. The SCC*mec* of HDE288, which carried class B *mec* and type 4 *ccr*, was designated type IV. Independently, we have reported another type IV SCC*mec* carried by community-acquired MRSA (CA-MRSA) strains isolated in the University of Chicago Children's Hospital *(5)*. These strains carried class B *mec* and type 2 *ccr*. Since MRSA strains carrying the latter elements are distributed mostly in CA-MRSA strains isolated in the United States or Europe, many researchers have regarded these elements as type IV SCC*mec* elements. At present, a consensus on a naming of these two type IV SCC*mec* has been established. The former was re-designated as type VI. A type V SCC*mec* carrying class C2 *mec* and type 5 *ccr* (*ccrC* and surrounding open reading frames) has been found in the CA-MRSA strain WIS[WGB8318] isolated in Australia *(6)*.

Diversity among the J regions in SCC*mec* elements has also been reported. Oliveira et al. *(4)* identified the type I SCC*mec* carrying an integrated copy of plasmid pUB110 and designated it type IA. Similarly, they identified a type III SCC*mec* that did not carry the integrated copy of plasmid pT181 downstream of *mecA* and called it type IIIA *(4)*. We have reported that SCC*mec* elements could be adequately subtyped further by differences found in the J1 region (formerly called the L-C region) such as IIa, IIb, IVa, IVb, IVc, and IVd *(5,7)*. In 2005, Shore et al. *(8)* reported seven novel variants of type II and type IV

SCC*mec* elements, including a type II SCC*mec* carrying a J1 region similar to the regions of type IVb, or type IV SCC*mec* elements carrying a J2 or J3 region different from previously described type IV SCC*mec* elements. Kwon et al. *(9)* reported a novel SCC*mec* subtype IVg.

As different types of SCC*mec* elements have been discovered, it has become evident that two MRSA clones are different if they carry different SCC*mec* elements, even if they belong to the same multilocus sequence typing type or the same pulsotype. Determination of the type of SCC*mec* element carried by an MRSA clone (SCC*mec* typing) has been an essential aspect of the epidemiology of MRSA since arriving at the consensus that MRSA clones should be defined both by the type of SCC*mec* element and by the type of *S. aureus* chromosome in which the SCC*mec* element is integrated.

Here, we describe the basic polymerase chain reaction (PCR) strategy used to identify *mec* and *ccr* genotypes in order to assign types of SCC*mec* elements as well as to identify the genes located in the J region to assign subtypes of SCC*mec* elements. We have listed the primers used in our laboratory, but other primers have been used successfully as well. Any primers that could identify specific *mec* and *ccr* genes, and genes in the J region, would be suitable.

2. Materials (*see* Note 1)

2.1. Genomic DNA Extraction (see Note 2)

1. Medium for the cultivation of cells: Heart-infusion broth (Eiken Kagaku, Tokyo, Japan) or L-broth: For the latter, dissolve 10 g of Bacto-tryptone (Difco, Detroit, MI), 5 g of Bacto–yeast extract (Difco), and 5 g of NaCl in 1 L of H_2O. Adjust the pH to 7.2–7.4 with NaOH. Sterilize by autoclaving.
2. Achromopeptidase (50,000 U/mL in 10 mM NaCl) (Wako Pure Chemical, Osaka, Japan) or lysostaphin (2 mg/L in 20 mM sodium acetate, pH 4.8) (Sigma, St. Louis, MO).
3. 10% Sodium dodecyl sulfate (SDS): Dissolve 10 g of SDS in 100 mL of H_2O. Sterilize by autoclaving.
4. Proteinase K: 10 mg/mL in diethylpyrocarbonate [DEPC]-H_2O (Wako Pure Chemical).
5. Tris-HCl saturated phenol: Melt phenol in a bottle in a water bath at 65°C. Add an equal volume of 0.5 M Tris HCl (pH 8.0) and 8-hydroxyquinoline to a final concentration of 0.1%. Stir the mixture on a magnetic stirrer for 15 min, and then turn off the stirrer. When the two phases have separated, remove the aqueous (upper) phase. Add an equal volume of 0.5 M Tris HCl (pH 8.0), and mix well with stirring for 15 min. After the two phases have separated, remove the aqueous phase. Repeat the extraction with 0.1 M Tris HCl (pH 8.0) until the pH of the aqueous phase is >7.6. Remove 90% of the aqueous phase, relative to the volume of the phenol phase. Store in a light-protected bottle at 4°C.
6. Chloroform:isoamyl alcohol (24:1). Mix 24 vol of chloroform and 1 vol of isoamyl alcohol.

7. 5 M NaCl (see **Note 1**): Dissolve 292.2 g of NaCl in 800 mL of H_2O. Adjust the volume to 1 L. Sterilize by autoclaving.
8. Ethanol.
9. 80% (v/v) Ethanol.
10. 1 M Tris-HCl (pH 8.0) (see **Note 1**): Dissolve 121.1 g of Tris base in 800 mL of H_2O. Adjust the pH to 8.0 by adding concentrated HCl and then adjust the volume to 1 L. Sterilize by autoclaving.
11. 0.5 M EDTA (pH 8.0) (see **Note 1**): Add 186.1 g of disodium ethylene diamine tetraacetate·$2H_2O$ to 800 mL of H_2O. Stir on a magnetic stirrer. Adjust the pH to 8.0 with NaOH (~20 g of NaOH pellets), and adjust the volume to 1 L. Sterilize by autoclaving.
12. T10E10 (pH 8.0): Mix 1 M Tris-HCl (pH 8.0), 0.5 M EDTA (pH 8.0), and DEPC-H_2O to make a final concentration of 10 mM Tris-HCl and 10 mM EDTA.
13. T10E1 (pH 8.0): Mix 1 M Tris-HCl (pH 8.0), 0.5 M EDTA (pH 8.0), and DEPC-H_2O to make a final concentration of 10 mM Tris-HCl and 1 mM EDTA.
14. DEPC-H_2O: Add DEPC (Sigma) to a final concentration of 0.02% to deionized (MilliQ) water in a bottle and shake vigorously. Keep at room temperature overnight and then sterilize by autoclaving.

2.2. PCR Amplification

1. dNTPs (2.5 mM): Prepare dNTPs as a 10X stock solution of 2.5 mM dATP, 2.5 mM dCTP, 2.5 mM dGTP, and 2.5 mM dTTP.
2. Reaction buffer (10X): Mix 5 M KCl, 1 M Tris-HCl (pH 8.3), 1 M $MgCl_2$, 2% (w/v) gelatin, and DEPC-H_2O to make final concentrations of 500 mM KCl, 100 mM Tris-HCl, 15 mM $MgCl_2$, and 0.01% (w/v) gelatin. Alternatively, a commercially available 10X stock solution can be used.
3. DEPC-H_2O.
4. Primers: Prepare primers at a concentration of 10 pmol/µL. Primers used for the assignment of SCC*mec* element types and their subtypes are given in **Tables 1** and **2**, respectively.
5. *Taq* DNA polymerase: Extaq (Takara Shuzo, Kyoto, Japan).
6. Tris-acetate (TAE) buffer (50X stock solution)* (see **Note 1**): Dissolve 242 g of Tris base in 800 mL of H_2O. Add 57.1 mL of glacial acetic acid and 100 mL of 0.5 M EDTA (pH 8.0). Adjust the volume to 1 L.
7. 0.8% Agarose gel: Melt 0.8 g of agarose in 100 mL of 1X TAE buffer by heating in a microwave oven. Cool the solutions to 50°C, and pour the agarose solution into a minigel mold. After the gel is completely set, carefully remove the comb.
8. Size markers: 1-kb ladder (Invitrogen) and λ*Hin*dIII (Takara Shuzo).
9. Loading buffer.
10. Control DNAs: Ideally, DNAs from characterized MRSA strains in which the entire SCC*mec* region has been sequenced should be used as controls. If no such strains are available, DNAs from MRSA strains, of which SCC*mec* region has been confirmed by amplifying with long-range PCR, are recommended. Strains carrying sequenced SCC*mec* elements of each type are as follows: type I (NCTC10442,

Table 1
Primers Used for Identification of *mec* and *ccr*

Detected gene(s) or gene alleles	Primer name	Nucleotide sequence (5'→3')	Expected sized of products (kb)
mecA	mA1	TGCTATCCACCCTCAAACAGG	0.28
	mA2	AACGTTGTAACCACCCCAAGA	
ccr gene complex			
ccr typing based on *ccrA*			
ccrB	βc	ATTGCCTTGATAATAGCCITTCT*	
ccrA1	α1	AACCTATATCATCAATCAGTACGT	Type 1; 0.7
ccrA2	α2	TAAAGGVATVAATGCACAAAACACT	Type 2; 1
ccrA3	α3	AGCTCAAAAGCAAGCAATAGAAT	Type 3; 1.6
common to three *ccr* genes	βc	ATTGCCTTGATAATAGCCITTCT*	0.56
	αc	ATCTATTTCAAAAATGAACCA	
ccrC	γ1	AGCCCAATTTTGATGGTTATTGA	0.52
	γ2	TGGAGAAGTACTCGTTACAATGT	
ccrA4, ccrB4	α4.1	TACTTACAAGGTCTAGGCTA	1.25
	β4.1	GTAGACATTCTCTACAATCG	
mec gene complex			
mecI-mecR1 (class A)	mI4	CAAGTGAATTGAAACCGCCT	0.64
	mcR5	CAGGGAATGAAAATTATTGGA	
IS1372-*mecA* (class B)	IS5	AACGCCACTCATAACATATGGAA	2
	mA6	TATACCAAACCGACAAC	
IS431*mec* L-*mecA* (class C)	mA2	AACGTTGTAACCACCCCAAGA	2
	IS2	TGAGGTAATTCAGATATTTCGATGT	

*I signify Inosine.

Table 2
Primers Used for Subtyping of SCC*mec*

Detected gene(s) or gene alleles	Primer name	Nucleotide sequence (5'–3')	Expected size of products (kb)
Primers for subtyping based on J1 region			
IIa	2a1	ATGTCAGAGCTTTCTAACTTAGGTCA	0.46
	2a2	TGAAATGAAAGCCGTGCCG	
IIb	2b1	AGCAATTTTTTCTCCTTCTGCTA	0.85
	2b2	TTATTAGATCAAGAGCCAAGTG	
IVa	4a1	TTTGAATGCCCTCCATGAATAAAAT	0.45
	4a2	AGAAAAGATAGAAGTTCGAAAGA	
IVb	4b1	AGTAATTTATCTTTGCGTA	1
	4b2	AGTCATCTTCAATATCGAGAAAGTA	
IVc	4c1	TCTATTCAATCGTTCTCGTATT	0.67
	4c2	TCGTTGTCATTTAATTCTGAACT	
IVd	4d1	TTTGAGAGTCCGTCATTATTTCTT	1.0
	4d2	AGAATGTGGTTATAAAGATAGCTA	
MLEP (previously called MREP)	cR4	GTTCAAGCCCAGAAGCGATGT	Type i; 1.6
	mR5	ATGCTCTTTGTTTTGCAGCA	Type ii; 1.7
	mR6	ATATTCTAGATCATCAATAGTTG	Type iii; 1
Primers for identification of integrated plasmids, and transposons			
ant(4')-1 in SCC*mec*	mD1	TAAGTGCTGAGCGCCTGAGGGAAT	2.02
	ant1	CAGACCAATCAACATGGCACC	
mer operon	merA2	TCTTCACAGCCTGTGCATGTCATGCCT	1.55
	merG	TGATACCGCGAATGAATCAAAGGT	

COL), type II (N315, Mu50, E-MRSA 252), type III (85/2082), type IVa (CA05, MW2, JCSC4744), type IVb (8-6/3P), type IVc (81/108, JCSC4788), type IVd (JCSC4469), type V (WIS[WBG8318]). For detailed information, access http://www.staphylococcus.net.

3. Methods

3.1. DNA Extraction (Small Scale)

1. Inoculate 4 mL of HI-broth or L-broth with a single colony of *S. aureus*. Incubate at 37°C overnight with shaking.
2. Pour 0.8 mL of the culture into an Eppendorf tube (1.5 mL). Centrifuge for 2 min at 20,400 g. Remove as much of the medium as possible using a Pipetman.
3. Resuspend the cell pellet in 400 µL of T10E10.
4. Add 5 µL of lysostaphin solution (2 mg/mL), and place the tube at 37°C for 60 min in a water bath or an incubator until the suspension becomes transparent. If the suspension does not become transparent, add an additional aliquot of lysostaphin and keep the tube at 37°C until it becomes transparent. Add 8 µL of achromopeptidase solution (50,000 U/mL), and place the tube at 55°C for 30 min in a water bath until the suspension becomes transparent.
5. Add 4 µL of proteinase K (10 mg/mL) to a final concentration of 100 µg/mL and 24 µL of 10% SDS. Place the tube in the incubater or the water bath at 37°C for more than 1 h or overnight.
6. Add an equal volume of Tris-Cl saturated phenol. Mix well by inverting the tube several times. Separate the phases by centrifuging at 20,400 g for 3–5 min at room temperature, and transfer the aqueous phase to a fresh tube.
7. Extract the aqueous phase twice with an equal volume of chloroform:isoamyl alcohol.
8. Transfer the aqueous phase to a fresh tube. Add 5 M NaCl to a final concentration of 0.2 M and 2 vol of 100% ethanol. Mix thoroughly.
9. Centrifuge at 20,400 g for 1 min and discard the supernatant. The chromosomal DNA sediments to the bottom of the tube.
10. Discard the supernatant. Wash the DNA pellet with 80% ethanol twice. Air-dry the pellet for 20–30 min, and redissolve the pellet of chromosomal DNA in 50 µL of T10E1.
11. Estimate the concentration of DNA using an ND-1000 UV/Vis spectrophotometer (NanoDrop Technologies, Wilmington, DE) or estimate roughly by running an agarose gel.

3.2. PCR Amplification

1. Prepare the premixture and aliquot to each tube. The premixture should contain the following components in a volume of 49 µL/aliquot: 5 µL of 10X reaction buffer, 4 µL of dNTPs, two to four oligonucleotide primers (1 to 2 µL), 1 U of *Taq* DNA polymerase, and DEPC-H$_2$O to adjust the final volume to 49 µL.

2. Add 1 µL of template DNA to each tube and subject to PCR. The PCR conditions are as follows: denaturation (94°C, 1 min), 30 cycles of denaturation (94°C, 1 min), annealing (50°C, 1 min), and extension (72°C, 2 min). Keep samples at 4°C following PCR.
3. Mount the agarose gel in the electrophoresis tank, and add enough 1X TAE buffer to cover the gel.
4. Mix 4 µL of sample from each PCR reaction with 1 µL of loading buffer, and load the samples into the wells of the gel. The gel is usually run at high voltage (100 V). Stop running at the appropriate time. Usually, we stop the procedure when the bromophenol blue has run two-thirds of the gel length.
5. Stain the DNA fragments in the agarose gel by soaking in 0.01% ethidium bromide solution for 20 min.
6. Take a photograph using transmitted UV light with Fas II (UV sample camera; Toyobo, Tokyo, Japan).

3.3. Analysis of Products of PCR Experiments

3.3.1. Assignment of mec

Figure 1 illustrates the structure of five types of SCC*mec* elements, their subtypes based on differences in the J1 region, and the location of the representative sets of primers used for PCR amplification. **Table 1** provides the primers used for the identification of *mec* and *ccr* and the expected sizes of the DNA fragments amplified by those sets of primers. It is important to estimate the sizes of amplified DNA fragments, and to confirm that DNA fragments of the appropriate sizes were generated.

Strains are first examined for the presence of the *mecA* gene to confirm that the strain is an MRSA strain (**Fig. 2A**). The four classes of *mec* gene complex are as follows: class A, IS*431*-*mecA*-*mecR1*-*mecI*; class B, IS*431*-*mecA*-Δ*mecR1*-IS*1272*; class C, IS*431*-*mecA*-Δ*mecR1*-IS*431*; and class D, IS*431*-*mecA*-Δ*mecR1* (10). Since the insertion sequence IS*431* was located downstream of *mecA* in all strains tested thus far, we omitted the identification of IS*431*. Class A *mec* strains were identified by PCR amplification of a DNA fragment using one primer in the *mecI* gene and the other in the *mecR1* gene (**Fig. 2B**). Similarly, the strain carried class B *mec* if a DNA fragment could be amplified from one primer in the *mecA* gene and the other in IS*1272*. PCR identified class C *mec* if a DNA fragment was amplified with one primer in IS*431* and the other in *mecA*. The class C *mec* gene complex was further classified into C1 and C2 by differences in the location of IS*431*.

It is important to confirm that the IS*1272* element is located in the vicinity of *mec*, since IS*1272* is sometimes distributed in MRSA strains. The identification of IS*1272* itself did not mean that the strain carried class B *mec*. When IS*1272* and IS*431* were not found by PCR, including long-range PCR (*see* **Note 3**), the strain was designated class D.

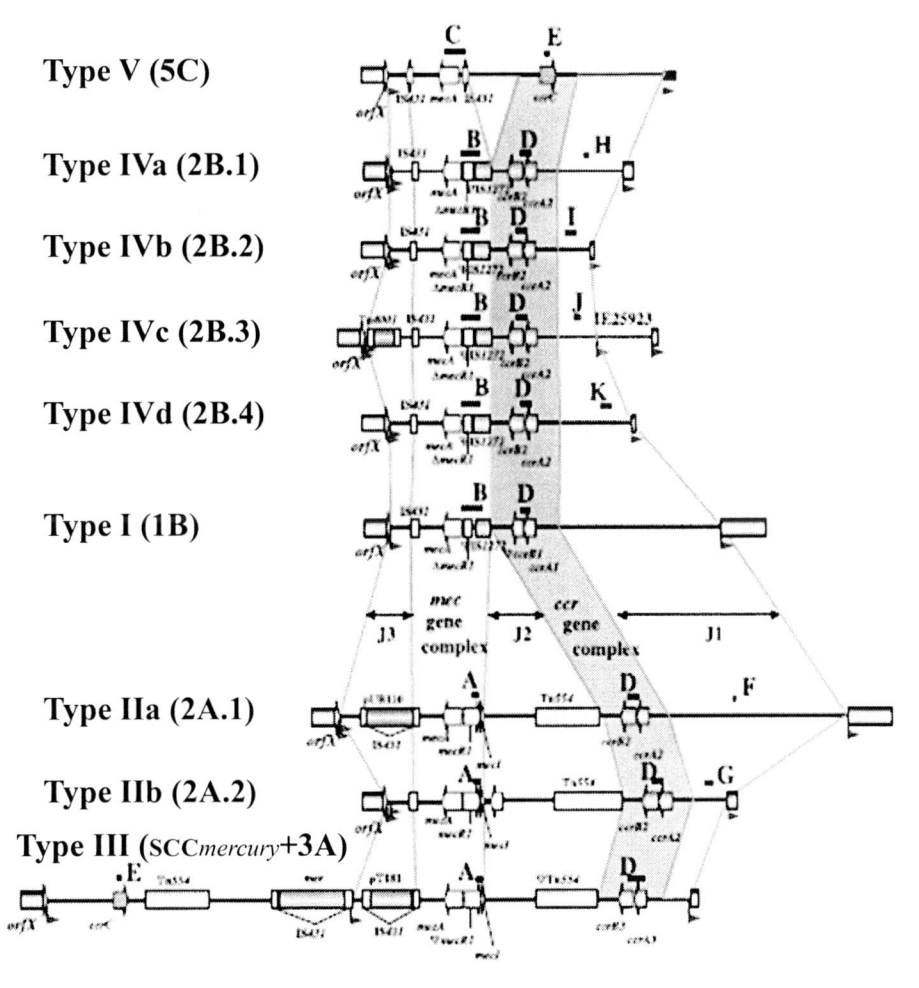

Fig. 1. Structures of SCC*mec* elements and location of primers. The structures of SCC*mec* elements are illustrated based on the nucleotide sequences deposited in the DDBJ/EMBL/GenBank databases under the following accession nos.: AB033763 (type I), D86934 (type II.1), AB127982 (type II.2), AB037671 (type III), AB063172 (type IV.1), AB063173 (type IV.2), AB096217 (type IV.3), AB097677 (type IV.4), and AB121219 (type V). Bars indicate the locations of representative sets of primers. A, *mecI-mecR1*; B, IS*1272-mecA*; C, IS*431*L-*mecA*; D, *ccrA1*, *ccrA2*, or *ccrA3* gene with *ccrB*; E, *ccrC*; F, J1 region of type IIa (2A.1) SCC*mec*; G, J1 region of type IIb (2A.2) SCC*mec*; H, J1 region of type IVa (2B.1) SCC*mec*; I, J1 region of type IVb (2B.2) SCC*mec*; J, J1 region of type IVc (2B.3) SCC*mec*; K, J1 region of type IVd (2B.4) SCC*mec*.

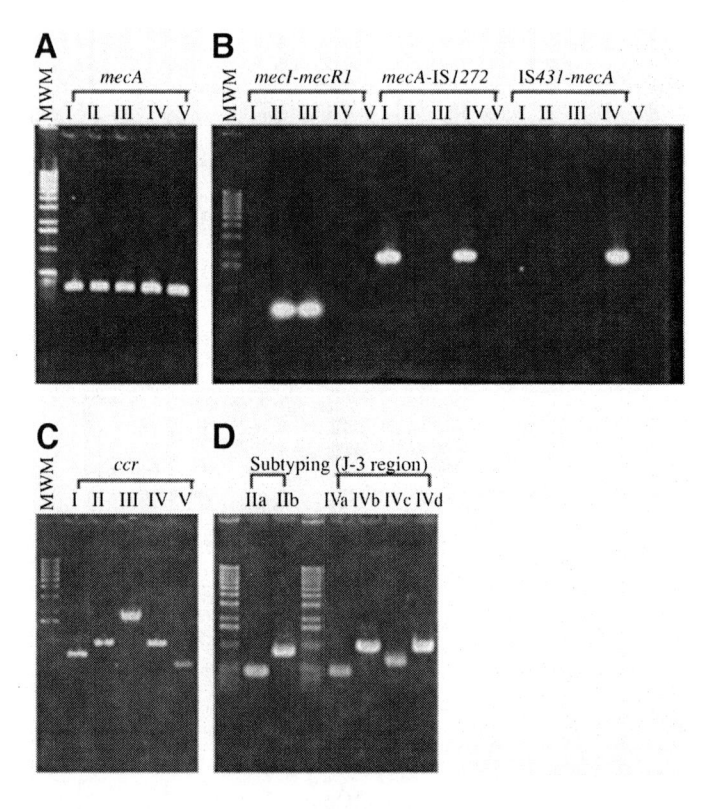

Fig. 2. Agarose gel electophoresis of amplified DNA fragments using sets of primers indicated above the gel. **A**, identification of *mecA*; **B**, identification of *mec* classes by amplifying gene lineage *mecI-mecR1* for class A *mec*, *mecA*-IS *1272* for class B *mec* and IS*431*-*mecA* for class C *mec*; **C**, identification of *ccr* genes, *ccrA1B1*, *ccrA2B2*, *ccrA3B3*, and *ccrC*; **D**, identification of specific genes in J1 regions of type II and type IV SCC*mec* elements. Chromosomal DNAs were extracted from representative strains and used as templates as follows: I, NCTC10442 (type I); II, N315 (type II); III, 85/2082 (type III); IV, CA05 (type IV); V, WIS[WBG8318] (type V); IIa, N315 (type IIa [II.1]); IIb, JCSC3063 (type IIb [II.2]); IVa, CA05 (type IVa [IV.1]); IVb, 8/6-3P (type IVb [IV.2]); IVc, 81/108 (type IVc [IV.3]); IVd, JCSC4469 (type IVd [IV.4]). A 1-kb ladder was used for molecular size markers.

3.3.2. Assignment of ccr

The type of *ccr* gene complex could be assigned by a multiplex PCR using four primers: cβ, a primer constructed using inosine at the nineteenth nucleotide position so that it recognizes three ccr genes, *ccrB1*, *ccrB2*, and *ccrB3*; α1, a primer specific to *ccrA1*; α2, a primer specific to *ccrA2*; and α3, a primer specific to *ccrA3*. The *ccr* gene type can be determined by the size of the amplified DNA fragment, which depends on which *ccr* genes are present: type 1 *ccr*, 0.7 kb; type 2 *ccr*, 1 kb; type 3 *ccr*, 1.6 kb (*see* **Table 1** and **Fig. 2C**). When no DNA fragment was amplified with the aforementioned primers (*see* **Note 4**), the strains were examined for the presence of type 4 *ccr* and type 5 *ccr*.

3.3.3. Assignment of J Regions

The SCC*mec* element J regions were further classified into three subregions: J1 (formerly called the L-C region), J2 (formerly called the C-M region), and J3 (formerly called the M-R region). Many differences in the J regions of SCC*mec* types have been reported, and they have been used to classify an SCC*mec* type in detail.

Table 2 provides the primers used to identify the J region differences. Oliveira and de Lencastre *(11)* developed a multiplex PCR that identifies J region differences specific to type I–III SCC*mec* elements. Zhang et al. *(12)* developed a multiplex PCR to identify *mec*, *ccr*, and type I–V–specific J1 regions.

Plasmids or transposons, which are primarily integrated in the J2 or J3 region, can be used as targets to differentiate SCC*mec* elements for epidemiological study. Oliveira et al. *(4)* found a type I SCC*mec* element that carried an integrated copy of pUB110, which they refer to as a type IA SCC*mec* element, and a type III SCC*mec* element that did not carry an integrated copy of pT181, which they refer to as a type IIIA SCC*mec* element. *mec* left extremity polymorphism (MLEP) typing (formerly called *mec* right extremity polymorphism typing [MREP]) is a PCR strategy to identify differences in the region of SCC*mec* flanking *orfX*. This multiplex PCR is composed of three primers: cR4, a primer in the chromosomal region flanking SCC*mec*; mR5, a primer for the extremities of the type I and type II SCC*mec* elements, which were called the downstream constant region (*dcs* region) by Oliveira et al. *(13)*; and mR6, a primer for the extremity of the type III SCC*mec* element.

3.3.4. Assignment of SCCmec Elements and a Proposal for a New Nomenclature

Table 3 provides the main SCC*mec* elements reported at present. The type of SCC*mec* can be defined based on the combination of mec and ccr, as given in **Table 3**. The differences in the J regions were identified by PCR with sets of primers that identify genes in the J regions. It should be noted that identification of *ccrC* is not definitive for the type V SCC*mec* element. If an MRSA carried a class A *mec* and was positive in PCR reactions to identify both the *ccrC* and *ccrA3ccrB3* genes, the strain could be type 3A SCC*mec*, since many type 3A SCC*mec* strains carry *ccrC*.

Table 3
Assingment for Types of SCC*mec* Elements

Type of SCC*mec*		Type of *ccr*	Class of *mec*	Characteristic features in J regions		Sequenced strains and sources of their sequences
Reported name	Proposed names			J1 region	J2 and J3 regions	
I	I.1.1.1	1	B	Subtype 1 specific ORFs		NCTC10442 (AB033763), COL www.tigr.org
IA	I.1.1.2	1	B	Subtype 1 specific ORFs	Carry pUB110	
IIa	II.1.1.1	2	A	kdp operon		N315 (D86934, http://www.bio.nite.go.jp/dogan/MicroTop?GENOME_ID=n315, MRSA252 http://www.sanger.ac.uk/Projects/S_aureus/)
IIb	II.2.1.2	2	A	Subtype 2 specific ORFs		JCSC3063 (AB127982)
IIA	II.3.1.1	2	A	Same as that of IVb	Insertion of IS*1182*	
IIB	II.3.2.1	2	A	Same as that of IVb		(AJ810123)
IIC	II.3.3.1	2	A	Same as that of IVb	Insertion/deletion of IS*1182*	
IID	II.3.1.2	2	A	Same as that of IVb	Insertion of IS*1182*, does not carry pUB110	

IIE	II.3.3.2	3	A	Same as that of IVb	AR13.1/3330.2 (AJ810120)
III	III.1.1.1	3	A		85/2082 (B037671)
IIIA	III.1.1.2	3	A		HU25 (AF422651-422696)
IVa	IV.1.1.1	2	B	Subtype 1 specific ORFs	CA05 (AB063172), MW2 www.bio.nite.go.jp/dogan/ MicroTop?GENOME_ID=mw2
IVb	IV.2.1.1	2	B	Subtype 2 specific ORFs	8/6-3P (AB063173)
IVc	IV.3.1.1	2	B	Subtype 3 specific ORFs	81/108 (AB096217)
IVE	IV.3.1.4	2	B	Same as that of IVc	AR43/3330.2 (AJ810121)
IVF	IV.2.1.4	2	B	Left extremity is different from that of IVc	
IVd	IV.4.1.1	2	B	Left extremity is different from that of IVb	JCSC4469 (AB097677)
IV	V.1	4	B	Subtype 4 specific ORFs	HDE288 (AF411935)
IVg	V1.5.1.1	2	B		M03-68 (DQ106887)
V	V.1	5	C		WIS (AB121219)

Insertion/deletion of IS1182, does not carry pUB110

Does not carry pT181

The first column in **Table 3** provides the reported names of SCC*mec* elements. It can be difficult to judge the differences among the elements by these names. We have classified SCC*mec* elements according to differences in the J1 region and named them IIa, IIb, IVa, IVb, IVc, and IVd, with the J1 type indicated by the lowercase letter. Oliveira et al. found differences in the presence or absence of an integrated plasmid and designated elements carrying the plasmid IA and IIIA *(4)* Shore et al. *(8)* found seven novel variants based on differences in the J regions and designated them IIA, IIB, IIC, IID, IIE, IVE, and IVF, with the J region differences indicated by capital letters.

The second column in **Table 3** gives our suggestions for renaming the SCC*mec* elements which are revised by the suggestions of many researchers. The names are based on the definition of SCC*mec* elements. First, the type of SCC*mec* is described by roman numerals, signifying the combination of the type of *ccr* indicated by a number and the class of *mec* indicated by an uppercase letter; and these designations are followed by numbers representing differences in the J1, J2, and J3 regions, each separated by a period (e.g., II.1.1.1). The type of *ccr* or differences in J regions could be numbered only in chronological order, according to the time of their identification.

By using this nomenclature, the type IV SCC*mec* element found in HDE288 would be called VI. Since the strain carries the sixth type of *ccr* and *mec* combination, type 4 *ccr* and classB *mec*. We hope that this nomenclature for SCC*mec* elements will prove useful in avoiding confusion in MRSA typing.

4. Notes

1. Reagents marked by an asterisk were prepared following previously described protocols *(14)*. All reagents whose manufacturers were not indicated were reagent grade.
2. In the preparation of chromosomal DNAs, other commercially available kits could be used unless they interfere with the later PCR amplification steps.
3. When no DNA fragment was amplified with sets of primers to identify class A to C *mec*, further PCR reactions to find the localization of *mecI* and *mecR1* genes with three sets of primers could be carried out, since this region sometimes has mutations or deletions *(15)*.
4. When no DNA fragments could be amplified with the set of primers to identify *ccr*1–5, those strains were judged unable to be typed.

Acknowledgments

This work was supported by a Grant-in-Aid for 21st Century COE Research and a Grant-in-Aid for Scientific Research on Priority Areas (13226114) from The Ministry of Education, Science, Sports, Culture and Technology of Japan.

References

1. Ito, T., Katayama, Y., and Hiramatsu, K. (1999) Cloning and nucleotide sequence determination of the entire *mec*DNA of pre-methicillin-resistant *Staphylococcus aureus* N315. *Antimicrob. Agents Chemother.* **43,** 1449–1458.
2. Katayama, Y., Ito, T., and Hiramatsu, K. (2000) A new class of genetic element, staphylococcal cassette chromosome *mec*, encodes methicillin resistance in *Staphylococcus aureus. Antimicrob. Agents Chemother.* **44,** 1549–1555.
3. Ito, T., Katayama, Y., Asada, K., et al. (2001) Complete structure of three types of staphylococcal cassette chromosome *mec*(SCC*mec*) integrated in the chromosome of MRSA strains in the world. *Antimicrob. Agents Chemother.* **45,** 1323–1336.
4. Oliveira, D. C., Tomasz, A., and de Lencasre, H. (2001) The evolution of pandemic clones of methicillin-resistant *Staphylococcus aureus*: identification of two ancestral genetic backgrounds and the associated *mec* elements. *Microb. Drug Resist.* **7,** 349–361.
5. Ma, X., Ito, T., Tiensasitorn, C., et al. (2002) A novel type of staphylococcal cassette chromosome *mec* (SCC*mec*) identified in community-acquired methicillin-resistant *Staphylococcus aureus* strains. *Antimicrob. Agents Chemother.* **46,** 1147–1152.
6. Ito, T., Ma, X. X., Takeuchi, F., Okuma, K., Yuzawa, H., and Hiramatsu, K. (2004) Novel type V staphylococcal cassette chromosome *mec* driven by a novel cassette chromosome recombinase, *ccrC. Antimicrob. Agents Chemother.* **48,** 2637–2651.
7. Hisata, K., Kuwahara-Arai, K., Yamanoto, M., et al. (2005) Dissemination of methicillin-resistant staphylococci among healthy Japanese children. *J. Clin. Microbiol.* **43,** 3364–3372.
8. Shore, A., Rossney, A. S., Keane, C. T., Enright, M. C., and Coleman, D. C. (2005) Seven novel variants of the staphylococcal chromosomal cassette *mec* in methicillin-resistant *Staphylococcus aureus* isolates from Ireland. *Antimicrob. Agents Chemother.* **49,** 2070–2083.
9. Kwon, N. G., Park, K. T., Moon, J. S., et al. (2005) Staphylococcal cassette chromosome *mec* (SCC*mec*) characterization and molecular analysis for methicillin-resistant *Staphylococcus aureus* and novel SCC*mec* subtype IVg isolated from bovine milk in Korea. *J. Antimicrob. Chemother.* **56,** 624–632.
10. Katayama, Y., Ito, T., and Hiramatsu, K. (2001) Genetic organization of the chromosome region surrounding *mecA* in clinical staphylococcal strains: role of IS*431*-mediated *mecI* deletion in expression of resistance in *mecA*-carrying, low-level methicillin-resistant *Staphylococcus haemolyticus. Antimicrob. Agents Chemother.* **45,** 1955–1963.
11. Oliveira, D. C. and de Lencastre, H. (2002) Multiplex PCR strategy for rapid identification of structural types and variants of the mec element in methicillin-resistant *Staphylococcus aureus. Antimicrob. Agents Chemother.* **46,** 2155–2161.
12. Zhang, K., McClure, J. A., Elsayed, S., Louie, T., and Conly, J. M. (2005) Novel multiplex PCR assay for characterization and concomitant subtyping of staphylococcal cassette chromosome *mec* types I to V in methicillin-resistant *Staphylococcus aureus. J. Clin. Microbiol.* **43,** 5026–5033.

13. Oliveira, D. C., Wu, S. W., and Lencastre, H. (2000) Genetic organization of the downstream region of the *mecA* element in methicillin-resistant *Staphylococcus aureus* isolates carrying different polymorphisms of this region. *Antimicrob. Agents Chemother.* **44,** 1906–1910.
14. Sambrook, J., Fritsch, E. F., and Maniatis, T. (1989) *Molecular Cloning,* Cold Spring Harbor Laboratory Press, Cold Spring Harbor, NY.
15. Okuma, K., Iwakawa, K., Turnidge, J. D., et al. (2002) Dissemination of new methicillin-resistant *Staphylococcus aureus* clones in the community. *J. Clin. Microbiol.* **40,** 4289–4294.

8

Targeted Gene Disruption for the Analysis of Virulence of *Staphylococcus aureus*

J. Ross Fitzgerald

Summary

In recent years, molecular genetic approaches to the study of the disease pathogenesis of *Staphylococcus aureus* have resulted in many new biological insights. I describe methods used for targeted disruption of staphylococcal genes leading to loss of gene function, important for studies of staphylococcal proteins and their role in virulence.

Key Words: *Staphylococcus aureus*; allele replacement; pathogenesis; virulence.

1. Introduction

The ability to generate stable mutations in staphylococcal genes leading to loss of gene function allows analysis of the contribution of individual loci to disease pathogenesis, bacterial metabolism, and survival.

Construction of stable gene knockouts requires the exchange by homologous recombination of a chromosomal wild-type (WT) gene with a mutated or inactivated allele. The established method for allele replacement requires the use of an antibiotic resistance cassette to mark the mutated allele or gene deletion *(1)*. Because of the small number of antibiotic resistance determinants that can be used successfully as selective markers in *Staphylococcus aureus*, this approach limits the number of genes that can be inactivated in a single strain. However, a novel method has recently been developed for disruption of staphylococcal genes that avoids the requirement for antibiotic resistance cassettes *(2)*. Using this method, an unlimited number of mutations can, in theory, be introduced into *S. aureus* without the requirement for antibiotic resistance markers (*see* **Note 1**). Both the standard and novel methods involve the construction of a plasmid that is temperature sensitive for replication and that contains an attenuated form of the target gene of interest. The plasmid is constructed in *Escherichia coli*, which is transformed into a primary electrocompetent *S. aureus* host before transduction to the *S. aureus* strain of interest and integration into the chromosome at the target locus. In this chapter, I describe the experimental details of both approaches.

2. Materials

2.1. Polymerase Chain Reaction for Plasmid Construction

1. DNA polymerase: *Taq* polymerase is typically used, although enzymes with proofreading activity such as *Pfu* and *Pwo* may be used to prevent incorporation of errors introduced by *Taq* polymerase that may affect expression of antibiotic resistance markers. Store at −20°C.
2. Reaction buffer: This is supplied with most commercial DNA polymerases. A typical buffer is composed of 10 mM Tris-HCl, pH 8.3, 50 mM KCl, and 1.5 mM MgCl$_2$. Optimal Mg^{2+} concentration in the final reaction mixture is usually in the range of 0.5–5.0 mM and is determined empirically. Store at −20°C.
3. Oligonucleotide primers: These are designed using DNA sequence data. They should be 20–30 nucleotides in length, avoid polybase sequences and regions of predicted secondary structure, contain a G/C clamp at the 3′ end, and have 40–60% G/C content. Store at −20°C.
4. Reagents: A typical polymerase chain reaction (PCR) will contain the following reagents: 50 mM KCl, 10 mM Tris-HCl, pH 8.3, 1.5 mM MgCl$_2$, 0.01% gelatin, 200 µM each of dATP, dGTP, dTTP, and dCTP, 20 pmol of each of the primers, and 2.5 U of *Taq* DNA polymerase (Promega, Madison, WI). The thermocycling program will depend on the melting temperature (*Tm*) of the primers and the size of template to be amplified.

2.2. Cloning and Electrotransformation

1. Competent bacterial strains: *E. coli* cloning hosts such as JM101, *S. aureus* restriction-modification mutant host such as RN4220 *(3)*, *S. aureus* WT strain containing target gene for disruption. Store at −70°C.
2. Shuttle plasmids (can replicate in *E. coli* and *S. aureus*): There are several including pCW59 *(4)* or pSK265, a derivative of pC194 with a multiple cloning site *(5)*. Store at −20°C.
3. Temperature-sensitive plasmid vectors, such as pTS1 and pTS2 derived from pE194 ts *(6)*, or pG+Host vectors *(7)*.
4. Antibiotic resistance determinants: A limited number of useful markers are currently available, including tetracycline (TetK) derived from pCW59 *(6)*, chloramphenical encoded by pTS2, erythromycin encoded by pE194 (ErmC) *(8)*, and kanamycin (pPQ126) *(9)*.
5. Appropriate restriction enzymes, ligase enzyme (NEB, Promega). Store at −20°C.
6. Medium: Luria broth (LB), Tryptic soya agar (TSA), and Tryptic soya broth (TSB) (Oxoid) containing the following appropriate antibiotics: erythromycin (10 µg/mL), tetracycline (2 µg/mL), chloramphenical (10 µg/mL), kanamycin (50 µg/mL), and 5-bromo-4-chloro-3-indolyl-β-D-galactopyranoside (X-gal) (150 µg/mL).

2.3. Preparation of Phage Stock and Transduction of Plasmid

1. Transducing phage, including θ11, 79, 80, 85 *(10)*. This is stable at 4°C for several months.
2. Phage broth: Oxoid nutrient broth no. 2 (20 g/L).
3. Phage base: phage broth and 7 g/L of Oxoid agar no.1.
4. Phage top agar: phage broth and 3.5 g/L of Oxoid agar no. 1.
5. 1 M CaCl$_2$.
6. 0.02 M Sodium citrate solution.
7. 10% Sodium citrate solution.
8. LB: standard Luria broth (Ca^{2+} precipitates in TSB).
9. Filters (0.45-μm) (Millipore, Bedford, MA).

2.4. Allele Replacement by Temperature Shifting with Enrichment

1. TSB broth with and without appropriate antibiotic.

2.5. Verification of the Presence of a Stably Disrupted Gene

1. Restriction enzyme *Hin*dIII and buffer (NEB). Store at −20°C.
2. 0.8% Agarose (Promega) gel, 0.5X TBE running buffer (10X TBE is 0.89 M Tris, 0.89 M boric acid, and 0.02 M EDTA, pH 8.4).
3. Nylon membrane: Hybond N+ (GE Healthcare).
4. Depurination solution: 0.2 M HCl.
5. Denaturation solution: 1.5 M NaCl, 0.5 M NaOH.
6. Neutralization solution: 5 M Tris-HCl, 3 M NaCl, pH 7.4.
7. 20X Saline sodium citrate (SSC): 3.0 M NaCl, 0.3 M Na-citrate, pH 7.0.
8. Hybridization solution: 5X SSC, 0.1% *N*-lauroylsarcosine, 0.02% sodium dodecyl sulfate (SDS), 1X blocking reagent (Roche).
9. Wash buffer: 0.1 M maleic acid, 0.15 M NaCl, pH 7.5, 0.3% Tween-20.
10. PCR Dig-labeling system (Roche).
11. Alkaline phosphatase–labeled Fab anti-Dig antibodies (Roche).
12. CDP-star solution (Roche).

3. Methods

3.1. PCR for Plasmid Construction

Genomic DNA isolation in our laboratory is carried out using a modification of the Edge Biosystems Genomic DNA isolation kit that includes an incubation at 37°C in the presence of lysostaphin (100 μg/mL; Sigma). PCR is used to amplify regions of the chromosome from 0.5 to 1 kb flanking each side of the target gene of interest. Primers are designed to incorporate appropriate restriction sites for insertion of the amplified fragments into the multiple cloning site of a vector such as pUC18/19, and for ligation of an antibiotic resistance cassette between the DNA fragments (**Fig. 1**). For example, a *Cla*I site would facilitate the cloning of a *Taq*I fragment containing the erythromycin resistance determinant from pE194, as described previously *(11)*.

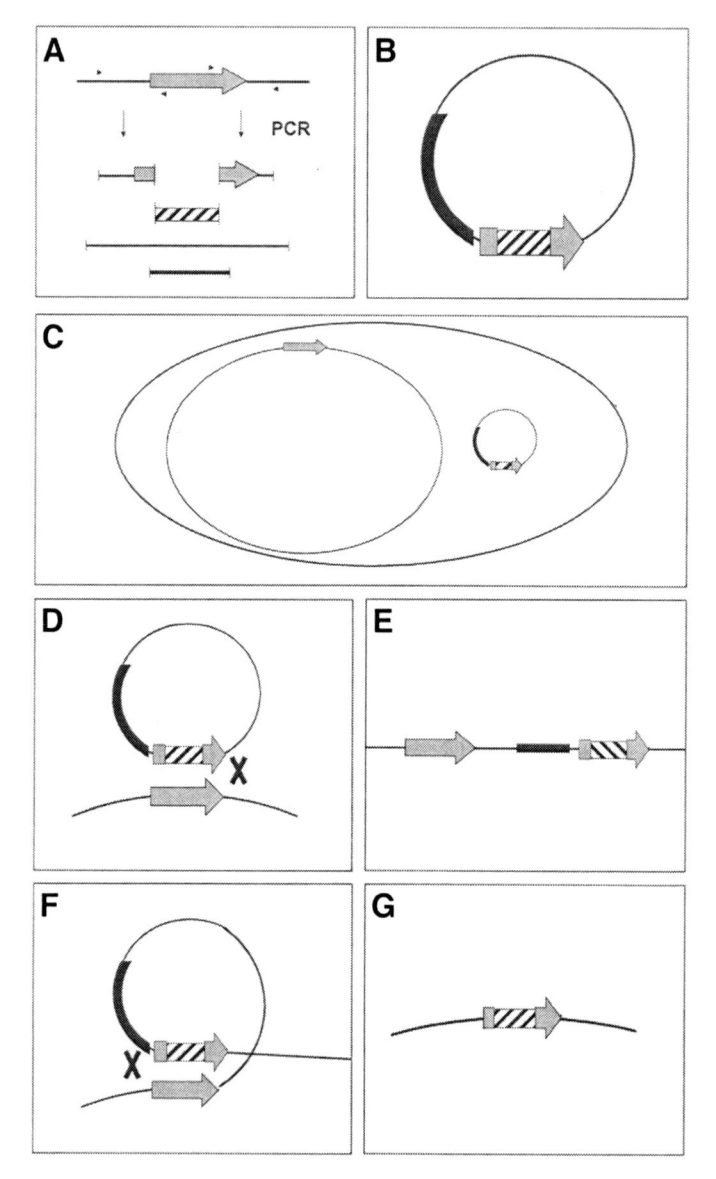

Fig. 1. Schematic representation of allele replacement. (**A**) Genetic components required for generation of a *ts* plasmid for allele replacement. PCR is used to generate DNA fragments specific for regions flanking the gene of interest. Small black arrows indicate the position of PCR primers relative to the target gene (solid gray arrows). The hatched box represents an antibiotic resistance cassette with ends compatible for ligation with the PCR-generated fragments. Restriction-digested plasmid vectors pUC18/19 (thin black line) and temperature-sensitive plasmid pTS2 (thick black line) are indicated. (**B**) Complete plasmid construct after ligation. (**C**) Representation of an *S. aureus* strain

3.2. Cloning and Electrotransformation

1. Digest PCR-generated fragments and plasmid vectors with the appropriate restriction enzymes, purify, and ligate with DNA ligase (Promega) at room temperature overnight. Transform the ligation mixture into an electrocompetent *E. coli* cloning host such as strain JM101 before plating on LB agar containing ampicillin (100 µg/mL) and incubating overnight at 37°C. Subculture positive transformants into LB broth with ampicillin (100 µg/mL), and purify the plasmid with a Qiagen plasmid miniprep kit (Qiagen).
2. Clone the antibiotic resistance cassette and a temperature-sensitive plasmid vector such as pTS2 into appropriate restriction sites to create the final temperature-sensitive plasmid construct (**Fig. 1**).
3. Because electrotransformation of many *S. aureus* strains is poorly efficient, an initial restriction-modification negative host such as RN4220 *(3)* is commonly used before transduction to the strain of interest. Make up a 1/25 dilution of an overnight culture of the strain to be transformed (e.g., RN4220) into 25 mL of fresh TSB and incubate with shaking at 200 rpm at 37°C until the midexponential growth phase (i.e., OD_{600} = 0.8–1.0).
4. Wash the harvested bacterial cells in an equal volume of distilled water at room temperature and pellet by centrifuging at 8000 rpm (12,000 ×g) for 10 min at 20°C. Wash the cells in 1/5 vol of 10% glycerol, harvest, and then resuspend in 1/10 vol of 10% glycerol and incubate at 20°C for 15 min. Pellet the cells once more and resuspend in a final volume of 800 µL of 10% glycerol. At this stage, cells may be snap-frozen in liquid nitrogen and stored at −70°C until required.
5. For each electrotransformation, mix a 70-µL aliquot of competent cells with 1 µg of plamid DNA. Transfer 60 µL of the mix to a 0.1-cm-gap electroporation cuvet and electroporate at 20°C, 100 Ω, 25 µF, and 2.3 kV (23 kV/cm). Immediately, transfer the cells to 390 µL of TSB broth and incubate with shaking at 37°C for 1 h. Plate 100-µL volumes of different dilutions onto TSA incorporating the appropriate antibiotic for the plasmid and incubate at 37°C for 16–24 h. Screen transformants for the presence of the plasmid with a Qiagen plasmid mini kit (Qiagen).

3.3 Preparation of Phage Stock and Transduction of Plasmid

1. Of an overnight TSB culture of *S. aureus* RN4220 containing the plasmid of interest, dilute 20 µL in 2 mL of phage broth and incubate at 37°C with shaking at 200 rpm for 4 h.
2. Add 200 µL of transducing phage (e.g., phage φ85; *see* **Note 2**) at dilutions of 10^{-3}–10^{-7} in phage broth to 300 µL of *S. aureus* cells, and incubate at room

---

with chromosomal WT gene of interest (gray arrow on large circle) and plasmid with mutant allele (small circle). (**D,E**) A single crossover integration event results in both WT and mutant copies of the target gene in the chromosome. (**F,G**) A second crossover event results in a single (mutant) copy of the gene in the chromosome.

temperature for 30 min. Add 10 mL of molten phage top agar (10 m*M* CaCl$_2$) at 55°C to each tube, mix, and pour immediately over the surface of two phage base plates (10 m*M* CaCl$_2$). Incubate the plates overnight at 37°C in an upright position in a plastic bag, to prevent drying of the agar. Choose the dilution of phage that results in just confluent lysis (i.e., the plaques are joining up), remove the top layer from both plates with a plate spreader, and centrifuge at 15,000 rpm for 10 min. Remove the supernatant and filter sterilize through a 0.45-µm filter (Millipore). The phage stock can now be stored at 4°C for several months.

3. Grow an overnight culture of the recipient *S. aureus* strain in 20 mL of TSB in a 250-mL flask at 37°C with shaking at 200 rpm. Harvest the cells by centrifuging at 10,000 rpm for 10 min and resuspend in 1 mL of TSB (i.e., 1/20 culture volume). Add 500 µL of bacterial cells to 1 mL of LB broth with 10 m*M* CaCl$_2$ and 500 µL of phage stock (*see* **Note 2**). In parallel, prepare a control reaction without phage consisting of 500 µL of bacteria and 1.5 mL of LB with 10 m*M* CaCl$_2$. Incubate both samples at 37°C for 25 min in a water bath, followed by incubation at 37°C for 15 min with shaking at 200 rpm. Transfer the samples to ice before adding 1 mL of ice-cold 0.02 *M* sodium citrate, harvest by spinning at 10,000 rpm for 10 min, resuspend in 1 mL of ice-cold sodium citrate (*see* **Note 3**), and leave on ice for 2 h. Finally, spread 100-µL aliquots onto TSA (0.05% sodium citrate) containing the appropriate selective antibiotic and incubate for 2 d at 28°C, the permissive temperature for plasmid replication (*see* **Note 4**). Single colony transductants that grow in the presence of selective antibiotic should contain the transduced plasmid (*see* **Note 5**).

3.4. Allele Replacement by Temperature Shifting with Enrichment

To generate single crossover chromosomal integrants, the *S. aureus* strain containing the transduced plasmid is grown overnight at the permissive temperature of 28°C in 2 mL of TSB in the presence of appropriate antibiotic. One hundred–microliter volumes of dilutions from 10^{-2} to 10^{-8} are plated in duplicate onto TSA containing appropriate antibiotic and incubated at either the restrictive temperature of 42°C or at 28°C (permissive temperature control). At the restrictive temperature, the plasmid cannot replicate and must integrate into the chromosome to survive (*see* **Note 6**). Colonies growing at 42°C represent single crossover integrants and should be subcultured twice to single colonies on TSA containing 0.05% sodium citrate and appropriate antibiotic at 42°C to ensure a pure culture (*see* **Note 7**).

To generate and enrich for double crossover excisants that have lost the backbone of the plasmid but have retained the disrupted allele of the gene, several rounds of temperature shifts from 28 to 42°C are carried out. Single crossover integrants are cultured in 20-mL volumes of TSB (no antibiotic) at 28°C for 12 h with shaking at 200 rpm, followed by incubation at 42°C for a further 12 h with shaking. Then a 1/100 dilution of each culture is made into fresh TSB and the temperature shifts are repeated. This process is continued until three rounds of dilutions and temperature shifts have been carried out over the course of 72 h. One hundred–microliter volumes of dilutions from 10^{-4} to 10^{-8} are then plated onto TSA containing appropriate antibiotic. The double crossover strains will retain the antibiotic resistance encoded by the cassette marking the mutant gene/gene deletion but will have lost the antibiotic resistance that was encoded by the plasmid backbone.

3.5. Verification of the Presence of a Stably Disrupted Gene

To confirm that a successful allelic exchange event has occurred resulting in a stably inherited mutated form of the gene, PCR amplification across the mutated gene will reveal whether the size of the template is consistent with the presence of the mutant allele. However, for absolute confirmation of the integrity of the mutant, Southern blot analysis of the WT and mutant strains should be carried out with a probe specific for the gene of interest.

1. Isolate genomic DNA from the parent and mutant strains as previously described. Digest 2 µg of genomic DNA from each strain with *Hin*dIII restriction endonuclease overnight at 37°C, and resolve by 0.8% agarose gel electrophoresis in 0.5X TBE.
2. Soak the agarose gel in depurination solution (0.2 M HCl) for 5 min, and rinse in distilled H$_2$O before denaturation for 45 min in 1.5 M NaCl, 0.5 M NaOH. Then soak the gel in neutralization buffer (5 M Tris-HCl, 3 M NaCl, pH 7.4) for 1 h.
3. Conduct Southern transfer of the restriction-digested DNA to nylon membrane (Millipore) overnight with transfer buffer (20X SSC) *(12)*, and crosslink the DNA to the nylon membrane by exposing each side of the membrane to UV light from a transilluminator for 1 min.
4. Conduct prehybridization for 3 h in 30 mL of standard buffer (5X SSC, 0.1% *N*-lauroylsarcosine, 0.02% SDS, 1X blocking reagent; Roche) at 65°C. Prepare the probe by PCR amplification of the target sequence with incorporation of digoxigenin (Roche), denaturation by boiling for 5 min, and chilling rapidly on ice before adding to the prehybridization solution and incubating overnight at 65°C.
5. After overnight hybridization, wash the membrane twice for 10 min each time in 2X SSC/0.1% SDS at room temperature followed by two 5-min washes in maleic acid buffer (0.1 M maleic acid, 0.15 M NaCl, pH 7.5) with 0.3% Tween-20.
6. Preincubate the membrane in buffer 2 (1X blocking reagent in maleic acid buffer) for 30 min before adding anti-Dig antibody (1/10,000 dilution) (Roche) and incubating for 30 min.
7. Wash the membrane twice for 10 min each time in maleic acid buffer with 0.3% Tween-20, and develop using CSPD solution (Roche) according to the manufacturer's instructions.

When using a probe that is specific for regions flanking the insertion site of the antibiotic marker, the sum of the sizes of the hybridizing restriction fragments in the mutant strain should be equal to the sum of the sizes of hybridizing fragments in the parent strain and the size of the antibiotic resistance cassette (**Fig. 2**). If an antibody for the encoded protein of interest is available, Western blot comparative analysis of the parent and mutant strains can be used to confirm the loss of phenotype in the mutant strain (**Fig. 2**).

4. Notes

1. A recently developed variation of this method allows the disruption of genes without the requirement that an antibiotic resistance marker be left in the chromosome *(2,13)*. This results in the potential for generation of many gene mutations in a single strain without the constraint of a limited number of useful antibiotic markers. A deletion is constructed in the gene of interest by overlapping PCR. In brief, PCR products of 500–1000 bp are amplified representing the left (oligonucleotides A and B) and the right (oligonucleotides C and D) of the sequence targeted for deletion. The oligonucleotides B and C are designed to have at least a 16-base complementary region to allow the products of the first PCR to anneal at their overlapping region during a second PCR. The second PCR is performed including 1 µL of both purified products of the first PCR as template with primers A and D to obtain a single amplicon. This fragment is purified and cloned into the pGEM-T easy vector (Promega). The fragment is then subcloned into the appropriate restriction site of the shuttle plasmid pMAD *(2)*, and the resulting plasmid is transformed into *S. aureus* RN4220 by electroporation as described previously. pMAD contains a temperature-sensitive origin of replication, a β-galactosidase gene expressed constitutively, and an erythromycin resistance gene. After overnight culture at the permissive temperature of 30°C, dilutions are plated onto TSA plates, and the plasmid is forced to integrate into the chromosome when grown at the restrictive temperature (43.5°C) in the presence of erythromycin (10 µg/mL) and X-gal (150 µg/mL). Light blue colonies indicate integration of a single plasmid into the chromosome. Dark blue colonies are indicative of the presence of multiple copies of the plasmid. Light blue colonies are inoculated into 10 mL of TSB and incubated for 24 h at 30°C with shaking, and 10-fold serial dilutions of the cultures are plated onto TSA (no antibiotic) containing X-gal (150 µg/mL). White colonies no longer contain the pMAD plasmid and should be erythromycin sensitive. These colonies are screened to confirm the allele replacement event by PCR and/or Southern blotting as described in this chapter.
2. Different strains of *S. aureus* will vary in their susceptibility to phage infection. Some strains may not be susceptible to infection by φ85, and additional transducing phage such as φ11, φ13, and φ42 should be tested.
3. Sodium citrate solution can go off after several months' storage at room temperature.
4. Very occasionally strains that have previously been recipients in transduction experiments may become lysogenized with phage, resulting in strains that are resistant to superinfection. In such a case, transduction will not be successful unless a different phage or strain is used for the experiments.

Analysis of Virulence of S. aureus

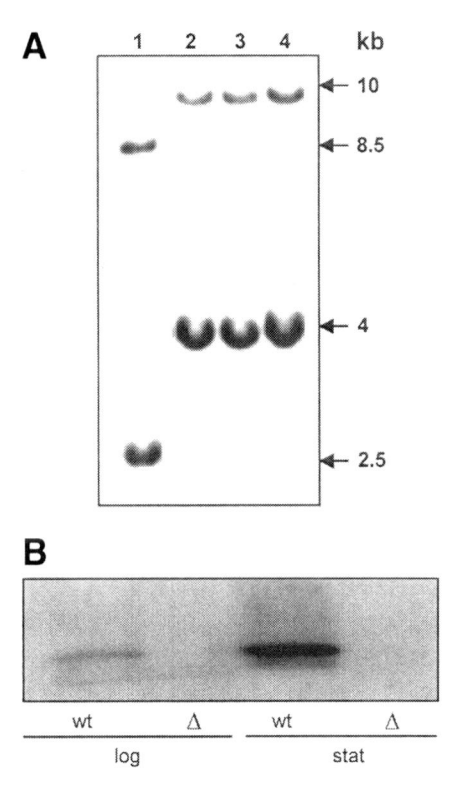

Fig. 2. Verification of integrity of disrupted gene. (**A**) Southern blot analysis to confirm replacement of a WT staphylococcal protein A (*spA*) gene with a mutated allele containing a kanamycin resistance cassette. Genomic DNA from parent and mutant strains was restriction digested with *Hin*dIII, resolved by agarose gel electrophoresis, transferred to a nylon membrane, and hybridized with a probe specific for a region of *spa* that includes the site of insertion of the kanamycin resistance cassette. Lane 1, *S. aureus* strain Newman WT; lanes 2–4, putative Newman::Δ*spA* mutants. The sum of the sizes of the hybridizing restriction fragments in the mutant strain is equal to the sum of the sizes of hybridizing fragments in the parent strain and the size of the kanamycin resistance cassette, indicating the existence of a single, disrupted allele of the *spA* gene in mutant strains. (**B**) Western blot analysis of Newman WT and mutant strains from exponential and stationary phases of growth with IgG antibody indicates the loss of SpA expression in the mutant strain.

5. Transductants should be subcultured twice to a single colony on TSA with sodium citrate to ensure the loss of transducing phage.
6. The restrictive temperature for temperature-sensitive replication may vary by 1–3°C depending on the strain and will need to be tested empirically.
7. *ts* replication mutations revert to temperature independence at a significant frequency. This should be borne in mind when screening survivors after temperature shifting.

Acknowledgments

I wish to thank Dr. José Penades for providing technical information and Prof. Tim Foster for introducing me to the molecular genetics of staphylococci.

References

1. Foster, T. (1998) Molecular genetic analysis of staphylococcal virulence, in *Methods in Microbiology, Bacterial Pathogenesis*, vol. 27 (Williams, P., Ketley, J., and Salmond, G. P. C., eds.), Academic, San Diego, pp. 433–454.
2. Arnaud, M., Chastanet, A., and Debarbouille, M. (2004) New vector for efficient allelic replacement in naturally nontransformable, low-GC-content, gram-positive bacteria. *Appl. Environ. Microbiol.* **70,** 6887–6891.
3. Kreiswirth, B. N., Lofdahl, S., Betley, M. J., O'Reilly, M., Schlievert, P. M., Bergdoll, M. S., and Novick, R. P. (1983) The toxic shock syndrome exotoxin structural gene is not detectably transmitted by a prophage. *Nature* **305,** 709–712.
4. Wilson, C. R., Skinner, S. E., and Shaw, W. V. (1981) Analysis of two chloramphenicol resistance plasmids from Staphylococcus aureus: insertional inactivation of Cm resistance, mapping of restriction sites, and construction of cloning vehicles. *Plasmid* **5,** 245–258.
5. Jones, C. L. and Khan, S. A. (1986) Nucleotide sequence of the enterotoxin B gene from Staphylococcus aureus. *J. Bacteriol.* **166,** 29–33.
6. Greene, C., McDevitt, D., Francois, P., Vaudaux, P. E., Lew, D. P., and Foster, T. J. (1995) Adhesion properties of mutants of Staphylococcus aureus defective in fibronectin-binding proteins and studies on the expression of fnb genes. *Mol. Microbiol.* **17,** 1143–1152.
7. Maguin, E., Prevost, H., Ehrlich, S. D., and Gruss, A. (1996) Efficient insertional mutagenesis in lactococci and other gram-positive bacteria. *J. Bacteriol.* **178,** 931–935.
8. O'Reilly, M., de Azavedo, J. C., Kennedy, S., and Foster, T. J. (1986) Inactivation of the alpha-haemolysin gene of Staphylococcus aureus 8325-4 by site-directed mutagenesis and studies on the expression of its haemolysins. *Microb. Pathog.* **1,** 125–138.
9. Luchansky, J. B., Benson, A. K., and Atherly, A. G. (1989) Construction, transfer and properties of a novel temperature-sensitive integrable plasmid for genomic analysis of Staphylococcus aureus. *Mol. Microbiol.* **3,** 65–78.
10. Novick, R. P. (1991) The Staphylococcus as a molecular genetic system, in *Molecular Biology of the Staphylococci* (Novick, R P., ed.), VCH, New York, pp. 1–37.
11. Fitzgerald, J. R., Monday, S. R., Foster, T. J., Bohach, G. A., Hartigan, P. J., Meaney, W. J., and Smyth, C. J. (2001) Characterization of a putative pathogenicity island from bovine Staphylococcus aureus encoding multiple superantigens. *J. Bacteriol.* **183,** 63–70.
12. Southern, E. M. (1992) Detection of specific sequences among DNA fragments separated by gel electrophoresis. *Biotechnology* **24,** 122–139.
13. Ubeda, C., Maiques, E., Knecht, E., Lasa, I., Novick, R. P., and Penades, J. R. (2005) Antibiotic-induced SOS response promotes horizontal dissemination of pathogenicity island-encoded virulence factors in staphylococci. *Mol. Microbiol.* **56,** 836–844.

9

Molecular Analysis of Staphylococcal Superantigens

Patrick M. Schlievert and Laura C. Case

Summary

Staphylococcal superantigens (SAgs) comprise a large family of exotoxins produced by *Staphylococcus aureus* strains. These exotoxins are important in a variety of serious human diseases, including menstrual and nonmenstrual toxic shock syndrome (TSS), staphylococcal pneumonias, and a recently described staphylococcal purpura fulminans. In addition, these SAg exotoxins are being increasingly recognized for their possible roles in many other human diseases, such as atopic dermatitis, Kawasaki syndrome, nasal polyposis, and certain autoimmune disorders. To clarify the full spectrum of human diseases caused by staphylococcal SAgs, it is necessary to have assays for them. At present there are 17 well-characterized, serologically distinct SAgs made by *S. aureus*: TSS toxin-1; staphylococcal enterotoxins (SEs) A, B, C (multiple minor variant forms exist), D, E, and I; and SE-like G, H, J, K, L, M, N, O, P, and Q. In addition, SE-like proteins R, S, T, and U have been identified but remain poorly characterized. The most straightforward way to analyze *S. aureus* strains for the well-characterized SAgs is through polymerase chain reaction for their genes; we provide here our method for this analysis. Although it would be ideal to confirm that all of the same SAgs are produced by *S. aureus* strains that have the genes, antibody reagents for SAg detection are only available for TSS toxin-1; SEs A–E; and enterotoxin-like proteins G, H, and Q. We provide a Western immunoblot procedure that allows in vitro quantification of these SAgs.

Key Words: Superantigens; *Staphylococcus aureus*; toxic shock syndrome; necrotizing pneumonia; purpura fulminans.

1. Introduction

Staphylococcus aureus causes a large number of human diseases, ranging from benign pimples and boils to life-threatening toxic shock syndrome (TSS), necrotizing pneumonia, and purpura fulminans *(1–7)*. The organism has an array of cell-surface and secreted virulence factors that allow it to cause illnesses *(6)*. The surface virulence factors allow the organism to colonize the host, through

From: *Methods in Molecular Biology: MRSA Protocols*
Edited by: Y. Ji © Humana Press Inc., Totowa, NJ

adhesion to mucosal surfaces and resistance to phagocytosis. The secreted factors, including exoenzymes and exotoxins, allow the organism to interfere with normal immune system function, spread into surrounding tissues, and access nutrients through cell damage. Among the secreted virulence factors that have known roles in serious human diseases, are the superantigens (SAgs) *(7,8)*.

SAgs are simple protein exotoxins secreted by *S. aureus* strains as variable traits, meaning that some strains make the proteins whereas others do not *(7,8)*. The SAg family made by *S. aureus* includes toxin-1 (TSST-1), the cause of nearly all menstrual TSS and half of nonmenstrual TSS *(2,9–11)*; staphylococcal enterotoxin (SE) serotypes A, B, Cn (in which n denotes that multiple variant forms exist), D, E, and I; and SE-like serotypes G, H, and J–U *(12)*. Incidentally, there is no enterotoxin F, because this protein was renamed TSST-1 in 1984. Of these SAgs, SEs A–E and I have the ability to induce emesis in monkeys and are thus correctly referred to as SEs *(12)*. The remaining SAgs either have not been tested for emetic activity or lack emetic activity and are thus correctly referred to as enterotoxin-like proteins (SE-like G, H, and J–U) and TSST-1 *(12)*. All SEs and SE-like SAgs are able to cause nonmenstrual TSS illnesses *(7,8)*, but in addition to TSST-1, only SEG appears to be associated with menstrual TSS *(13)*.

For purposes of this chapter, we provide methods to study TSST-1; SEs A–E and I; and SE-like proteins G, H, and J–Q. Their genes have been well characterized, and specific DNA primers may be made for their detection *(12)*. We believe similar methods may be used for detection of the genes of enterotoxin-like proteins T and U when more information is available for these SAgs. We also provide methods for quantifying production of selected SAg proteins as measures to confirm gene function. These include tests for TSST-1; SEs A–E; and SE-like proteins G, H, and Q. These proteins appear to be made in higher concentrations in vitro than the other SAgs, and they have been associated with human diseases *(2,7,8,10,11,13,14)*. Assays for other SAgs, as they become recognized as more important, can be developed based on the techniques provided.

TSS is defined by the presence of fever, hypotension/shock, a red rash, desquamation of the skin on recovery, and a variable multiorgan component *(15–17)*. Many TSS patients do not meet all five of these defining criteria, which has led to the establishment of categories of illness including probable TSS and toxin-mediated disease *(18,19)*. Thus, although it is clear that TSS is caused by SAgs, the spectrum of TSS illnesses remains to be defined completely. It is relatively easy to identify menstrual TSS, but there are large numbers of categories of nonmenstrual TSS-like illnesses, many of which are only now being associated with SAg production and *S. aureus*. Nonmenstrual TSS-like illnesses include illness associated with wounds, boils, and abscesses *(19)*; postsurgical *(19)*; post–respiratory virus TSS *(5,20)*, purpura fulminans *(5)*,

Molecular Analysis of Staphylococcal SAgs

and necrotizing pneumonia *(1,14)*; and recalcitrant erythematous desquamating disorder of patients with acquired immunodeficiency syndrome *(21)*. Other illnesses that may be associated with staphylococcal SAg production include atopic dermatitis and anaphylactic reactions *(22,23)*, nasal polyposis *(3)*, mycosis fungoides (cutaneous T-cell lymphoma) *(24)*, Kawasaki-like illness *(25)*, and some cases of acute onset rheumatoid arthritis. To understand the role of staphylococcal SAgs in human illnesses, assays have been developed for their measurement. These include first polymerase chain reaction (PCR)–based techniques to demonstrate the presence of the genes, followed by antibody-based assays for protein production. These latter antibody assays were previously unnecessary for measures of TSST-1 and SEs B and C, because there was a one-to-one correlation between the presence of the genes and production of high levels of protein by the strains *(8,26)*. This is not the case with other SAgs, where, e.g., SEA-positive strains may make undetectable protein (<5 pg/mL to 5 µg/mL) in vitro, depending on the strain analyzed.

SAgs have an unusual mechanism of action in their activation of the immune system, and thus, they are often used as probes of immune system function. Such studies are likely to become even more important as the spectrum of SAg illnesses expands, including into autoimmunity. SAgs act by causing T-lymphocyte proliferation independent of the antigenic specificity of the T-cell; rather, such proliferation is dependent on the composition of the variable region of the β-chain of the typical α-β T-cell receptor (TCR) *(7,27)*. Thus, TSST-1 stimulates all human T-cells with the β-chain TCR variable region 2 (VβTCR-2), but not other T-cells. The consequence of this stimulation is that VβTCR-2 T-cells may expand to become 60–70% of all of a TSS patient's T-cells during acute TSS, as opposed to 10% normally. The ability of the SAgs to stimulate T-cells also depends on the proteins' abilities to bind to major histocompatibility complex II (MHC II) molecules on the surface of antigen-presenting cells, most importantly macrophages *(7)*. The consequence of the cross bridging of VβTCR on T-cells with MHC II on macrophages is massive cytokine release by macrophages (tumor necrosis factor-α [TNF-α] and interleukin-1β) and T-cells (TNF-β and interferon-γ) and consequent TSS illness *(7)*. Presumably, variations in amounts and types of SAgs made by *S. aureus* strains control the severity and type of illness produced.

2. Materials
2.1. S. aureus *Strains*

1. Control *S. aureus* strains for detection of SAg genes: TSS strain MN8 for TSST-1 and atopic dermatitis strain MN22 for all others; the latter strain is a naturally occurring clinical isolate with the genes for all SAgs except TSST-1 (*see* **Note 1**).

2. Control *S. aureus* strains for detection of SAg production: strain MN8 for production of TSST-1; CDC 11 for SEA; nonmenstrual TSS isolate MNHO for SEB; community-associated methicillin-resistant isolate MW2 for SEC and SEH (strain represents CDC USA 400 necrotizing pneumonia clones); isolate KSI1410 for SED; FRI 918 for SEE; MN6 for SE-like G; and MNNJ for SE-like Q (*see* **Note 2**). Production of the remaining SAgs is not well characterized; thus, we do not present data related to detection of their proteins.
3. SAgs SEA–E, SE-like H, and TSST-1: These may be purchased from Toxin Technology (Sarasota, FL), to serve as control proteins. The Schlievert laboratory makes all necessary SAgs to serve as controls. A detailed description of their production is beyond the scope of this chapter but has been provided elsewhere *(4)*.
4. Todd Hewitt medium (Becton, Dickinson, and Company, Sparks, MD), for growth of *S. aureus* strains.

2.2. Polymerase Chain Reaction

1. Bacteria grown overnight in 25 mL of Todd Hewitt broth.
2. DNeasy Tissue Kit (Qiagen, Valencia, CA), for DNA extraction.
3. ReddyMix PCR Master Mix with loading dye (Abgene, Rochester, NY).
4. Primers (*see* **Table 1**) from Sigma Genosys (St. Louis, MO). They are ordered DESALT and on a 0.05-μmol scale. Once received, resuspend the primers in water to a final concentration of 100 pmol/μL (*see* **Note 3**).
5. TempAssure 8-tube strips (0.8 mL), to hold the reaction mixtures (USA Scientific, Ocala, FL).
6. PCR machine such as the Thermal Mastercycler (which is what we use) with gradient #5331 (Eppendorf, Hamburg, Germany).
7. DNA gel:
    a. Type 1 agarose (Sigma-Aldrich, St. Louis. MO).
    b. Tris acetate EDTA (TAE) buffer made at a 10X concentration. The ingredients for the 10X buffer are 96.9 g of Tris base, 13.6 g of sodium acetate, 6.05 g of EDTA (disodium), and 2 L of distilled water.
8. DNA gel apparatus QS-710 (International Biotechnologies, New Haven, CT).
9. exACTGene 1 kb Plus DNA Ladder (Fisher, Pittsburgh, PA).
10. Ethidium bromide solution made to a final concentration of 1 μg/μL using powdered ethidium bromide dissolved in distilled water. This chemical is very hazardous and should only be weighed out under a chemical hood and handled with gloves.

2.3. Antisera

1. Polyclonal hyperimmune antisera against the SAgs SEA–E, SE-like H, and TSST-1 (Toxin Technology, Sarasota, FL) (*see* **Note 4**). Hyperimmune antisera against SE-like G and Q are not commercially available, but we produce antisera in rabbits.
2. Antisera made by immunization of rabbits with the various purified SAgs. It is our experience that injection of 25 μg of SAg emulsified in incomplete Freund's adjuvant into the nape of the neck of rabbits every other week for three to four injections results in high-level serum antibody titers.

Table 1
PCR Primers for SAgs Presented from 5N to 3N

Primer name	Primer sequence	Approximate size (bp)
SEA Forward	ATTGTTTTGGGGGAGTTTGAAGTT	400
SEA Reverse	TACATTGCGTTTTATTGGTTGCTC	
SEB Forward	GTATGATGATAATCATGTATCAGCAATA	640
SEB Reverse	CGTAAGATAAACTTCAATCTTCACATC	
SEC Forward	GAGTCAACCAGACCCTATGCC	610
SEC Reverse	CGCCTGGTGCAGGCATC	
SED Forward	GAGACTAGCCGCAATCTATCC	650
SED Reverse	GCTGCATTTAGTAATGCTGGCTG	
SEE Forward	GGTAGCGAGAAAAGCGAAG	450
SEE Reverse	GCCTTGCCTGAAGATCTAGCTC	
SEG Forward	TGAATGCTCAACCCGATCCTAAAT	580
SEG Reverse	CAAACCAAAAACTTGTATTGTTCTTTTCA	
SEH Forward	TTCACATCATATGCGAAAGCAGAA	620
SEH Reverse	CAGATTTTAAAGTTTTATTGTCTTCA	
SEI Forward	CGTATGCTCAAGGTGATATTGGTG	580
SEI Reverse	AAAAACTTACAGGCAGTCCATCTCC	
SEJ Forward *Bam*HI	TTTAGGATCCCTACAGAACCAAAGG	900
SEJ Reverse *Nco*I	GTTTCCATGGATAGCAAAAATGAAAC	
SEK Forward	GTGTCTCTAATAATGCCAGCGCTC	650
SEK Reverse	TTTGGTAGCCCATCATCTCC	
SEL Forward	CACCAGAATCACACCGCTTA	450
SEL Reverse	TCCCCTTATCAAAACCGCTAT	
SEM Forward	TTTTGCTATTCGCAAAATCATATCGCA	800
SEM Reverse	TCAACTTTCGTCCTTATAAGATATTTCTAC	
SEN Forward	TGAGATTGTTCTACATAGCTGCAA	720
SEN Reverse	AATTAGATGAGCTAACTGTTCTATTATCAC	
SEO Forward	TAGTGTAAACAATGCATATGCAAATG	950
SEO Reverse	ATTATGTAAATAAATAAACATCAATATGATA	
SEP Forward	GGAAGCTAAAGCAGAGACAC	660
SEP Reverse	CCCGTTTCATATGAAGTGCCACC	
SEQ Forward	GCTTCAAGGAGTTAGTTCTGG	500
SEQ Reverse	CTCTCTGCTTGACCAGTTCCGGTG	
TSST-1 Forward	GAAATTTTTCATCGTAAGCCCTTTGTTG	625
TSST-1 Reverse	TTCATCAATATTTATAGGTGGTTTTTCA	

2.4. Sodium Dodecyl Sulfate-Polyacrylamide Gel Electrophoresis

1. 30% Acrylamide aqueous solution (29 g of acrylamide, 1 g of bis-acrylamide/100 mL). Acrylamide is neurotoxic, so care should be taken when working with this compound.
2. TEMED.
3. Lower (separating) buffer: 1 M Tris-HCl, pH 8.8, 20% (w/v) sodium dodecyl sulfate (SDS). Store at room temperature.
4. Upper (stacking) buffer: 1 M Tris-HCl, pH 6.8, 20% (w/v) SDS. Store at room temperature.
5. Ammonium persulfate: Prepare fresh each time a 10% solution in water.
6. Running buffer: 0.05 M Tris base, 0.4 M glycine, 0.1% (w/v) SDS. Store at room temperature.
7. Sample buffer (2X): 0.125 M Tris-HCl, pH 6.8, 4% (w/v) SDS, 20% (v/v) glycerol, 2% (v/v) 2-mercaptoethanol, 0.001% (w/v) bromophenol blue in water.
8. Prestained molecular weight markers (SeeBlue® Plus2; Invitrogen, Carlsbad, CA).
9. SDS-polyacrylamide gel electrophoresis (PAGE) and Western transfer apparatus: We use a Bio-Rad mini PROTEAN® 3 Cell and Western transfer apparatus (Hercules, CA).

2.5. Western Immunoblotting

1. Transfer buffer: 0.025 M Tris base (do not adjust the pH), 0.05 M glycine.
2. Immunoblot polyvinyl difluoride (PVDF) membrane (Bio-Rad) and methanol to equilibrate the PVDF membrane.
3. Tris-buffered saline with Tween-20 (TBST): 0.01 M Tris-HCl, pH 7.5, 0.05 M NaCl, 0.5% (v/v) Tween-20.
4. Blocking solution: 1% (w/v) bovine serum albumin (BSA) in TBST.
5. Primary antibody: 10 µL/30 mL of TBST.
6. Secondary antibody (10 µL/30 mL of TBST): Alkaline phosphatase–conjugated antirabbit IgG (Sigma-Aldrich).
7. Substrate: Buffer (20 mL) containing 0.05 M Tris-HCl, pH 9.6, 0.004 M $MgCl_2$, 66 µL of nitroblue tetrazolium dye (2 mg/mL in water) and 33 µL of 5-bromo-4-chloro-3 indolyl phosphate (5 mg/mL in dimethyl formamide).

2.6. Density Scans

1. Image J 1.34S program from NIH (this can be downloaded from http://rsb.info.nih.gov/ij/).

3. Methods

For assessing the presence of the genes for all known SAgs and the proteins for SEA–E; SE-like G, H, and Q; and TSST-1, strains to be analyzed plus controls are cultured aerobically (with shaking at 200 rpm) at 37°C until the organisms are well into stationary phase. Most SAgs are made either during the exponential phase (from an initial absorbance of 0 to 1.0 at 600 nm wavelength; examples are SEA and SE-like K) or post exponential phase (from 1.0 to approx 3.0 absorbance at 600 nm; examples are SEB, SEC, and TSST-1). We use Todd Hewitt broth as the growth medium.

3.1. Polymerase Chain Reaction

1. Grow bacteria overnight in 25 mL of Todd Hewitt medium at 37°C with shaking at 200 rpm. Add the grown culture (1.5 mL) to a microcentrifuge tube and spin down at 14,000 rpm for 2 min. Aspirate the supernate, and extract the DNA from the pellet according to the protocol found on page 33 of the DNeasy Tissue Handbook provided by the supplier of the kit. There are two deviations from the DNeasy method:
   a. Add 20 µL of lysostaphin to the DNeasy enzymatic lysis buffer (*see* **Note 5**). Lysostaphin can be purchased from Sigma-Aldrich and is resuspended in distilled water to a final concentration of 1 mg/mL (it can be stored frozen with repetitive freeze/thaws at −20°C until used). The enzymatic lysis buffer is made fresh for each batch of bacteria grown.
   b. Instead of performing elutions twice with 200 µL of EB buffer each time, elute the DNA twice with 100 µL of water each time.
2. After the DNA is extracted in a final volume of 200 µL, make the PCR reaction mixture. The ReddyMix PCR Master Mix must stay on ice during the preparation. To perform PCR of the genes, add the following reagents to a 0.2-mL tube to make a 50-µL total volume reaction: 47 µL of ReddyMix PCR Master Mix, 1 µL of the forward primer of the gene to be amplified, 1 µL of the reverse primer of the gene to be amplified, 1 µL of bacterial DNA. These values may be scaled down by half if needed. Also include a positive control bacterial DNA sample and a no-template negative control.
3. After all reaction mixtures are made, tightly close the caps of the tubes, and insert the tubes into an Eppendorf Thermal Mastercycler. The cycle settings are as follows: 94°C for 1 min, 52°C for 1 min, and 72°C for 2 min for 35 cycles; 72°C for 3 min; and a hold at 4°C until the DNA gel is prepared and ready to be subjected to electrophoresis.
4. To make the DNA gel, dilute 10X TAE buffer to 1X with distilled water. Add 3.3 g of agarose to 300 mL of 1X TAE buffer to give a final gel concentration of 1.1%. Microwave the mixture until it becomes clear. Allow the hot gel mixture to cool on the bench for 5 min. Pour the gel into the gel tray until the bottom of the tray is covered with 0.75–1 cm of gel. Immediately add 10 µL of ethidium bromide to the mixture to give a final concentration of approx 0.5 µg/mL and stir in before the gel solidifies. Add a comb to the gel with the appropriate number of wells (samples + ladder).
5. Once the gel has completely solidified, remove the comb and place the gel in the gel-running apparatus. Then completely submerge the gel in the same 1X TAE buffer that was used to prepare the DNA gel.

6. Add DNA ladder (10 μL) to the leftmost lane, and add 10 μL of PCR products individually to additional wells. Attach the gel apparatus cathode and anode to a power supply, and electrophorese the samples at approx 100 V for 1 h.
7. Take the gel to an ultraviolet (UV) light box and expose to UV light. The UV light will cause the DNA to luminesce, allowing detection of the PCR products. Products under 100 bp in length are assumed to be degraded DNA or remnants of primers. **Figure 1** shows the results of testing a strain of *S. aureus* from a recently emergent clone (referred to as USA 300 by Centers for Disease Control) that caused necrotizing pneumonia for the presence of SAg genes. The strain is positive only for the SE-like Q gene (*see* **Note 6**).

3.2. Growth and Preparation of S. aureus for SAg Protein Assays

1. Inoculate *S. aureus* strains into 10 mL Todd Hewitt broth in 125-mL Erlenmeyer flasks, and culture aerobically until stationary phase.
2. Place the samples in 50-mL conical centrifuge tubes (Falcon polypropylene; Becton, Dickinson, and Company), and add 40 mL of absolute ethanol (4 vol) to the cultures to precipitate SAgs while killing the *S. aureus* cells. We have found that this 80% final concentration of ethanol will precipitate 100% of SAgs in as little as 2 h; the treated cells may remain at room temperature or 4°C for months without losing the ability to be resolubilized.
3. Collect the precipitates by centrifuging at 1000g for 15 min. Pour off the ethanol, and place the samples flat under a laminar flow hood for 30 min to dry off the remaining ethanol (*see* **Note 7**).
4. Resuspend the samples in 1 mL of distilled water (10 times concentrated relative to the original supernate concentration), and centrifuge in a microfuge (14,000 ×g for 5 min).
5. Treat a small amount (100 μL) of the clarified supernates with 100 μL of SDS-PAGE sample buffer, and leave at room temperature overnight to denature the proteins.

3.3. Sodium Dodecyl Sulfate-Polyacrylamide Gel Electrophoresis

As stated previously, we use the Bio-Rad mini PROTEAN 3 Cell and Western transfer apparatus. The gel apparatus is equipped with a minigel system that is 0.75 mm thick. The glass slides are carefully washed with soap and water, then rinsed with ethanol, and dried. The apparatus is assembled and the gels are poured.

1. Prepare the lower separating gel by mixing 1.75 mL of acrylamide solution, 1.9 mL of lower buffer, 1.35 mL of distilled water, 10 μL of TEMED, and 70 μL of ammonium persulfate for each gel (*see* **Note 8**). Pipet the solution into the gel apparatus, leaving space for the stacking gel. We overlay the gel with distilled water. This gel takes approx 30 min to polymerize.
2. Pour off the overlay solution and prepare the stacking gel.
   To make the stacking gel solution, mix 0.18 mL of acrylamide, 0.24 mL of upper Tris buffer, 0.9 mL of distilled water, 5 μL of TEMED, and 10 μL of ammonium persulfate per each gel.

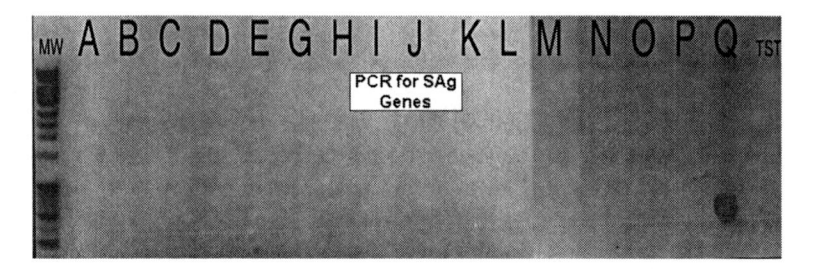

Fig. 1. PCR analysis of DNA from *S. aureus* strain that caused necrotizing pneumonia.

3. Pour the gel, put the comb in place, and allow the gel to polymerize for 30 min.
4. Carefully remove the comb, and place the gel in the apparatus for electrophoresis.
5. Add running buffer to the upper and lower chambers, and add 20 µL of each sample to the wells using a pro-pipet equipped with extended tips. We electrophorese control SAg samples of 10.0, 1, 0.1, 0.01, and 0.001 µg/mL, molecular weight markers, as recommended by the supplier, and then 20-µL amounts of the unknown samples (these samples can also be diluted 10- and 100-fold to ensure that the SAg concentrations are in the range of appropriate quantification).
6. Completely assemble the apparatus with electrodes and turn on the apparatus.
7. Electrophorese the samples at 80 V until the bromophenol blue marker dye reaches the bottom of the separating gel (*see* **Note 9**). Then turn off the apparatus. The gels are now ready to be used for Western immunoblotting.

3.4. Western Immunoblotting for SAgs

The samples that have been separated by SDS-PAGE, including unknowns, prestained molecular weight markers, and purified standard SAgs, are transferred to PVDF membranes for Western immunoblotting. These directions assume use of the Bio-Rad mini PROTEAN 3 Cell and Western transfer apparatus.

1. Wet the PVDF membrane in absolute methanol and then place in a glass dish containing Western transfer buffer. At the same time, place two pieces of foam (included with the apparatus) and two pieces of 3MM paper cut to the size of the PVDF membrane in the glass dish.
2. Remove the SDS-PAGE gel from the SDS apparatus, separate the glass plates containing the gel with a spatula, remove the stacking gel with a spatula, and place the plate with the gel attached in the glass dish with transfer buffer. Remove the gel from its glass plate using gentle agitation and a spatula, and then place on a piece of wetted 3MM paper. In order, place the following on the transfer cassette: one piece of wetted foam, the piece of 3MM paper with attached gel, the PVDF membrane, the second piece of 3MM paper, and the second piece of wetted foam. Be careful not to allow air bubbles to become trapped between the 3MM paper with gel, the PVDF membrane, and the second piece of 3MM paper, to ensure uniform transfer. We set up the cassette with the

black side down and add the stack toward the clear side. When inserted into the transfer apparatus, the black side is toward the cathode (which is marked as black).
3. Fill the transfer tank with transfer buffer and insert the closed cassette.
4. Allow electrotransfer of the samples onto the PVDF membrane for about 1 h, beginning with 100 V and monitoring the amperage until it shifts from approx 250 mA at the start and rises to 350 mA.
5. Disassemble the apparatus and remove the PVDF membrane. Nitrocellulose and Nytran membranes can also be used, but we use PVDF membranes because they are durable and useful for protein sequencing (*see* **Note 10**).
6. Place the PVDF membrane in 30 mL of TBS (two times) to wash the membrane. There is no incubation with the TBS; add it immediately and then pour off.
7. Block the PVDF membrane for 30 min by adding 30 mL of TBST containing 1% BSA and gently agitating on a shaking platform (*see* **Note 11**).
8. Directly add the primary antibody, and agitate the membrane for 30 min.
9. Pour off the primary antibody and agitate the gel two times for 5 min each with TBST.
10. Add secondary antibody in 30 mL of TBST, and incubate with agitation for 30 min.
11. Pour off the secondary antibody, and wash the PVDF membrane twice (5 min each) with TBST and then once with TBS (5 min).
12. Add alkaline phosphatase buffer with added substrate (30 mL total), and agitate the PVDF membrane until standard SAg samples have reacted, turning the membrane purple. The time to stop the reaction is just when the background also begins to turn purple.
13. Remove the PVDF membrane from the substrate solution, rinse with distilled water, and air-dry.
14. Photograph the PVDF membrane and visualize the image on a computer.

3.5. Determination of Density

For measuring SAgs, we use the NIH imaging program to capture the density of the protein bands. To do this, we use the program to draw around each stained band and determine its density. The background density is determined from three regions of the PVDF membrane that do not contain protein and subtracted from the SAg values.

It is important to note that *S. aureus* makes protein A, which has the ability to bind IgG nonspecifically and thus will also stain the PVDF membranes. The protein A bands are not present in the same molecular weight range as the SAgs and, thus, do not interfere.

Standard amounts of SAg are used to prepare a standard curve by plotting band density vs the logarithm of the SAg concentration. This typically gives a straight line with an R^2 value >95. The amount of SAg made by the unknown samples is derived from the graph.

4. Notes

1. *S. aureus* strains may be stored indefinitely lyophilized or frozen at −20°C. To store strains, grow the organisms in Todd Hewitt broth until stationary phase; usually overnight will suffice. Collect the cells by centrifugation (1000g, 15 min), and suspend them in 1/10 vol with Todd Hewitt broth supplements with 10% rabbit blood (for lyophilization) or Todd Hewitt broth containing 10% glycerol (for freezing). The *S. aureus* cells may then be lyophilized in ampoules or frozen.
2. If the investigator wishes to assess only whether or not SAgs are produced by the strains and not quantified, double immunodiffusion may be used instead of Western immunoblotting. It is our experience that the antisera and control SAgs available from Toxin Technology are suitable for such analyses. Double immunodiffusion is performed after growing the strains in Todd Hewitt broth and concentrating the culture fluids by treatment with 4 vol of ethanol and resuspension to one-tenth the original volume. Microscope slides are used to prepare the double immunodiffusion gels. These are coated with 4 mL of 0.75% agarose melted in phosphate-buffered saline (0.005 M sodium phosphate, pH 7.2; 0.15 M NaCl). After the slides have solidified, wells 4 mm in diameter and 4 mm from each other are punched in the agarose in a hexagonal pattern. Control SAg (20 µL of 50–100 µg/mL) and samples to be tested (20 µL) are added to adjacent wells. The slides are incubated for 4 h at 37°C in a humidified chamber or overnight at room temperature. The slides are read for precipitation lines forming lines of identity with control SAgs. In this assay, staphylococcal protein A does not interfere with the reactions, because rabbit IgG does not precipitate with protein A, even though it will not specifically react with rabbit IgG. It is our experience that this assay can be used with 100% effectiveness to detect SEB, SEC, SED, SEE, SE-like H, and TSST-1 qualitatively.
3. When PCR is performed, the investigator may wish to use multiplex analysis in which multiple assays are performed in the same reaction tube, with agarose gel electrophoresis used to separate the different samples based on size. We do not routinely perform such assays. However, such analyses are possible if the researcher uses primers of different sizes. Some of the primer combinations given in **Table 1** can be used for this purpose.
4. It is our experience that the antisera and SAgs purchased from Toxin Technology are stable when stored at 4°C as long as they do not become contaminated. We have retained such preparations for up to 1 yr without loss of activity. If the researcher wishes to store samples frozen, the preparations should not be repetitively frozen and thawed.
5. Lysostaphin is required for lysis of *S. aureus* cells. Despite the fact that many textbooks state that Gram-positive bacteria can be easily lysed with lysozyme, this is not the case with staphylococci.
6. In our experience, it is always best to determine the SAg gene composition of the strains to be tested prior to performing assays for the proteins. This allows conservation of expensive antisera and control SAgs (purchased from Toxin Technology). If the investigator wishes to test only for SEB, SEC, and TSST-1,

only assay for either the genes or SAg proteins needs be performed. We find that there is a one-to-one correlation with the presence of the genes and their production. These proteins typically are made in concentrations of 5–100 µg/mL of culture fluid. Other SAg testing requires assays for both the genes and the proteins: the proteins may or may not be made by the strains, and the proteins may be made in very low concentrations by the strains.

7. It is unnecessary to dry the samples completely. We allow air to blow across the tubes until excess ethanol is removed and the surface of the residual pellet has a dull finish.
8. Ammonium persulfate is inactivated by oxygen; thus, vigorous mixing is strongly discouraged. If the gels do not polymerize, it is possible that the ammonium persulfate is too old (we purchase new ammonium persulfate every 6 mo) or that the compound is inactivated by exposure to air. In the latter case, this can be overcome by mixing the gel in a side-arm flask and de-aerating the solution while mixing. This can be done by connecting the side-arm flask to a vacuum for about 2 min while mixing.
9. The molecular weights of all SAgs are between 22,000 and 30,000. This corresponds to a gel region with minimal other contaminating staphylococcal proteins. It is worth noting that the SAgs separated by this method will be near the gel bottom. Thus, the researcher may wish to electrophorese the samples until the bromophenol blue dye is only three-fourths of the way to the bottom.
10. We use PVDF membranes for Western immunoblotting for two reasons: first, they are sturdy and resist tearing; and, second, we often electrophorese second samples with the intent of staining the PVDF membrane with Coomassie brilliant blue dye to allow visualization of all proteins. In this case we may wish to cut out and partially destain proteins for submission for direct protein sequence analysis. This is easily accomplished with the use of PVDF membranes.
11. Three percent gelatin may also be used in place of 1% BSA. In that case, be sure to incubate with blocking solution at 37°C to prevent the gelatin from solidifying.

Acknowledgment

This work was supported by USPHS research grant HL36611 from the National Heart, Lung, and Blood Institute.

References

1. (1999) From the Centers for Disease Control and Prevention. Four pediatric deaths from community-acquired methicillin-resistant *Staphylococcus aureus*—Minnesota and North Dakota, 1997–1999. *JAMA* **282,** 1123–1125.
2. Bergdoll, M. S., Crass, B. A., Reiser, R. F., Robbins, R. N., and Davis, J. P. (1981) A new staphylococcal enterotoxin, enterotoxin F, associated with toxic-shock-syndrome *Staphylococcus aureus* isolates. *Lancet* **1,** 1017–1021.
3. Bernstein, J. M., Ballow, M., Schlievert, P. M., Rich, G., Allen, C., and Dryja, D. (2003) A superantigen hypothesis for the pathogenesis of chronic hyperplastic sinusitis with massive nasal polyposis. *Am. J. Rhinol.* **17,** 321–326.

4. Blomster-Hautamaa, D. A. and Schlievert, P. M. (1988) Preparation of toxic shock syndrome toxin-1. *Methods Enzymol.* **165,** 37–43.
 5. Kravitz, G., Dries, D. J., Peterson, M. L., and Schlievert, P. M. (2005) Purpura fulminans due to *Staphylococcus aureus. Clin. Infect. Dis.* **40,** 941–947.
 6. Lowy, F. D. (1998) *Staphylococcus aureus* infections. *N. Engl. J. Med.* **339,** 520–532.
 7. McCormick, J. K., Yarwood, J. M., and Schlievert, P. M. (2001) Toxic shock syndrome and bacterial superantigens: an update. *Annu. Rev. Microbiol.* **55,** 77–104.
 8. Dinges, M. M., Orwin, P. M., and Schlievert, P. M. (2000) Exotoxins of *Staphylococcus aureus. Clin. Microbiol. Rev.* **13,** 16–34.
 9. Schlievert, P. M. (1986) Staphylococcal enterotoxin B and toxic-shock syndrome toxin-1 are significantly associated with non-menstrual TSS. *Lancet* **1,** 1149, 1150.
10. Schlievert, P. M., Shands, K. N., Dan, B. B., Schmid, G. P., and Nishimura, R. D. (1981) Identification and characterization of an exotoxin from *Staphylococcus aureus* associated with toxic-shock syndrome. *J. Infect. Dis.* **143,** 509–516.
11. Schlievert, P. M., Tripp, T. J., and Peterson, M. L. (2004) Reemergence of staphylococcal toxic shock syndrome in Minneapolis-St. Paul, Minnesota, during the 2000–2003 surveillance period. *J. Clin. Microbiol.* **42,** 2875–2886.
12. Lina, G., Bohach, G. A., Nair, S. P., Hiramatsu, K., Jouvin-Marche, E., and Mariuzza, R. (2004) Standard nomenclature for the superantigens expressed by *Staphylococcus. J. Infect. Dis.* **189,** 2334–2336.
13. Jarraud, S., Cozon, G., Vandenesch, F., Bes, M., Etienne, J., and Lina, G. (1999) Involvement of enterotoxins G and I in staphylococcal toxic shock syndrome and staphylococcal scarlet fever. *J. Clin. Microbiol.* **37,** 2446–2449.
14. Fey, P. D., Said-Salim, B., Rupp, M. E., et al. (2003) Comparative molecular analysis of community- or hospital-acquired methicillin-resistant *Staphylococcus aureus. Antimicrob. Agents Chemother.* **47,** 196–203.
15. Davis, J. P., Chesney, P. J., Wand, P. J., and LaVenture, M. (1980) Toxic-shock syndrome: epidemiologic features, recurrence, risk factors, and prevention. *N. Engl. J. Med.* **303,** 1429–1435.
16. Shands, K. N., Schmid, G. P., Dan, B. B., et al. (1980) Toxic-shock syndrome in menstruating women: association with tampon use and *Staphylococcus aureus* and clinical features in 52 cases. *N. Engl. J. Med.* **303,** 1436–1442.
17. Todd, J. K., Kapral, F. A., Fishaut, M., and Welch, T. R. (1978) Toxic shock syndrome associated with phage group 1 staphylococci. *Lancet* **2,** 1116–1118.
18. Parsonnet, J. (1998) Case definition of staphylococcal TSS: a proposed revision incorporating laboratory findings. *Int. Congress Symp. Ser.* **229,** 15.
19. Reingold, A. L., Hargrett, N. T., Dan, B. B., Shands, K. N., Strickland, B. Y., and Broome, C. V. (1982) Nonmenstrual toxic shock syndrome: a review of 130 cases. *Ann. Intern. Med.* **96,** 871–874.
20. MacDonald, K. L., Osterholm, M. T., Hedberg, C. W., et al. (1987) Toxic shock syndrome: a newly recognized complication of influenza and influenza like illness. *JAMA* **257,** 1053–1058.
21. Cone, L. A., Woodard, D. R., Byrd, R. G., Schulz, K., Kopp, S. M., and Schlievert, P. M. (1992) A recalcitrant, erythematous, desquamating disorder associated with toxin-producing staphylococci in patients with AIDS. *J. Infect. Dis.* **165,** 638–643.

22. Hofer, M. F., Harbeck, R. J., Schlievert, P. M., and Leung, D. Y. (1999) Staphylococcal toxins augment specific IgE responses by atopic patients exposed to allergen. *J. Invest. Dermatol.* **112,** 171–176.
23. Hofer, M. F., Lester, M. R., Schlievert, P. M., and Leung, D. Y. (1995) Upregulation of IgE synthesis by staphylococcal toxic shock syndrome toxin-1 in peripheral blood mononuclear cells from patients with atopic dermatitis. *Clin. Exp. Allergy* **25,** 1218–1227.
24. Jackow, C. M., Cather, J. C., Hearne, V., Asano, A. T., Musser, J. M., and Duvic, M. (1997) Association of erythrodermic cutaneous T-cell lymphoma, superantigen-positive *Staphylococcus aureus*, and oligoclonal T-cell receptor V beta gene expansion [published erratum appears in Blood 1997 89:3496]. *Blood* **89,** 32–40.
25. Leung, D. Y., Meissner, H. C., Fulton, D. R., Murray, D. L., Kotzin, B. L., and Schlievert, P. M. (1993) Toxic shock syndrome toxin-secreting *Staphylococcus aureus* in Kawasaki syndrome. *Lancet* **342,** 1385–1388.
26. Bohach, G. A., Fast, D. J., Nelson, R. D., and Schlievert, P. M. (1990) Staphylococcal and streptococcal pyrogenic toxins involved in toxic shock syndrome and related illnesses. *Crit. Rev. Microbiol.* **17,** 251–272.
27. Marrack, P. and Kappler, J. (1990) The staphylococcal enterotoxins and their relatives. *Science* **248,** 705–711.

10

Investigation of Biofilm Formation in Clinical Isolates of *Staphylococcus aureus*

James E. Cassat, Chia Y. Lee, and Mark S. Smeltzer

Summary

As with many other bacterial species, the most commonly used method to assess staphylococcal biofilm formation in vitro is the microtiter plate assay. This assay is particularly useful for comparison of multiple strains including large-scale screens of mutant libraries. When such screens are applied to the coagulase-negative staphylococci in general, and *Staphylococcus epidermidis* in particular, they are relatively straightforward by comparison with microtiter plate assays used to assess biofilm formation in other bacterial species. However, in the case of clinical isolates of *Staphylococcus aureus*, including methicillin-resistant *S. aureus*, we have found it necessary to employ specific modifications including precoating of the wells of the microtiter plate with plasma proteins and supplementation of the medium with both salt and glucose. In this chapter, we describe the microtiter plate assay in the specific context of clinical isolates of *S. aureus* and the use of these modifications. A second in vitro method, which also is generally dependent on coating with plasma proteins and supplementation of the growth medium, is the use of flow cells. In this method, bacteria are allowed to attach to a surface and then monitored with respect to their ability to remain attached to the substrate and differentiate into mature biofilms under the constant pressure of fluid shear force. Although flow cells are not applicable to large-scale screens, we have found that they provide a more reproducible and accurate assessment of the capacity of *S. aureus* clinical isolates to form a biofilm. They also provide a means of analyzing structural differences in biofilm architecture and isolating bacteria and/or spent media for analysis of physiological and metabolic changes associated with the adaptive response to growth in a biofilm. While a primary focus of this chapter is on the use of in vitro assays to assess biofilm formation in clinical isolates of *S. aureus*, it is important to emphasize two additional considerations. First, it has become increasingly evident that biofilm formation in *S. epiderimidis* and *S. aureus* is not equivalent. Additionally, to date, most studies with *S. aureus* have been done with a very limited number of strains, almost all of which are derived from the NCTC strain designated 8325, and we have found that these strains are not representative of the most relevant clinical isolates. As with the specific elements of our flow cell system, we have written this chapter

From: *Methods in Molecular Biology: MRSA Protocols*
Edited by: Y. Ji © Humana Press Inc., Totowa, NJ

to reflect our focus on clinical isolates of *S. aureus* and the specific methods that we have found most reliable in that context. Second, as is often the case, in vitro methods do not necessarily reflect events that occur in vivo. Several in vivo methods to assess biofilm formation have been described, and these generally fall into one of two categories. The first focuses directly on staphylococcal diseases that are generally thought to include a biofilm component (e.g., endocarditis, osteomyelitis, septic arthritis). A discussion of these models is also beyond the scope of this chapter, but examples are easily found in the staphylococcal literature. The second approach uses some form of implanted device in an attempt to focus more directly on implant-associated biofilms. We use a model in which a small piece of Teflon catheter is implanted subcutaneously in mice and used as a substrate for colonization. We have the advantage of using bioluminescent derivatives of *S. aureus* clinical isolates and the IVIS® imaging system. However, because this system is not generally available, we restrict technical comments in this chapter to our use of an implanted catheter model evaluated by direct microbiological analysis of explanted catheters *(2)*.

Key Words: Polysaccharide intercellular adhesin; poly-*N*-acetyl glucosamine; microbial surface components recognizing adhesive matrix molecules; flow cell; implant-associated biofilm.

1. Introduction

Staphylococcus aureus is among the most prominent of all bacterial pathogens. It is a commensal inhabitant of a significant proportion of the healthy population, but it also has the capacity to cause a diverse array of infections ranging from relatively superficial skin infections to serious, life-threatening infections including endocarditis, pneumonia, and osteomyelitis. Many forms of staphylococcal infection are associated with the formation of a bacterial biofilm on either native tissues (e.g., cartilage, bone) or implanted biomaterials (e.g., catheters, orthopedic devices). For reasons that are not completely understood, this biofilm significantly impairs antimicrobial therapy even in those cases caused by strains that are not resistant to the relevant antibiotics *(1,2)*. For this reason, considerable effort has been expended to define the specific staphylococcal factors that promote biofilm formation and/or persistence within a biofilm. The two most common in vitro methods are the microtiter plate assay and flow cells, while the most common in vivo method is the use of an implanted biomaterial that is either inoculated directly or preinoculated prior to implantation *(3,4,5)*. Our application of these three methods with respect to clinical isolates of *S. aureus* is the specific focus of this chapter.

In general, biofilm formation in all bacterial species involves four relatively distinct phases. The first phase is nonspecific interactions that promote the transient adherence to a substrate. These interactions are defined by general characteristics of the bacterium and the substrate (e.g., hydrophobicity). We have not investigated these interactions, but we have found that some microtiter plates work better than others, and this presumably reflects subtle differences in surface chemistry. With that in mind, we have included information regarding the specific components that we have found most reliable. The second phase is attachment to the substrate via specific bacterial adhesins. In *S. aureus* and many other Gram-positive pathogens, there is considerable evidence to suggest that this stage is mediated by the surface-exposed protein adhesins referred to as microbial surface components recognizing adhesive matrix molecules (MSCRAMMS) *(6,7)*. This is consistent with the need to coat the substrate used for in vitro studies with plasma proteins. It should be noted that, in our experience, plasma coating is not necessary with *Staphylococcus epidermidis*, and there are exceptions to the rule in *S. aureus (8)*. However, they are rare, and we have written this chapter to emphasize the rule rather than the exception. The third phase is the accumulation phase, in which bacteria adhere to each other in a fashion that ultimately results in the formation of a mature biofilm. In *S. epidermidis*, this is closely correlated with the presence and expression of the *ica* (intercellular adhesin) operon and the consequent production of the polysaccharide intercellular adhesin *(9,10)*. The *ica* operon is present in most *S. aureus* isolates, and in some cases, it is required for biofilm formation *(11)*. However, that is clearly not the case in all strains *(12,13)*. Whether this reflects the existence of an alternative means of accumulation or simply the fact that polysaccharide intercellular adhesin production is important under growth conditions that are not reflected in current biofilm models remains unclear, but it should be noted that there is evidence that *ica* is preferentially expressed under in vivo growth conditions *(14,15)*. This further emphasizes the need to verify the results of any in vitro biofilm assay using appropriate in vivo models, and it is for this reason that we have included a discussion of our murine model in this chapter. The fourth and final phase of biofilm formation is dispersal or release of bacteria from the biofilm. Although this can occur as a function of shear forces rather than any specific bacterial attribute, many bacteria also use specific mechanisms of dispersal, and in the case of *S. aureus*, there is evidence to suggest that production of phenol-soluble modulins (PSMs) may be important in that regard *(16)*. As an example, there is convincing evidence that induction of the accessory gene regulator (*agr*), which results in expression of the PSM δ-toxin, results in detachment of *S. aureus* from mature biofilms *(17)*. Indeed, studies from several laboratories have demonstrated that expression of *agr* is negatively correlated with biofilm formation in both *S. aureus* and *S. epidermidis (8,18)*.

2. Materials
2.1. Microtiter Plate Biofilm Assay
1. Biofilm media: tryptic soy broth (#211822; BD Biosciences) supplemented with 3.0% NaCl and 0.5% dextrose.

2. Carbonate-bicarbonate buffer (C3041; Sigma, St. Louis, MO).
3. Lyophilized human plasma (P9523; Sigma): Prepare a 20% suspension by resuspending 5 mL of lyophilized human plasma in 20 mL of filter-sterilized carbonate-bicarbonate buffer (*see* **Note 1**).
4. Phosphate-buffered saline (PBS) (10X PBS stock): 1.37 M NaCl, 27 mM KCl, 100 mM Na$_2$HPO$_4$, 18 mM KH$_2$PO$_4$. Adjust the pH to 7.4 with HCl if necessary, and autoclave before storing at room temperature. Prepare a working solution by diluting 1 part with 9 parts of water.
5. Flat-bottomed polystyrene 96-well tissue culture plates (#3596; Corning, Corning NY).

2.2. Flow Cell Biofilm Assay

1. Stovall Flow Cell Kit (#FLCAS0001; Stovall Life Sciences).
2. 10-L Polycarbonate media reservoir (#ACCFL0010 Stovall Life Sciences).
3. Lyophilized human plasma (P9523; Sigma).
4. Carbonate-bicarbonate buffer (C3041; Sigma).
5. Autoclavable tubing #EW-96429-42; Cole Parmer).
6. Luer-Lok™ syringes (#14-823-2B; Fisher).
7. Male and female Luer-Lok connectors.
8. Three-way stopcock (#K75; Baxter Pharmaseal).
9. Tabletop incubator (LabLine Thermal Rocker: 14-512-30; Fisher) (*see* **Note 2**).
10. Peristaltic pump (Ismatec Low Flow, High Accuracy, 12 channel: #ACCFL0013; Stovall Life Sciences).
11. Insulin syringes (#14-841-31; Fisher).

2.3. Catheter-Based Model of In Vivo Biofilm Formation

1. Six- to eight-week-old NIH-Swiss female mice (*see* **Note 3**).
2. 14-Gage Teflon iv catheters (#14-841-11 or similar; Fisher): Precut catheters into 1-cm segments and sterilize by autoclaving prior to surgery.
3. Vetbond™ tissue adhesive (#NC9259532; Fisher).
4. 2,2,2-Tribromoethanol (TBE) (#T48402; Sigma-Aldrich) (*see* **Note 4**): Prepare a stock solution of TBE by mixing 25 g of TBE with 15.5 mL of tert-amyl alcohol (#152463; Sigma-Aldrich) in a dark bottle. Stir for 12–24 h at room temperature until the TBE is completely dissolved. Wrap the stock solution in foil and keep at room temperature (the stock solution is both hydroscopic and photosensitive). Prepare a working solution of TBE prior to surgery by mixing 0.5 mL of the TBE stock with 39.5 mL of PBS or 0.9% saline. Stir in a dark bottle until complete resuspension has occurred (this may take several hours). Filter sterilize the resuspended working solution and then store in the dark at 4°C. The working solution, stored properly, can be used for several months.
5. Fisher Scientific Sonic Dismembrator model 500 (#15-338-550) with 1.2-in. tapped horn (#15-338-56) and 1/8-in.-diameter microtip (#15-338-67).
6. PBS: Prepare as described in **Subheading 2.1, item 4**.
7. Tryptic soy agar (#236950; BD Biosciences).

3. Methods
3.1. Microtiter Plate Biofilm Assay
3.1.1. Day 1
1. Add 200 µL of 20% human plasma into the required number of wells and incubate overnight at 4°C.
2. Start overnight cultures of each test strain in biofilm medium (tryptic soy broth supplemented with 0.5% dextrose and 3.0% NaCl).

3.1.2. Day 2
1. Remove plasma from the wells by gentle aspiration with a sterile pipet tip. Care must be taken to avoid forceful suction of plasma from the well. Slowly and gently move the vacuum tip down the side of the well until all fluid has been removed. Take care not to aspirate the contents from the bottom of the well.
2. After ensuring that all overnight cultures grew to a comparable extent, dilute overnight cultures 1:200 in sterile biofilm medium (*see* **Note 5**)
3. Inoculate microtiter plate wells with 200 µL of diluted cultures. Fill the desired number of replicate wells for each strain. Include control wells consisting of sterile biofilm medium alone. Incubate the plate at 37°C without shaking for 24 h.

3.1.3. Day 3
1. Aspirate bacterial cultures from each well using the method described in **Subheading 3.1.2., step 1**. Wash the wells gently three times with 200 µL of sterile PBS.
2. Fix with 200 µL of 100% ethanol. Immediately aspirate off the ethanol, and let the microtiter plate dry for 10 min with the lid off in a sterile hood.
3. Stain the biofilm by adding 200 µL of crystal violet to each well for exactly 2 min. Gently aspirate off the crystal violet from each well.
4. Gently wash the wells three times with 200 µL of sterile PBS. Allow the plate to dry overnight with the lid on.

3.1.4. Day 4
1. Elute crystal violet by filling the wells with 100 µL of 100% ethanol for 10 min.
2. Gently pipet the eluted stain from each well into a new microtiter plate. Read the absorbance using an enzyme-linked immunosorbent assay plate reader at an absorbance of 595 nm (*see* **Note 6**).

3.2. Flow Cell Biofilm Assay
Diagram of the Stovall flow cell (**Fig. 1**).

3.2.1. Plasma Coating
Plasma coating should be performed in a sterile environmental hood if possible.
1. Resuspend 5 mL of lyophilized human plasma in 20 mL of carbonate-bicarbonate buffer.

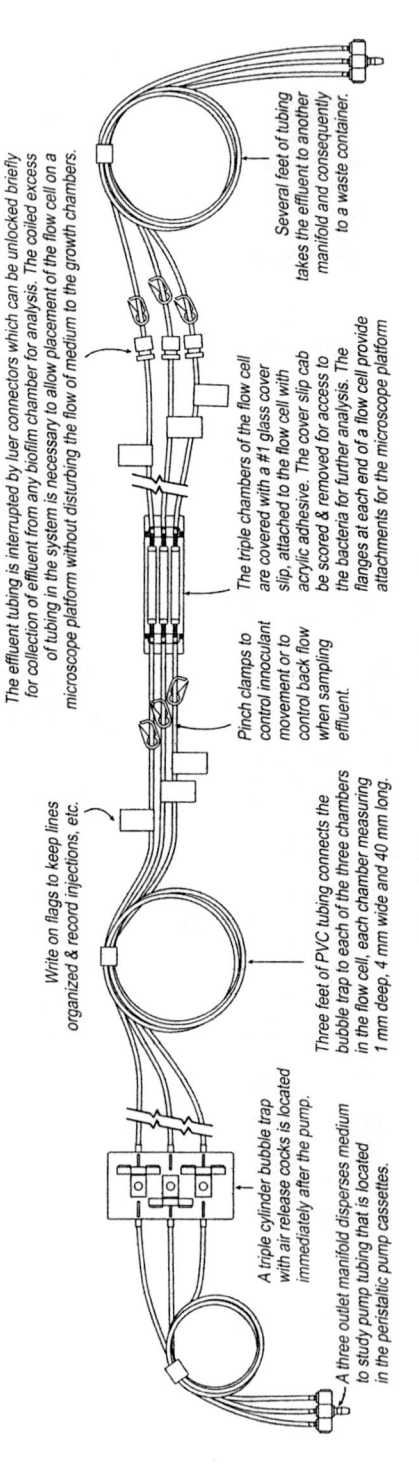

Fig. 1. Diagram of Stovall flow cell.

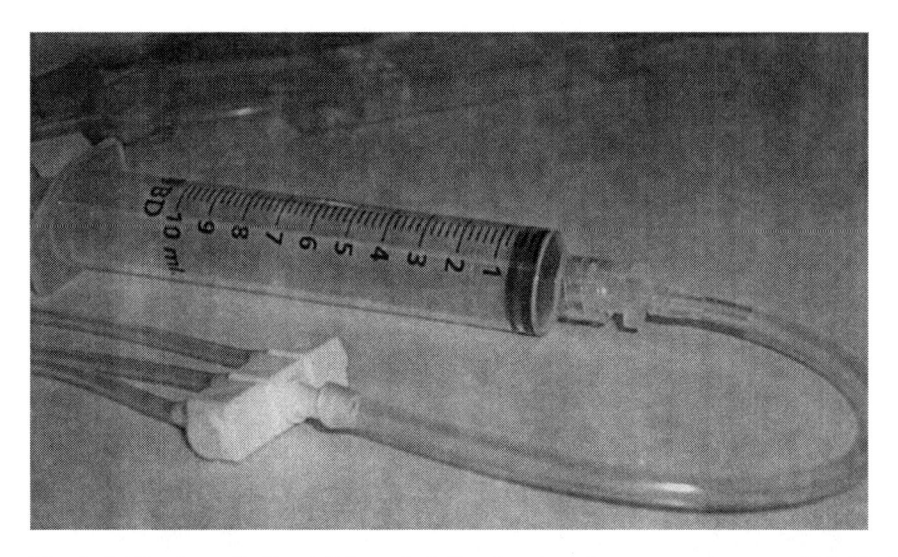

Fig. 2. A 20-mL sterile "male" Luer-Lok fitted syringe is connected to a small (~6 in.) section of sterile tubing by means of a "female" Luer-Lok adapter. The other end of the sterile tubing is subsequently attached to the flow cell output manifold. This apparatus is used to introduce plasma into the flow cell circuit.

2. Connect a sterile section of tubing fitted with a female Luer connector at one end to the flow cell output manifold. Connect a 20-mL Luer-Lok syringe to the female Luer connector (**Fig. 2**).
3. Connect a sterile section of tubing fitted with a female Luer connector at one end to the flow cell input manifold. Place this section of tubing into a sterile 50-mL beaker containing the resuspended 20% human plasma (**Fig. 3**).
4. Open all six pinch clamps on the flow cell apparatus. Slowly draw plasma into the flow cell tubing by exerting a slight pressure on the plunger of the syringe connected to the flow cell output manifold. Continue drawing plasma into the flow cell until each chamber is filled (*see* **Note 7**).
5. Close all six pinch clamps. To ensure sterility after drawing plasma into the flow cell, attach a 20-mL Luer-Lok syringe to the female Luer adapter connected to the flow cell input manifold. Rinse the connection with 70% isopropanol to remove residual plasma.
6. Incubate the entire flow cell apparatus at 4°C for 24 h.

3.2.2. Establishing Flow of Medium

1. Prior to sterilization of the media reservoir containing a sufficient quantity of biofilm medium (*see* **Note 8**), attach a three-way stopcock to the external tubing connected to the media reservoir (**Fig. 4**, inset). Wrap the stopcock in foil to ensure sterility once autoclaved. Confirm that the stopcock is in the "off" position such that the biofilm medium cannot exit the reservoir during sterilization.

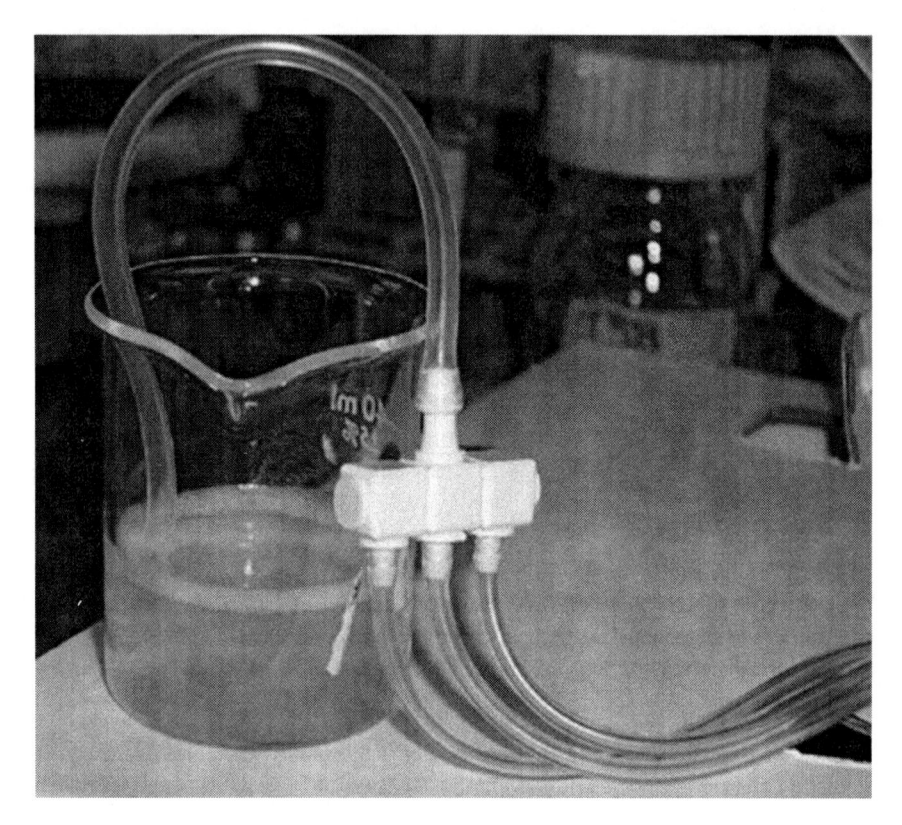

Fig. 3. A small (~6 in.) section of sterile tubing containing a "female" Luer-Lok adapter at one end is connected to the flow cell input manifold. This "female" Luer-Lok-equipped end of the sterile tubing is subsequently placed in a 50-mL beaker containing 25 mL of resuspended 20% human plasma.

2. Using aseptic technique, carefully remove the 20-mL Luer-Lok syringe and female Luer adapter from the tubing section connected to the flow cell input manifold. Replace the female Luer adapter with a threaded male adaptor. Connect the threaded male Luer adapter to the three-way stopcock attached to the sterile media reservoir (**Fig. 5**).
3. Remove the 20-mL Luer-Lok syringe from the section of tubing connected to the flow cell output manifold. Insert this section of tubing into a vessel suitable for collecting flow cell effluent waste (*see* **Note 9**). Connect the flow cell to a peristaltic pump by placing each of the three pieces of tubing between the flow cell input manifold and the bubble trap apparatus in adjacent pump channels (**Fig. 6**).
4. Open all six pinch clamps on the flow cell apparatus. Turn the three-way stopcock connected to the media reservoir to the "on" position, such that the medium can now enter the flow cell apparatus.

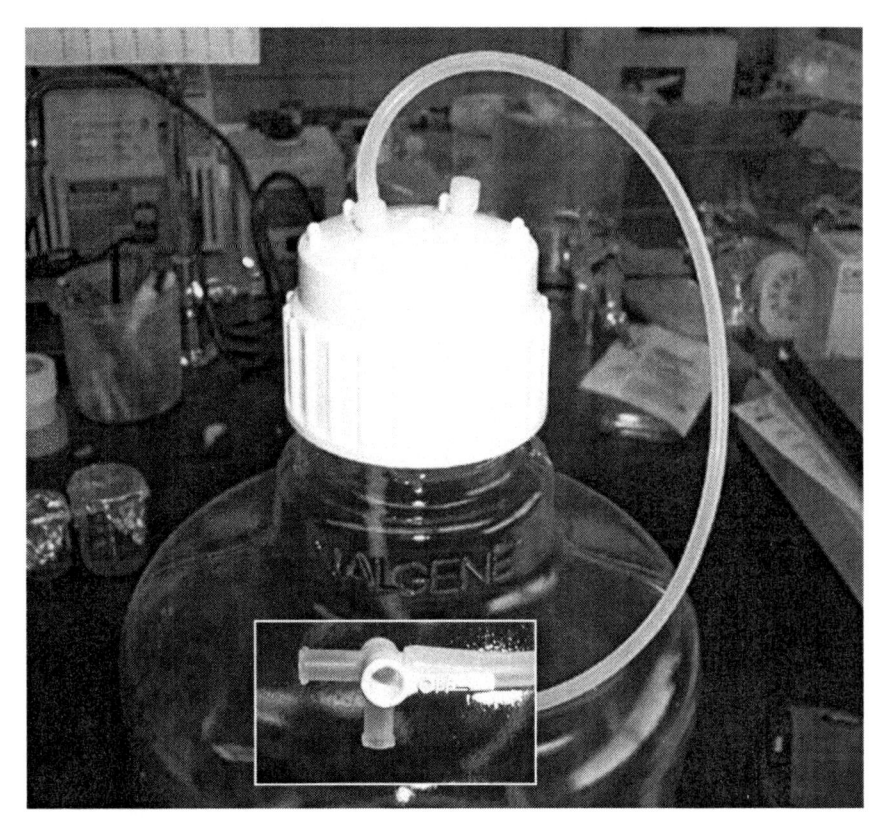

Fig. 4. Prior to sterilization of the media reservoir, a three-way stopcock is fitted to the external tubing and placed in the "off" position.

5. After ensuring that all tubing connections are intact, turn on the peristaltic pump at a rate of 1.5 mL/min (approx 0.5 mL/min per flow cell chamber). Prepare the bubble trap apparatus by turning one of the bubble trap stopcocks to the "on" (vertical) position until sterile medium has filled approximately half of the bubble trap (lower right). Repeat this process for the remaining two bubble traps and then return the bubble trap stopcock to the "off" (horizontal) position (**Fig. 7**).
6. Allow fresh biofilm medium to flow through the flow cell apparatus for 20–30 min to remove all the plasma. If bubbles have accumulated in the flow cell chambers, they can be removed by turning the chamber vertically and lightly tapping on the surface. Remove all bubbles in or near the flow cell chamber before inoculation.

3.2.3. Inoculation of Flow Cell Chambers

1. Prepare strains by setting up an overnight broth culture in biofilm medium. Prior to inoculation of the flow cell, standardize each culture to be tested based on spectrophotometer readings (*see* **Note 10**).

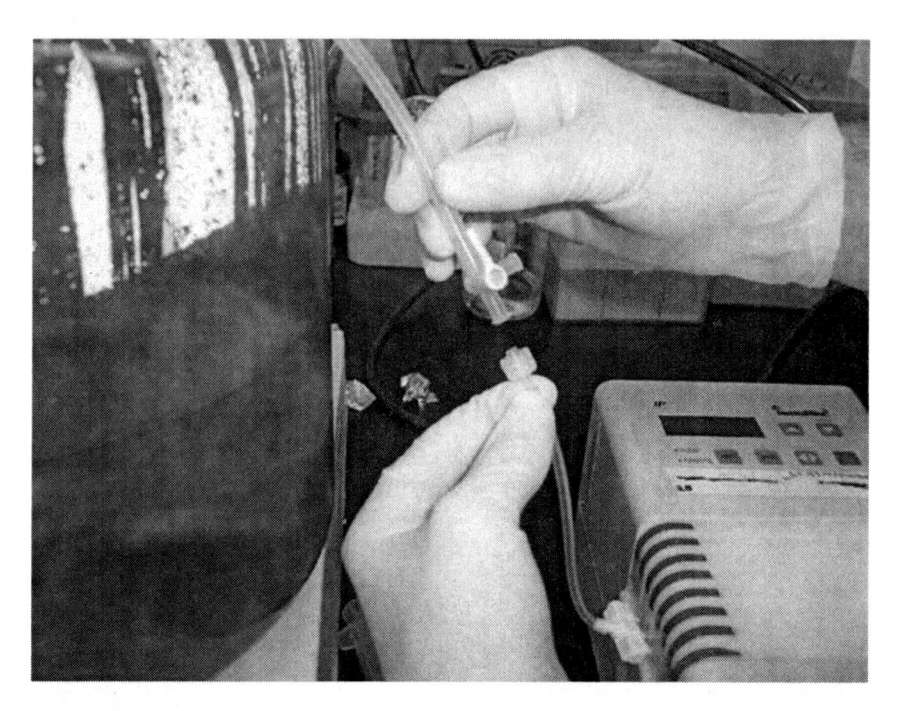

Fig. 5. The "female" Luer-Lok adapter on the section of tubing connected to the flow cell input manifold is removed and replaced with a "male" Luer-Lok adapter. The flow cell may now be aseptically connected to the sterilized media reservoir by means of the three-way stopcock.

2. Turn off the peristaltic pump and close all pinch clamps on the flow cell apparatus.
3. Clean the section of tubing between the upstream pinch clamps and the flow cell chamber with a sterile alcohol pad to prepare for inoculation. Apply a small piece of self-sealing tape (included in the Stovall Flow Cell Kit) to this section of tubing, and clean the tape with a sterile alcohol pad (**Fig. 8**).
4. Draw up 0.5 mL of each standardized bacterial culture into an insulin syringe. Working on one chamber at a time, carefully insert the needle through the self-sealing tape and into the lumen of the flow cell tubing (**Fig. 9**).
5. Open the downstream pinch clamp, and slowly inject the bacterial suspension making sure that the turbid suspension fills the flow cell chamber. Take care not to introduce bubbles into the chamber. After injection, carefully remove the needle and clean the self-sealing tape with a sterile alcohol pad once more. Close the downstream pinch clamp. Repeat this process for each strain in the respective flow cell chambers.
6. After inoculation, place the flow cell chamber upside down in an incubator to allow bacteria to attach (**Fig. 10**). Use a small weight to stabilize the flow cell in a flat position. Ensure that the upstream and downstream tubing is not pinched as it enters or exits the incubator (**Fig. 11**).

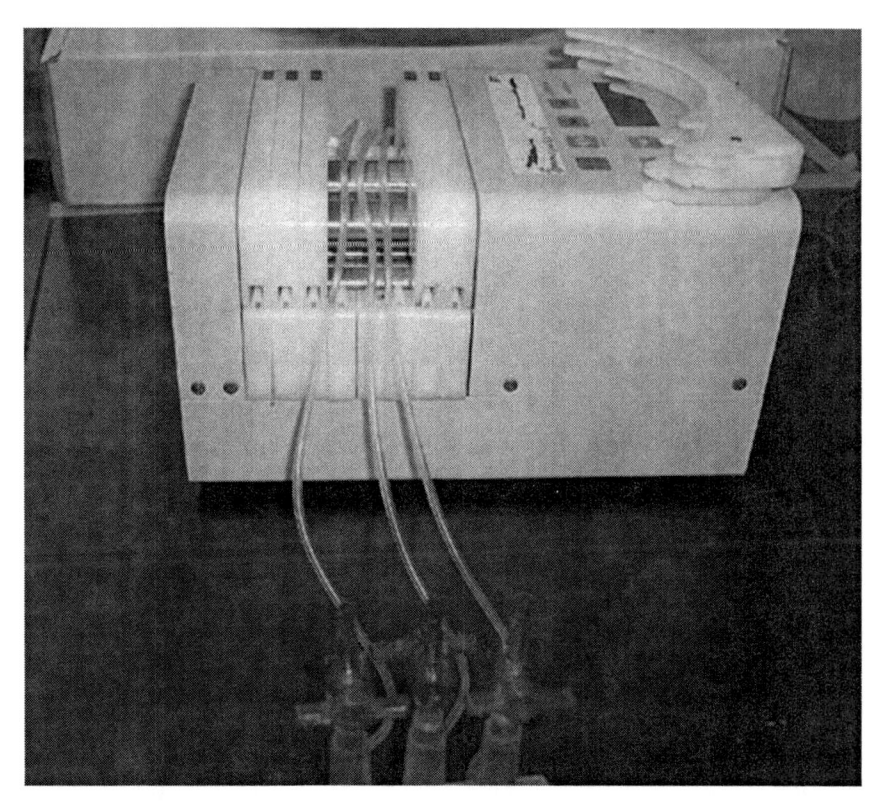

Fig. 6. The flow cell is connected to a peristaltic pump by placing each of the three pieces of tubing between the flow cell input manifold and the bubble trap apparatus into adjacent pump channels.

7. Incubate the inoculated flow cell with the flow off for 1 h at 37°C.
8. Return the flow chamber to the upright position and start the peristaltic pump at a flow rate of 1.5 mL/min (*see* **Note 11**). Incubate the flow cell at 37°C for the remainder of the experiment.
9. Observe the bubble trap periodically to ensure that it is approximately half full, adjusting the stopcocks as necessary to allow more medium to enter the cylinder.

3.3. Catheter-Based Model of In Vivo Biofilm Formation

3.3.1. Preparation of Bacterial Inocula

1. Grow each bacterial strain at 37°C with constant aeration to the desired concentration, as measured by optical density (OD) (*see* **Note 12**). Harvest bacterial cells by centrifugation and resuspend in PBS containing 10% dimethyl sulfoxide and 5% bovine serum albumin.
2. After determining viable colony counts by plating on suitable growth medium, store aliquots at −80°C.

Fig. 7. Bubble traps in the flow cell apparatus are sequentially filled by turning a bubble trap stopcock to the "on" position. Each bubble trap cylinder is filled approximately half full with medium, and then the bubble trap stopcock is returned to the "off" position. This process is repeated for each of three bubble traps per flow cell.

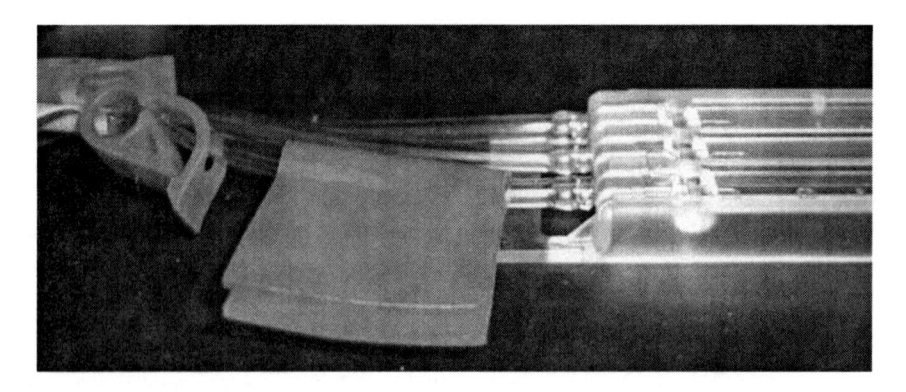

Fig. 8. Tubing just upstream of the flow cell chamber is prepared for inoculation by cleansing with a sterile alcohol pad followed by application of self-sealing tape (included in the Stovall Flow Cell Kit).

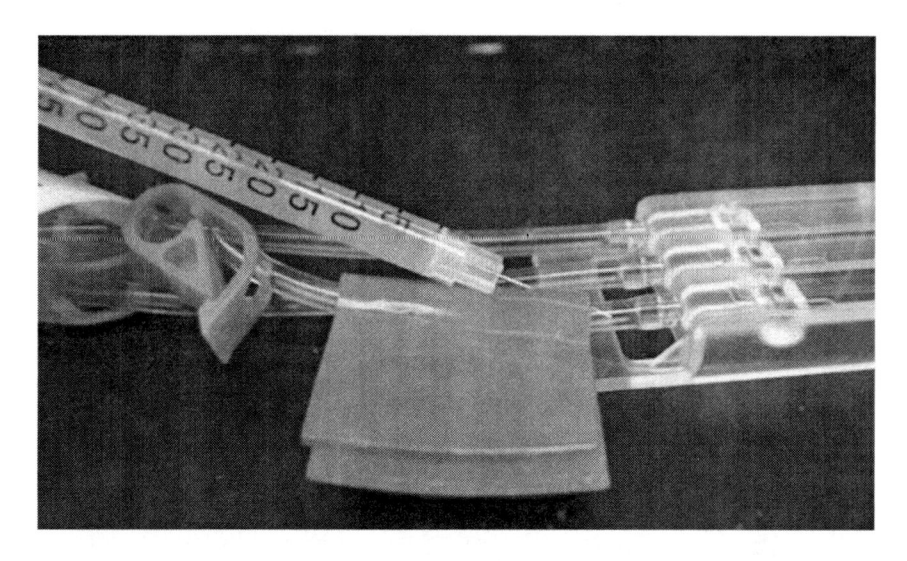

Fig. 9. The needle of an insulin syringe containing the flow cell inoculum is carefully inserted through the self-sealing tape and into the lumen of the tubing just upstream of the flow cell chamber. The bacterial inoculum (0.5 mL) is slowly introduced into the flow cell chamber while the upstream pinch clamp is closed and the downstream pinch clamp is open. After full injection of the inoculum, the injection site is cleaned with a sterile alcohol pad, and the downstream pinch clamp is closed.

3. Prior to injection, thaw the aliquots on ice and wash twice with sterile PBS.

3.3.2. Implantation of sc Catheter Segments

1. Anesthetize mice by injecting 0.5–0.7 cc of TBE intraperitoneally (approx 0.4–0.75 mg of TBE/g of body weight). Induction of anesthesia should occur within 5–15 min.
2. After ensuring adequate anesthesia, shave the dorsal flanks of each mouse. Clean the shaved areas first with Betadyne and then with alcohol. Allow the area to dry before making incisions.
3. Make a small (~1 cm) incision in the shaved area over one flank by lifting the skin and cutting with surgical scissors. Using forceps or a blunt probe, insert the catheter segment into the incision and approx 3 cm cephalad into the sc space. Ensure that the catheter does not shift back toward the incision site. Close the wound with surgical adhesive. Repeat this process for the other flank (i.e., two catheters per mouse).

3.3.3. Inoculation of Catheter Lumen

1. Prepare inocula by filling insulin syringes with 100 µL of bacterial suspension consisting of the desired number of colony-forming units.

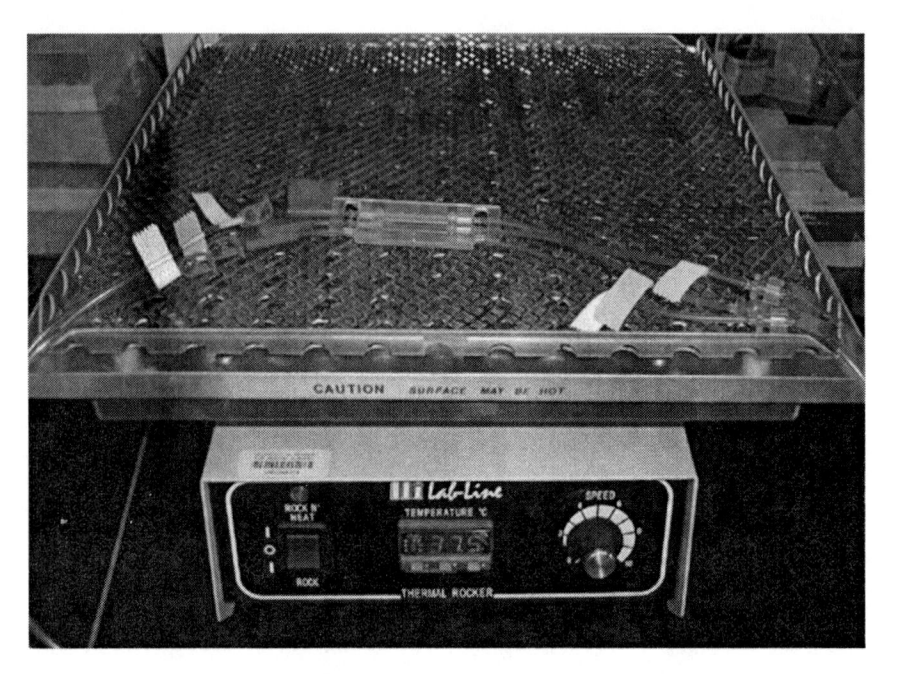

Fig. 10. After inoculation, the flow cell chamber is placed upside down in a 37°C incubator. Growth at 37°C is maintained without medium flow for 1 h, after which the flow cell is returned to an upright position and medium flow is resumed.

2. Ensure that the surgical wound is closed and that the adhesive is completely dry before inoculation (*see* **Note 13**).
3. Working from the cephalad side of the catheter, carefully insert the needle subcutaneously and into the lumen of the catheter. It is helpful to use one hand to secure the catheter and surrounding skin while manipulating the syringe with the other hand.
4. Slowly inject the bacterial suspension into the lumen of the catheter. Forceful injections will increase the chances of inoculating outside of the catheter.
5. Carefully remove the needle and gently clean the injection site with isopropanol.
6. Monitor infected mice for signs of distress until awake and mobile (*see* **Note 14**).

3.3.4. Assessment of Catheter Infection

1. At the desired time point(s), humanely euthanize mice according to the protocols approved at the researcher's institution.
2. Using aseptic technique, make a small incision with surgical scissors and carefully remove each catheter from the sc space using sterile forceps.
3. To remove nonadherent or loosely adherent bacteria from the catheter, carefully dunk the catheter into sterile PBS three times before placing it into a sterile container containing 5 mL of sterile PBS.

Biofilm Formation in Clinical Isolates of S. aureus

Fig. 11. The flow cell tubing should not be impacted as it enters or exits the incubator. The flow cell chamber should be level with the incubator surface at all times throughout the experiment (a small weight may be used for this purpose).

4. Sonicate the explanted catheters to remove adherent bacteria. We have found that 5 min of sonication (using the Fisher Sonic Dismembrator at a setting of 2) is sufficient to remove a prototypic clinical isolate of *S. aureus* from both 2- and 10-d-old catheter-associated biofilms (*see* **Note 15**).
5. Make serial dilutions of each sample and plate on an appropriate medium to obtain quantitative colony counts. Correct for the dilution factor and the volume plated to determine the total number of bacteria recovered per explanted catheter.

4. Notes

1. This formulation provides enough plasma to coat just over one full plate (125 wells). While this adds considerable expense to the protocol, we have tried alternative concentrations (as low as 5%) and have found that this increases variability between wells. Nevertheless, variability is unavoidable, and for this reason it is mandatory that all assays be done in replicates. We typically employ at least four and sometimes eight wells per test strain.
2. This incubator has a heated platform with a removable cover, which allows us to do our assays on the benchtop while maintaining 37°C in the flow cell itself. A standard laboratory incubator can be used assuming it can accommodate all

components of the flow cell system or has ports that can be used to extend the tubing from the media reservoir/pump to the spent-medium collection vessel.
3. There are a number of suitable choices for murine species. We chose to use NIH-Swiss mice based on studies indicating that these mice are also an appropriate choice for our other experiments investigating the pathogenesis of staphylococcal septic arthritis.
4. There are a number of options for murine anesthesia. We have found that administration of TBE results in a rapid and predictable anesthesia with a relatively low incidence of adverse reactions or overdose.
5. If cultures did not grow comparably, it may be necessary to make appropriate modifications to the starting dilution. Note that we have also tried alternative starting densities for these assays and have found that a 1:200 dilution, which corresponds to an OD (560 nm) of approx 0.05, yields the most reproducible results.
6. It may be necessary to dilute the eluted stain in PBS in order to obtain an absorbance value within the linear range of the plate reader. Results can be expressed in terms of absolute absorbance value, but we often express our results relative to a well-characterized reference strain. This is particularly appropriate when screening mutants generated in the reference strain.
7. To avoid wasting plasma, do not open the stopcocks on the bubble trap at this time. If bubbles accumulate during this step, ensure that they do not remain inside the flow chamber itself before incubating at 4°C. Bubbles inside the tubing are acceptable at this point.
8. A sufficiently sized sterile media reservoir must be used to ensure that the flow of medium is not compromised once established. The 10-L media reservoir from Stovall (**Fig. 4**) comes prefitted with tubing and is highly recommended. Ensure that the media reservoir is sufficiently sterilized by autoclaving, because the biofilm medium in the reservoir can become contaminated quite easily.
9. Alternatively, the flow cell output manifold may be removed, and each of the three flow cell chamber effluents may be collected individually for subsequent analyses.
10. With our prototype clinical isolate (UAMS-1) and its corresponding mutants, we typically use 0.5 mL of a standardized overnight culture to inoculate each flow cell *(8)*.
11. Flow rate is an experimentally defined parameter. Setting a flow rate that is too slow will result in planktonic growth within the flow chamber, owing to a failure to remove nonadherent cells. Setting a flow rate that is too fast will prevent biofilm growth, owing to the presence of high shear forces. It may be necessary to set up multiple experiments with varying flow rates, especially when comparing *S. aureus* mutants thought to be impaired in biofilm formation.
12. The relationship between OD and viable cell counts depends on many factors and should be determined empirically for each isolate of *S. aureus*.
13. It is highly recommended that infected mice be sufficiently anesthetized such that after inoculation the bacterial suspension can settle in the lumen of the catheter without being disturbed by movement of the mouse.

14. Signs of distress in mice include any of the following: rapid breathing rate; slow, shallow, or labored breathing; rapid weight loss; ruffled fur or rough hair coat; hunched posture; difficulty moving; hypothermia or hyperthermia; anorexia; diarrhea; or constipation.
15. Since strains of *S. aureus* have varying capacities for both adherence and persistence within a biofilm, it is highly recommended that additional mice be included in the experimental protocol for the purpose of determining the appropriate amount of sonication. Using aseptic technique, sonicate the catheter in 5 mL of PBS for 30 s (this volume of PBS allows sufficient space for the sonicating tip to operate without touching the catheter segment). Remove 100 µL for serial dilution and subsequent determination of colony-forming units. Repeat this step after another 30 s of sonication.

References

1. Lewis, K. (2001) Riddle of biofilm resistance. *Antimicrob. Agents Chemother.* **45,** 999–1007.
2. Keren, I., Kaldalu, N., Spoering, A., Wang, Y., and Lewis, K. (2004) Persister cells and tolerance to antimicrobials. *FEMS Microbiol. Lett.* **230,** 13–18.
3. Christensen, G. D., Simpson, W. A., Bisno, A. L., and Beachey, E. H. (1983) Experimental foreign body infections in mice challenged with slime-producing *Staphylococcus epidermidis*. *Infect. Immun.* **40,** 407–410.
4. Rupp, M. E., Ulphani, J. S., Fey, P. D., and Mack, D. (1999) Characterization of *Staphylococcus epidermidis* polysaccharide intercellular adhesin/hemagglutinin in the pathogenesis of intravascular catheter-associated infection in a rat model. *Infect. Immun.* **67,** 2656–2659.
5. Kadurugamuwa, J. L., Sin, L., Albert, E., et al. (2003) Direct continuous method for monitoring biofilm infection in a mouse model. *Infect. Immun.* **71,** 882–890.
6. Patti, J. M., Allen, B. L., McGavin, M. J., and Hook, M. (1994) MSCRAMM-mediated adherence of microorganisms to host tissues. *Annu. Rev. Microbiol.* **48,** 585–617.
7. Sillanpaa, J., Xu, Y., Nallapareddy, S. R., Murray, B. E., and Hook, M. (2004) A family of putative MSCRAMMs from *Enterococcus faecalis*. *Microbiology* **150,** 2069–2078.
8. Beenken, K. E., Blevins, J. S., and Smeltzer, M. S. (2003) Mutation of *sarA* in *Staphylococcus aureus* limits biofilm formation. *Infect. Immun.* **71,** 4206–4211.
9. Cafiso, V., Bertuccio, T., Santagati, M., et al. (2004) Presence of the *ica* operon in clinical isolates of *Staphylococcus epidermidis* and its role in biofilm production. *Clin. Microbiol. Infect.* **10,** 1081–1088.
10. Fitzpatrick, F., Humphreys, H., and O'Gara, J. P. (2005) The genetics of staphylococcal biofilm formation—will a greater understanding of pathogenesis lead to better management of device-related infection? *Clin. Microbiol. Infect.* **11,** 967–973.
11. Cramton, S. E., Gerke, C., Schnell, N. F., Nichols, W. W., and Gotz, F. (1999) The intracellular adhesin (*ica*) locus is present in *Staphylococcus aureus* and is required for biofilm formation. *Infect. Immun.* **67,** 5427–5433.

12. Beenken, K. E., Dunman, P. M., McAleese, F., et al. (2004) Global gene expression in *Staphylococcus aureus* biofilms. *J. Bacteriol.* **186,** 4665–4684.
13. Fitzpatrick, F., Humpreys, H., and O'Gara, J. P. (2005) Evidence for *icaADBC*-independent biofilm development mechanism in methicillin-resistant *Staphylococcus aureus* clinical isolates. *J. Clin. Microbiol.* **43,** 1973–1976.
14. Fluckiger, U., Ulrich, M., Steinhuber, A., et al. (2005) Biofilm formation, *icaADBC* transcription, and polysaccharide intercellular adhesin synthesis by staphylococci in a device-related infection model. *Infect. Immun.* **73,** 1811–1819.
15. McKenney, D., Pouliot, K. L., Wang, Y., et al. (1999) Broadly protective vaccine for *Staphylococcus aureus* based on an in vivo–expressed antigen. *Science* **284,** 1523–1527.
16. Yao, Y., Sturdevandt, D. E., and Otto, M. (2005) Genomewide analysis of gene expression in *Staphylococcus epidermidis* biofilms: insights into the pathophysiology of *S. epidermidis* biofilms and the role of phenol-soluble modulins in formation of biofilms. *J. Infect. Dis.* **191,** 289–298.
17. Yarwood, J. M., Bartels, D. J., Volper, E. M., and Greenberg, E. P. (2004) Quorum sensing in *Staphylococcus aureus* biofilms. *J. Bacteriol.* **186,** 1838–1850.
18. Vuong, C., Gerke, C., Somerville, G. A., Fischer, E. R., and Otto, M. (2003) Quorum-sensing control of biofilm factors in *Staphylococcus epidermidis*. *J. Infect. Dis.* **188,** 706–718.

11

Comparative Analysis of Staphylococcal Adhesion and Internalization by Epithelial Cells

Xudong Liang and Yinduo Ji

Summary

Multiple drug resistance to antibiotics is a major public health problem. Many mechanisms may be involved in such resistance. Increasing data have shown that *Staphylococcus aureus* can invade different types of nonphagocytic cells, which, in turn, may contribute to evasion of the toxicity of certain antibiotics. The fibronectin-binding proteins are required for *S. aureus* to adhere to and internalize into the host cells. We have shown that a two-component signal transduction system, SaeRS, is essential for bacterial adhesion and invasion of the epithelial cells.

Key Words: *Staphylococcus aureus*; adhesion; invasion; human epithelial cells.

1. Introduction

Staphylococcus aureus is a major human and animal pathogen. This organism can cause a wide variety of superficial and severe diseases, such as pneumonia, endocarditis, and toxic shock syndrome *(1)*. For *S. aureus* to initiate and establish infections, it must adhere to host cells *(2)*. *S. aureus* expresses a series of microbial surface components that recognize and interact with the extracellular matrix (ECM) components of the host *(3)*. Expression of these surface proteins is upregulated at the early stage of infection by different regulators, such as *agr*, *sae*, and *sar* *(4,5)*. Fibronectin-binding proteins are the main surface-associated proteins that function as adhesins by assembling the ECM protein Fn that bridges to the host cell receptors such as $\alpha_5\beta_1$-integrin *(6)*. The adherence of *S. aureus* to host cell integrins activates the Src family protein tyrosine kinase pathway, which triggers rearrangement of the actin cytoskeleton and is required for the host cells to uptake the bacteria *(7,8)*.

From: *Methods in Molecular Biology: MRSA Protocols*
Edited by: Y. Ji © Humana Press Inc., Totowa, NJ

Increasingly, studies have demonstrated that *S. aureus* can invade different types of host cell, which may allow the bacteria to evade the host immune system and decrease the potency of certain antibiotics *(9)*. This chapter describes a method that can be used for the comparative analysis of methicillin-resistant *S. aureus* adhesion and invasion of a variety of host cells.

2. Materials

1. RPMI-1640 medium (Invitrogen, Carlsbad, CA) supplemented with 10% fetal bovine serum (FBS) (Invitrogen) (equals RPMI-1640 medium/10% FBS).
2. Solution of sterile phosphate-buffered saline (PBS).
3. Trypsin-EDTA (Invitrogen).
4. Tryptic soy broth (TSB) (BD BioSciences, San Jose, CA).
5. Trypic soy broth (BD BioSciences) agar (Fluka) plates (TSA) (prepared according to the manufacturer's instructions).
6. Antibiotic/antimycotic (Invitrogen): Aliquot in 1.5-mL tubes at 1 mL.
7. RPMI-1640 medium supplemented with 10% FBS containing 100 µg/mL of gentamicin (Fisher Scientific, Pittsburgh, PA) and 5 µg/mL of lysostaphin (Sigma, St. Louis, MO).
8. Solution of Triton X-100 (0.025%) (Fisher Scientific).

3. Methods

3.1. Bacterial Culture

1. Incubate *S. aureus* isolates overnight (16 h) in 5 mL of TSB with shaking (225 rpm) at 37°C.
2. Harvest the bacterial cells by centrifuging for 10 min at 3200*g*.
3. Prior to infection, wash the bacterial cells once in PBS, adjust the cell density to approx 0.200 at OD_{600nm}, and keep the cells on ice.
4. Add 150 µL of diluted bacterial solution (OD_{600} = 0.200) to a tube containing 5 mL of RPMI-1640 medium/10% FBS.

3.2. Cell Culture (see Note 1)

1. Incubate A549 human epithelial cells in tissue culture plates (100 × 20 mm; SARSTED) in RPMI-1640 medium/10% FBS containing 1% antibiotic-antimycotic in a 37°C, 5% CO_2 incubator.
2. Subculture the epithelial cells every 2 to 3 d.

3.3. Invasion Assay (see Note 1)

3.3.1. Day 1

1. Grow A549 cells to 70% confluence in the culture plates.
2. Remove the old medium and wash the cells with 3.0 mL of warm PBS.
3. Add a solution of 400 µL of trypsin and incubate for 3–5 min at room temperature.
4. Add a total of 2 mL of warm RPMI-1640 medium/10% FBS containing 1% antibiotic/antimycotic and suspend all the cells.

5. Transfer the suspended cells (2 mL) into a new tube containing 8 mL of RPMI-1640 medium/10% FBS containing 1% antibiotic/antimycotic.
6. Mix the cells and transfer into each well (1 mL/well) of 24-well culture plate as needed.
7. Put the culture plates containing the cells in a 37°C, 5% CO_2 incubator.

3.3.2. Day 2

1. Remove the old medium from each well of the cell culture plate, and add 1 mL of warm fresh RPMI-1640 medium/10% FBS to each well. Incubate the epithelial cells overnight in a 37°C, 5% CO_2 incubator.
2. Incubate the bacterial cells overnight in 5 mL of TSB (with or without antibiotic) at 37°C with shaking (225 rpm).

3.3.3. Day 3

1. Harvest the bacterial cells from overnight culture by centrifuging for 10 min at 3200g.
2. Wash the collected bacterial cells once with PBS.
3. Spin down the bacterial cells and resuspend in 5 mL of ice-cold PBS.
4. Dilute the bacterial cells and adjust the OD to approx 0.20 at OD_{600nm}.
5. Add the total 150 µL of diluted bacterial solution (OD_{600} = 0.20) to a new sterile tube containing 5 mL of RPMI-1640 medium/10% FBS. This is called the original bacterial culture.
6. Remove the old medium from each well of the 24-well tissue culture plates, and add 1 mL of warm RPMI-1640 medium/10% FBS to each well.
7. Mix the original bacterial cultures by vortexing, and add 500 µL of the bacterial solution into the appropriate wells of the tissue culture plates.
8. Tilt the tissue culture plates back and forth six times and spin for 5 min at 100g.
9. Incubate the tissue culture plate for 2 h in a 37°C, 5% CO_2 incubator.
10. To calculate the amount of bacterial cells added in each well, dilute the original bacterial cultures by serial dilutions of 10^{-1}, 10^{-2}, 10^{-3}, and 10^{-4} with PBS, and plate out 25 µL four times on TSA plates (see **Note 2**).
11. Take the tissue culture plate from the incubator, collect the supernatants from the wells for a total bacterial count, and discard the supernatants from the invasion wells.
12. Wash the epithelial cells three times with 1 mL of warm PBS.
13. Add a total of 1 mL of RPMI-1640 medium/10% FBS to each well for a total bacterial count.
14. Add a total of 1 mL of RPMI-1640 medium/10% FBS containing 100 µg/mL of gentamicin and 5 µg/mL of lysostaphin to each invasion well.
15. Incubate the tissue culture plate for 1 h in a 37°C, 5% CO_2 incubator.
16. For a total bacterial count, collect the supernatants from each well; dilute by serial dilutions of 10^{-1}, 10^{-2}, 10^{-3}, and 10^{-4} with PBS; and plate out 25 µL four times on TSA plates (see **Note 3**).
17. Collect the supernatants from the total wells and discard the supernatants from the invasion wells, and wash the cells three times with 1 mL of warm PBS.

18. Add 150 μL of trypsin to each well and incubate for 3–5 min at room temperature.
19. Carefully resuspend the cells and transfer into 1.5-mL tubes on ice.
20. Add 400 μL of 0.025% Triton X-100 to each well to wash the wells, and then transfer into the corresponding 1.5-mL tubes.
21. Mix the collected cells by vortexing for 30 s; dilute by serial dilutions of 10^{-1}, 10^{-2}, 10^{-3}, and 10^{-4} with PBS; and plate out 25 μL four times on TSA plates (*see* **Note 3**).
22. Incubate the TSA plates overnight at 37°C.

3.3.4. Day 4

1. Count the colonies on all of the plates.
2. Calculate the average CFU, CFU/well, growth index, % invasion, corrected invasion, and relative invasion using the following formulas:

$$\text{CFU} = \text{number of colonies counted}$$

$$\text{CFU/well} = (\text{avg. CFU}/0.025 \text{ mL}) \times \text{dilution factor (DF)} \times \text{total volume in well}$$

$$\text{Growth index} = \text{total CFU/original CFU}$$

$$\% \text{ Invasion} = \text{internalized CFU/original CFU} \times 100$$

$$\text{Corrected invasion} = \% \text{ invasion/growth index}$$

$$\text{Relative invasion} = (\text{corrected invasion of mutant/corrected invasion of control}) \times 100\%$$

The figure of relative invasion can be created in an Excel file (**Fig. 1**).

3.4. Adhesion Assay (see Note 1)

3.4.1. Day 1

1. Grow A549 cells to 70% confluence in the culture plate.
2. Remove the old medium, and wash the cells with 3.0 mL of warm PBS.
3. Add a solution of 400 μL of trypsin and incubate for 3–5 min.
4. Add a total of 2 mL of warm RPMI-1640 medium/10% FBS containing 1% antibiotic/antimycotic, and suspend all the cells.
5. Transfer the suspended cells (2 mL) into a new tube containing 8 mL of RPMI-1640 medium/10% FBS containing 1% antibiotic/antimycotic.
6. Mix the cells and transfer into each well (1 mL/well) of the 24-well culture plate as needed.
7. Put the culture plate containing the cells in a 37°C, 5% CO_2 incubator (*see* **Note 3**).

3.4.2. Day 2

1. Remove the old medium from each well of the tissue culture plate, and add 1 mL of warm fresh RPMI-1640 medium/10% FBS to each well. Incubate the cells overnight in a 37°C, 5% CO_2 incubator.
2. Incubate the bacterial cells overnight in 5 mL of TSB (with or without antibiotic) at 37°C with shaking (225 rpm).

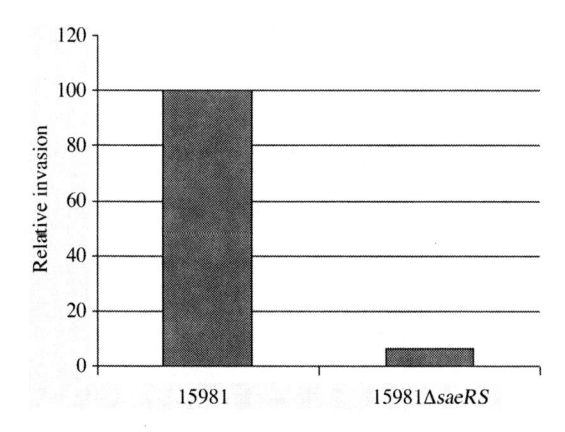

Fig. 1. Effects of SaeSR on internalization of *S. aureus* by epithelial cells. The bacterial cells were collected from overnight cultures (wild-type 15981, the *saeSR* null mutant strain15981Δ*saeRS*; *[10]*), washed with PBS, diluted, and resuspended in RPMI-1640 medium with 10% FBS, respectively, just prior to infection of monolayer cells. Relative invasion was calculated as described. The experiments were repeated three times, with similar results.

3.4.3. Day 3

1. Harvest the bacterial cells from overnight culture by centrifuging for 10 min at 3200*g*.
2. Wash the collected bacterial cells once with PBS.
3. Spin down the bacterial cells and resuspend in 5 mL of ice-cold PBS.
4. Dilute the bacterial cells, and adjust the OD to approx 0.20 at OD_{600nm}.
5. Add a total of 150 μL of diluted bacterial solution (OD_{600} = 0.20) to a new sterile tube containing 5 mL of RPMI-1640 medium/10% FBS (*see* **Note 4**). This will be the original bacterial culture.
6. Remove the old medium from each well of the 24-well tissue culture plate, and add 1 mL of warm RPMI-1640 medium/10% FBS (*see* **Note 4**) to each well.
7. Tilt the tissue culture plate back and forth six times and spin for 5 min at 100*g*.
8. Incubate the tissue culture plate for 1 h in a 37°C, 5% CO_2 incubator.
9. To calculate the number of bacterial cells added in each well, dilute the original bacterial cultures by serial dilutions of 10^{-1}, 10^{-2}, 10^{-3}, and 10^{-4} with PBS and plate out 25 μL four times on TSA plate (*see* **Note 5**).
10. Take the tissue culture plate out of the incubator after 1 h of incubation, and discard the supernatants.
11. Wash the epithelial cells three times with 1 mL of warm PBS.
12. Add a total of 150 μL of trypsin to each well, and incubate for 3–5 min at room temperature.
13. Carefully resuspend the cells and transfer into 1.5-mL tube, respectively, on ice.
14. Add 400 μL of 0.025% Triton X-100 to each well to wash the wells, and transfer into the corresponding 1.5-mL tubes.

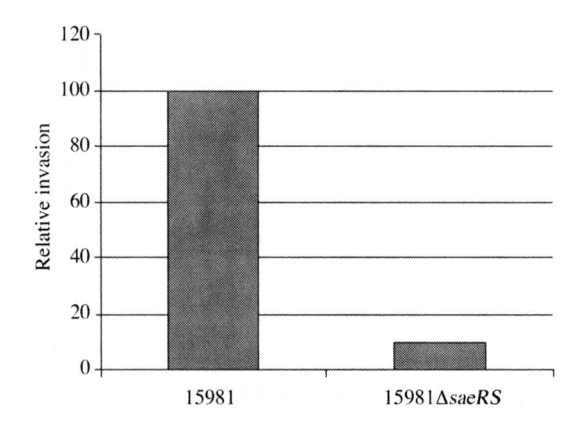

Fig. 2. Effects of SaeSR on adherence of *S. aureus* by epithelial cells. The bacterial cells were collected from overnight cultures (wild-type 15981, the *saeSR* null mutant strain15981Δ*saeRS*; *[10]*), washed with PBS, diluted, and resuspended in RPMI-1640 medium with 10% FBS, respectively, just prior to infection of monolayer cells. Relative adherence was calculated as described. The experiments were repeated three times, with similar results.

15. Mix the collected cells by vortexing for 30 s; dilute by serial dilutions of 10^{-1}, 10^{-2}, 10^{-3}, and 10^{-4} with PBS; and plate out 25 μL four times on TSA plates (*see* **Note 3**).
16. Incubate the TSA plates overnight at 37°C.

3.4.4. Day 4

1. Count the colonies on all plates.
2. Calculate the average CFU, CFU/well, % adhesion, and relative adhesion using the following formulas:

$$\text{CFU} = \text{number of colonies counted}$$

$$\text{CFU/well} = (\text{avg. CFU}/0.025 \text{ mL}) \times \text{dilution factor (DF)} \times \text{total volume in well}$$

$$\% \text{ Adhesion} = \text{adhered and internalized CFU/original inoculum's CFU} \times 100$$

$$\text{Relative adhesion} = (\text{adhesion of mutant/adhesion of control}) \times 100\%$$

The figure of relative adhesion can be created in an Excel file (**Fig. 2**).

4. Notes

1. A common problem in the adhesion and invasion assay is bacterial contamination. To avoid potential contamination, the adhesion and invasion assay should be performed in a biosafety hood and gloves should be worn.
2. The bacterial solution should be mixed by vortexing before it is inoculated into the wells of the tissue culture plates.

3. To avoid cross contamination between different samples, all plates should be dried before starting an assay, and all plates should be completely dry before placed in the 37°C incubator.
4. The RPMI-1640 medium/10% FBS should not contain antibiotic/antimycotic.
5. Let the plates sit until late afternoon and then put them in a 37°C incubator.

Acknowlegments

This work was supported in part by USPHS research grant AI057451 from the National Institute of Allergy and Infectious Disease. We thank Dr. Lasa for providing 15981 and 19581Δ*SaeRS* Strains.

References

1. Lowy, F. D. (1998) *Staphylococcus aureus* infections. *N. Engl. J. Med.* **339,** 520–532.
2. von Eiff, C., Becker, K., Machka, K., Stammer, H., and Peters, G. (2001) Nasal carriage as a source of *Staphylococcus aureus* bacteremia. *N. Engl. J. Med.* **344,** 11–16.
3. Foster, T. J. and Höök, M. (1998) Surface protein adhesins of *Staphylococcus aureus*. *Trends Microbiol.* **6,** 484–488.
4. Novick, R. P. (2003) Autoinduction and signal transduction in the regulation of staphylococcal virulence. *Mol. Microbiol.* **48,** 1429–1449.
5. Liang, X., Yu, C., Sun, J., Liu, H., Landwehr, C., Holmes, D., and Ji, Y. (2006) Inactivation of a two-component signal transduction system, SaeRS, eliminates adherence and attenuates virulence of *Staphylococcus aureus*. *Infect. Immun.* **74,** 4655–4665.
6. Sinha, B., Francois, P. P., Nusse, O., et al. (1999) Fibronectin-binding protein acts as *Staphylococcus aureus* invasin via fibronectin bridging to integrin alpha5beta1. *Cell. Microbiol.* **1,** 101–117.
7. Agerer, F., Michel, A., Ohlsen, K., and Hauck, C. R. (2003) Integrin-mediated invasion of *Staphylococcus aureus* into human cells requires Src family protein-tyrosine kinases. *J. Biol. Chem.* **278,** 42,524–42,531.
8. Wang, B., Yurecko, R., Dedhar, S., and Cleary, P. P. (2005) Integrin-linked kinase is an essential link between integrins and uptake of bacterial pathogens by epithelial cells. *Cell. Microbiol.* **8,** 257–266.
9. Alexander, E. H. and Hudson, M. C. (2001) Factors influencing the internalization of *Staphylococcus aureus* and impacts on the course of infections in humans. *Appl. Microbiol. Biotechnol.* **56,** 361–366.
10. Toledo-Arana, A., Merino, N., Vergara-Irigaray, M., Débarbouillé, M., Penadés, J. R., and Lasa, I. (2005) *Staphylococcus aureus* develops an alternative, *ica*-independent biofilm in the absence of the *arlRS* two-component system. *J. Bacteriol.* **187,** 5318–5329.

12

Comparative Analysis of MRSA

Fumihiko Takeuchi, Tadashi Baba, and Keiichi Hiramatsu

Summary

This chapter explains computer techniques for comparing genes, proteins, or genomes of methicillin-resistant *Staphylococcus aureus* (MRSA). In the principle methodology for comparative genomics, first researchers obtain the data of DNA sequences and phenotypes for various strains of interest, and from those they infer what difference/similarity in sequences results in what difference/similarity in phenotypes. Usually, the obtained hypothesis provides guidance for the succeeding biological experiments, such as producing knockout strains or conducting transcriptome or proteome analysis, which examine the hypothesis. Even for physicians and experimental researchers, these computer-aided researches would be needed in order to understand the physiological characteristics and pathogenic abilities of the MRSA that they deal with in this "genome era." This chapter involves no experiments and is confined to computer analysis. We explain methods for extracting the difference/similarity between sequences of nucleotide, amino acid, or even the whole genomes of bacteria. We also introduce how to compare the pathways between strains that possess different sets of genes.

Key Words: Comparative genomics; bioinformatics; bl2seq; GATA; MUMmer; ClustalW; TreeView; Multi-LAGAN; KEGG.

1. Introduction

Genomic concepts of the bioinformatics techniques dealt in this chapter can be found in *(1)* and theoretical background can be found in *(2)*. The computer programs introduced here are free software available from the World Wide Web as listed in the references. As for the platform to run these programs, we adopted Macintosh OSX (MacOSX). The availability of the programs for other platforms, such as Microsoft Windows, is provided in the Web pages of respective programs. Although we explain how to use the programs, the accompanying documents will need to be consulted for installation instructions.

Many of the bioinformatics tools, including some explained herein, are available only for the Unix platforms, which usually include Linux, Solaris, and MacOSX. We employed MacOSX here, because it is the most available to life science researchers among the platforms that include the Unix features. In MacOSX, the Unix command-line-interface is accessible by opening the "Applications" folder, then the "Utilities" folder there, followed by launching the "Terminal" application found there. In the examples given here, type in the command printed in bold typewriter font, and press the return key. The Unix features of MacOSX are explained in **ref. 3**, and general Unix knowledge adequate for bioinformatics can be found in **ref. 4**. We recommend these books for troubleshooting problems for these Unix-based software.

In the examples that we have provided, we used sequences registered in Genbank/EMBL/DDBJ available from **ref. 5** by the accession number.

2. Materials

2.1. Gene Finding and Annotation—Glimmer, RBSfinder, BLAST

For a DNA sequence, the program Glimmer *(6)* searches the region for protein coding genes. Afterward, RBSfinder *(7)* searches for regions in the vicinity of the gene start where the ribosome might bind and uses this information to adjust the gene start. The function of each gene is predicted by searching homologous genes already registered in the protein database by BLAST *(8)* (*see* **Note 1**). There are various software packages that perform the pipeline of these annotation tasks, either commercial *(9)* or noncommercial *(10)*. The easiest way might be to use Internet-based services, such as those by Nano+Bio-Center *(11)* (*see* **Fig. 1**).

The nucleotide sequence that one wishes to annotate is prepared in FASTA format. We provide an example using the chromosome sequence of *Staphylococcus aureus* strain N315 (accession no. NC_002745).

2.2. Local Alignment Between Two Sequences—bl2seq, GATA

The program bl2seq *(12)* finds local matches between two different sequences, or within one sequence. The extracted matches include repetitive sequences such as insertion sequences. The GATA *(13)* programs provide graphical user interface for bl2seq. The G-InforBIO program *(14)* also includes similar functionality.

A pair of nucleotide (or amino acid) sequences is prepared in FASTA format. We provide an example using the sequences for mobile genetic elements that confer methicillin resistance to staphylococci, SCC*mec* type II (accession no. D86934) and type III (accession no. AB037671) (*see* **Note 2**).

2.3. Maximum Unique Matches Among Multiple Sequences—MUMmer

For two or more sequences, the MUMmer *(15)* package extracts their matches that appear only once in each sequence (*see* **Notes 3** and **4**). For the case of two sequences, a tool for visualizing the matches is also provided.

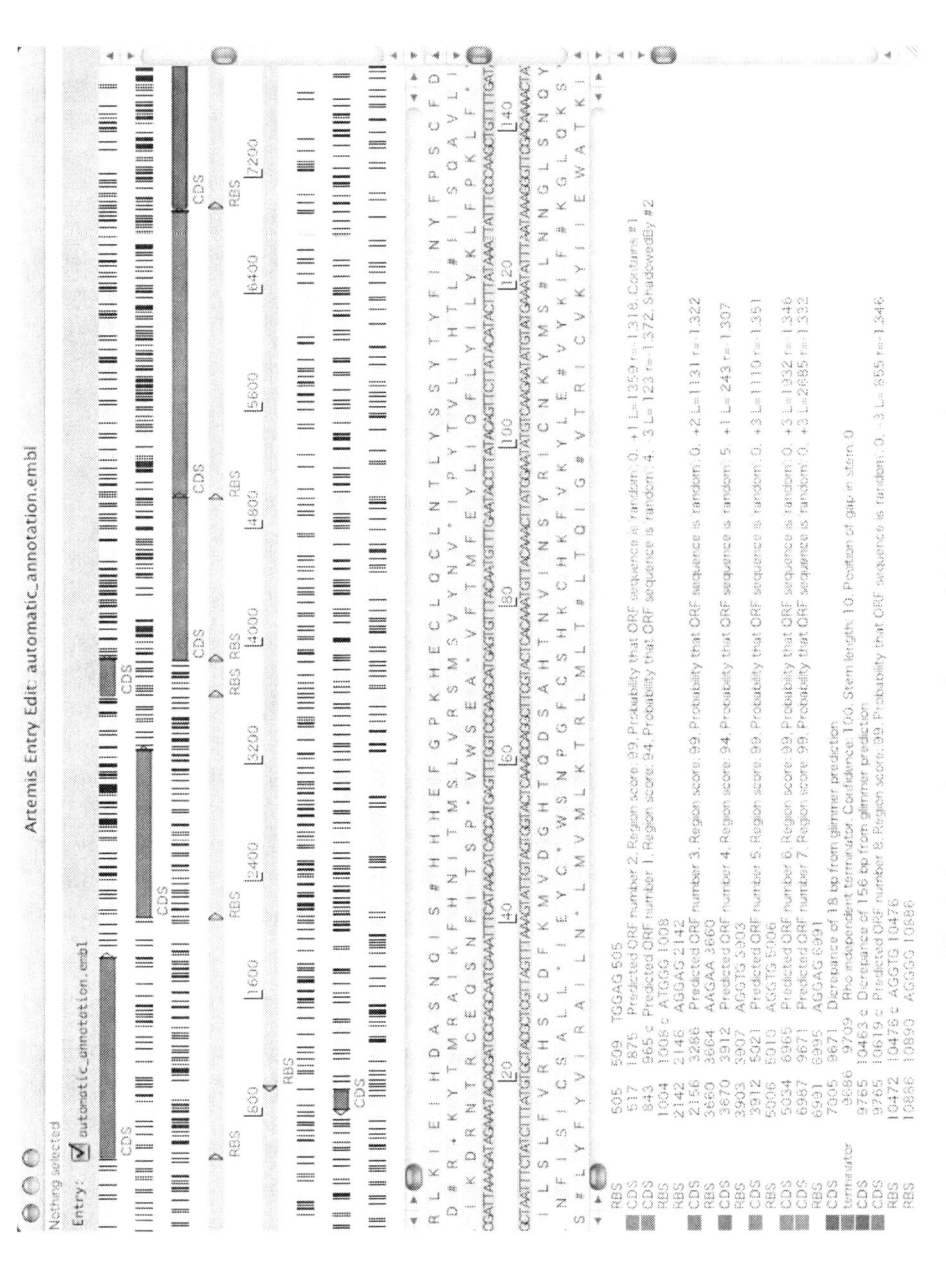

Fig. 1. Automatic annotation provided by Nano+Bio Center.

A pair of nucleotide (or amino acid) sequences in FASTA format is prepared. We provide an example using the whole chromosome sequences of *Staphylococcus haemolyticus* strain JCSC1435 (accession no. NC_007168) and *S. aureus* strain N315 (accession no. NC_002745).

2.4. Multiple Alignment and Phylogenetic Tree—ClustalW, TreeView

ClustalW *(16)* (and ClustalX *[17]* with graphical user interface) performs alignment of multiple nucleotide (or amino acid) sequences. ClustalW can also compute the phylogenetic tree for the aligned sequences, but TreeView *(18)* is needed for the visualization.

Multiple nucleotide (or amino acid) sequences are prepared in FASTA format. Before alignment, it must be confirmed that the sequences are collinear (i.e., not having translocations or inversions) using the programs bl2seq-GATA or MUMmer. We provide an example using *S. aureus* coagulase gene *coa* from seven strains (accession nos. AB158549, AB158550, AB158551, AB158552, AB158553, AB158554, AB158555).

2.5. Whole-Genome Alignment—Multi-LAGAN

Multi-LAGAN *(19)* is one of the recently developed programs that can align multiple sequences that are megabases long, such as whole bacterial genomes (*see* **Note 5**).

Nucleotide sequences are prepared in FASTA format. Before alignment, it must be confirmed that the sequences are collinear (i.e., not having translocations or inversions) by the programs bl2seq-GATA or MUMmer. We provide an example using the whole chromosomal sequences of *S. aureus* strains N315 (accession no. NC_002745) and Mu50 (accession no. NC_002758) (*see* **Note 6**).

2.6. Pathways—KEGG

Whole-genome determination of an organism can enable researchers to predict its complex and sophisticated cellular processes and physiological features. The KEGG, or Kyoto Encyclopedia of Genes and Genomes *(20)*, displays such cellular processes as interactive biochemical pathway maps as long as the organism's whole genome has been determined. The software, better called databases, does not need to be installed into individual local computers. KEGG consists of Web-based databases and, therefore, only its home page needs to be accessed (*see* **Note 7**).

Basically no materials are required to use the KEGG services. They provide very detailed and interactive information by identifying proteins involved in known biochemical pathways from whole-genome sequence and up-to-date annotation (*see* **Note 8**). KEGG includes KEGG PATHWAY Database, with

237 known biochemical, regulatory, and information pathways and so on, allowing users to view which organism and/or strains have the potential ability to accomplish a biochemical reaction.

3. Methods

3.1. Gene Finding and Annotation—Glimmer, RBSfinder, BLAST

1. Type in as follows to run Glimmer, which outputs the result to files "tmp.coord", "g2.coord" etc.
   % **run-glimmer2 NC_002745.fasta**
   Genome is NC_002745.fasta
   Find non-overlapping orfs in tmp.coord
   Longest orf = 20199
   1557 putative genes extracted
   (abbreviated)
   2564 putative genes extracted

2. Type in as follows to run RBSfinder, which outputs the result to the file "rbsfinder-out.txt":
   % **rbs_finder.pl NC_002745.fasta g2.coord rbsfinderout.txt 50**
   Summary:
   # of orfs that have RBS before original start codon loc= 562 -> 21.92%
   # of orfs that have RBS before new start codon loc= 1109 -> 43.25%
   # of orfs that have no RBS= 893 -> 34.83%
   Total # of orfs: 2564

3. For the extracted genes, query the translated amino acid sequence to protein-protein BLAST, which returns similar proteins and their functions.

3.2. Local Alignment Between Two Sequences—bl2seq, GATA

1. Launch GATAligner. Enter "D86934.fasta" as reference sequence and "AB037671.fasta" as comparative sequence; select a proper folder; and enter, e.g., "test" as file base name. Press "Align Sequences."
2. Launch GATAPlotter. Enter "test" as Alignment file. Press "Go." The result appears as in **Fig. 2** (*see* **Note 9**).

3.3. Maximum Unique Matches Among Multiple Sequences—MUMmer

1. Type in as follows to run MUMmer and mummerplot, which compute and plot the maximum unique matches:
   % **mummer -mum -b -c NC_007168.fasta NC_002745.fasta > mummer.mums**
   # reading input file "NC_007168.fasta" of length 2685015
   # construct suffix tree for sequence of length 2685015
   (abbreviated)
   # reading input file "NC_002745.fasta" of length 2814816
   # matching query-file "NC_002745.fasta"

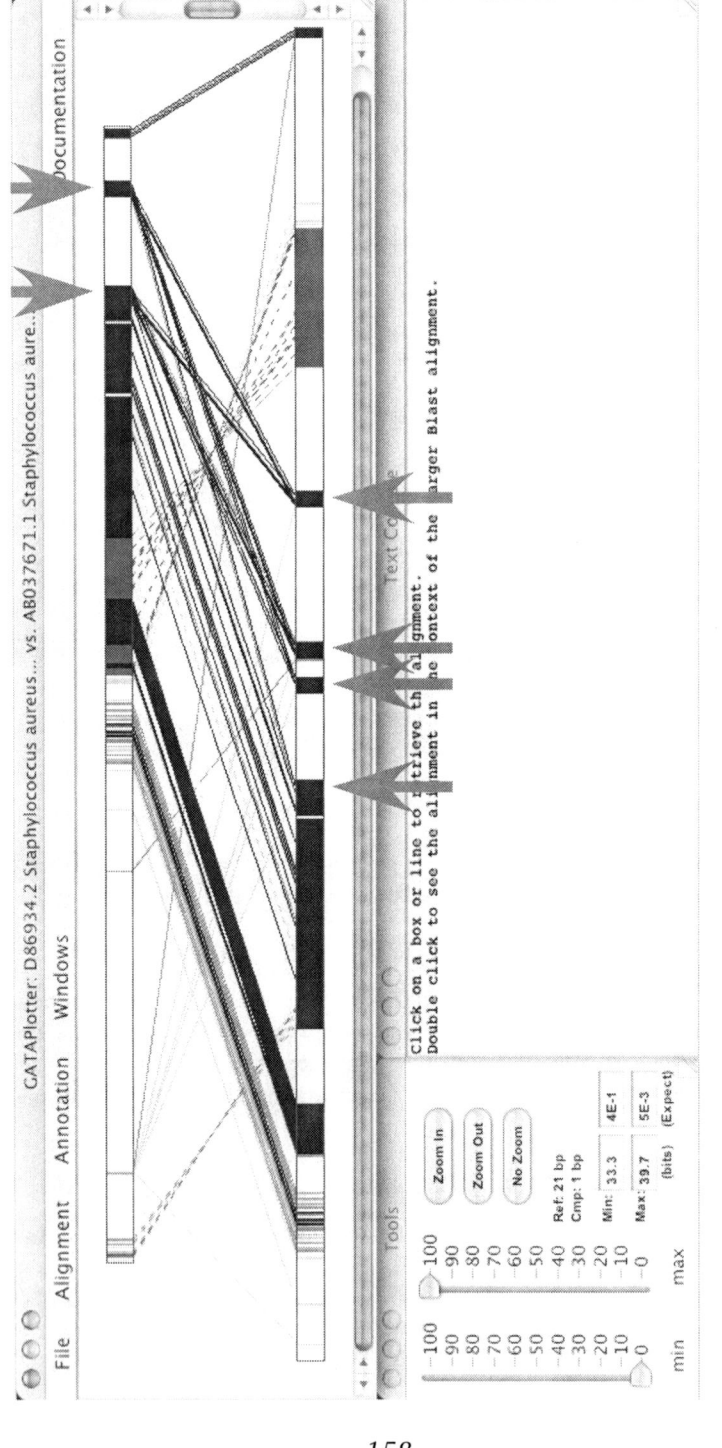

Fig. 2. Local alignments between two sequences computed by bl2seq with the graphical user interface of GATA.

Comparative Analysis of MRSA

```
# against subject-file "NC_007168.fasta"
# COMPLETETIME mummer NC_007168.fasta 15.69
# SPACE mummer NC_007168.fasta 5.36
mummer.mums
% mummerplot -x "[0,2685015]" -y "[0,2814816]" -postscript -p mummer
gnuplot 3.8j patchlevel 0
Reading mummer file mummer.mums (use mummer -c)
Writing plot files mummer.fplot, mummer.rplot
Writing gnuplot script mummer.gp
Rendering plot mummer.ps
```

2. View the generated file "mummer.ps", e.g., with Preview (**Fig. 3;** *see* **Note 10**).

3.4. Multiple Alignment and Phylogenetic Tree—ClustalW, TreeView

1. Launch ClustalX. From the menu "File" > "Load Sequences", add "AB158549.fasta". Then, from the menu "File" > "Append Sequences", add the other sequences one by one.
2. Perform alignment by menu "Alignment" > "Do Complete Alignment". The result appears as in **Fig. 4**.
3. Open the menu "File" > "Save Sequences as…". Save range from 2463 to 3282 to file "coa5prime.aln", which corresponds to the 5′ half of the *coa* gene (*see* **Note 11**).
4. Open the menu "File" > "Load Sequences". Click "Yes" and select "coa5prime.aln".
5. Open the menu "Trees" > "Draw N-J Tree" (*see* **Note 12**) and save the phylogenetic tree to file "coa5prime.ph".
6. Launch TreeViewPPC. This program runs in Classic environment of MacOSX. From the menu "File" > "Open…" open "coa5prime.ph". The phylogenetic tree is plotted as in **Fig. 5**. If this does not work, change the file name to "coa5prime.ph.txt" to make the program recognize it as a text file.

3.5. Whole-Genome Alignment—Multi-LAGAN

1. Type in as follows to run Multi-LAGAN, which outputs the sequences with gaps inserted as in the alignment to a multi-fasta file "alignment":
   % mlagan NC_002745.fasta NC_002758.fasta -tree "(N315 Mu50)" -out alignment
   using given phylogenetic tree:
   (N315 Mu50)
   outputting to: alignment
   Using NC_002745.fasta NC_002758.fasta (12, 0, 25, 0) x
   (abbreviated)
   final alignment...
   mlagan — end.

2. The result file can also be view by the mpretty utility. (Type space key for scroll.)
   % mpretty.pl alignment | more

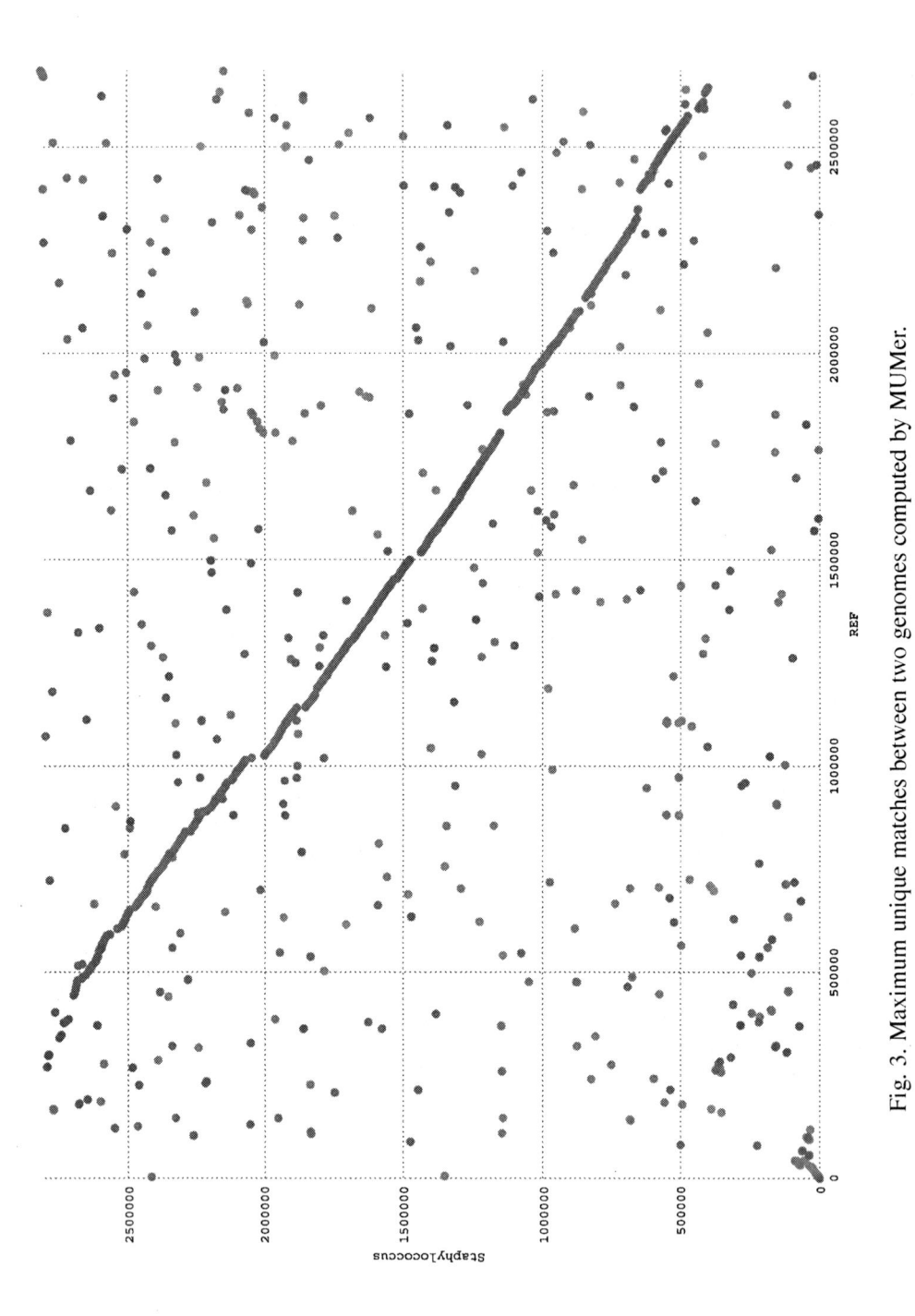

Fig. 3. Maximum unique matches between two genomes computed by MUMer.

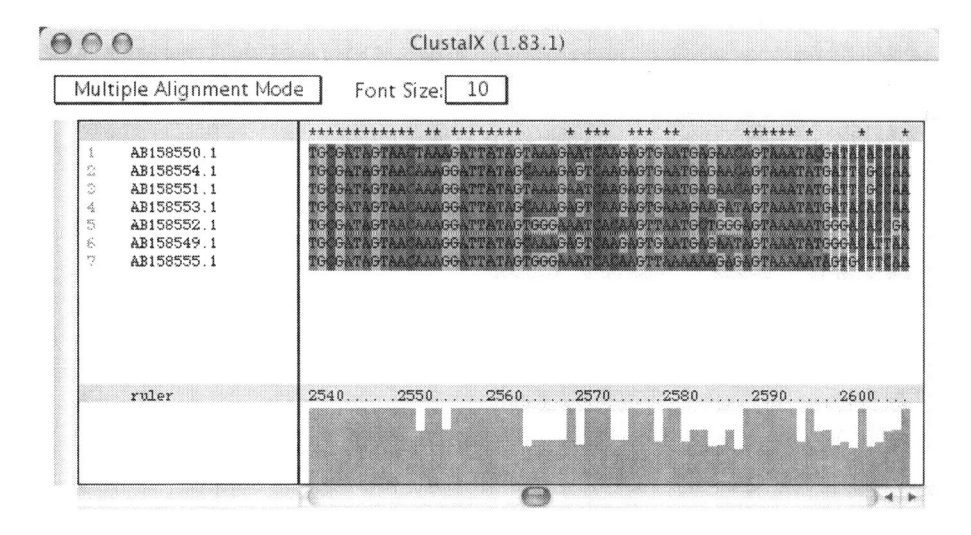

Fig. 4. Multiple sequence alignment computed by ClustalW.

N315_ :
AAGGTGTTTATCCACAGAAATGGGGATAGTTAT-CCAGAATTGTGTACAA @
33497/2814816
Mu50_ :
AAGGTGTTTATCCACAGAAATGGGGATAGTTATCCCAGAATTGTGTACCA @
33499/2878040
= 33451 33461 33471 33481 33491

N315_ :
TTTAAAGAGAAA–TACCCACAATGCCCACAGAGTTATCCACAAATACA @
33544/2814816
Mu50_ :
TTTAAAGGAGAAATTACCCAACAATGCCCACAGAGTTATCCACAAATACA @
33549/2878040
= 33501 33511 33521 33531 33541

N315_ :
CAAGTTATACACTAAAAATTGGGCATAAATGTCAGGAAAATATCAAAAAC @
33594/2814816
Mu50_ :
CAAGTTATACACTAAAAATTGGGCATAAATGTCAGGAAAATATCAAAAAC @
33599/2878040
= 33551 33561 33571 33581 33591

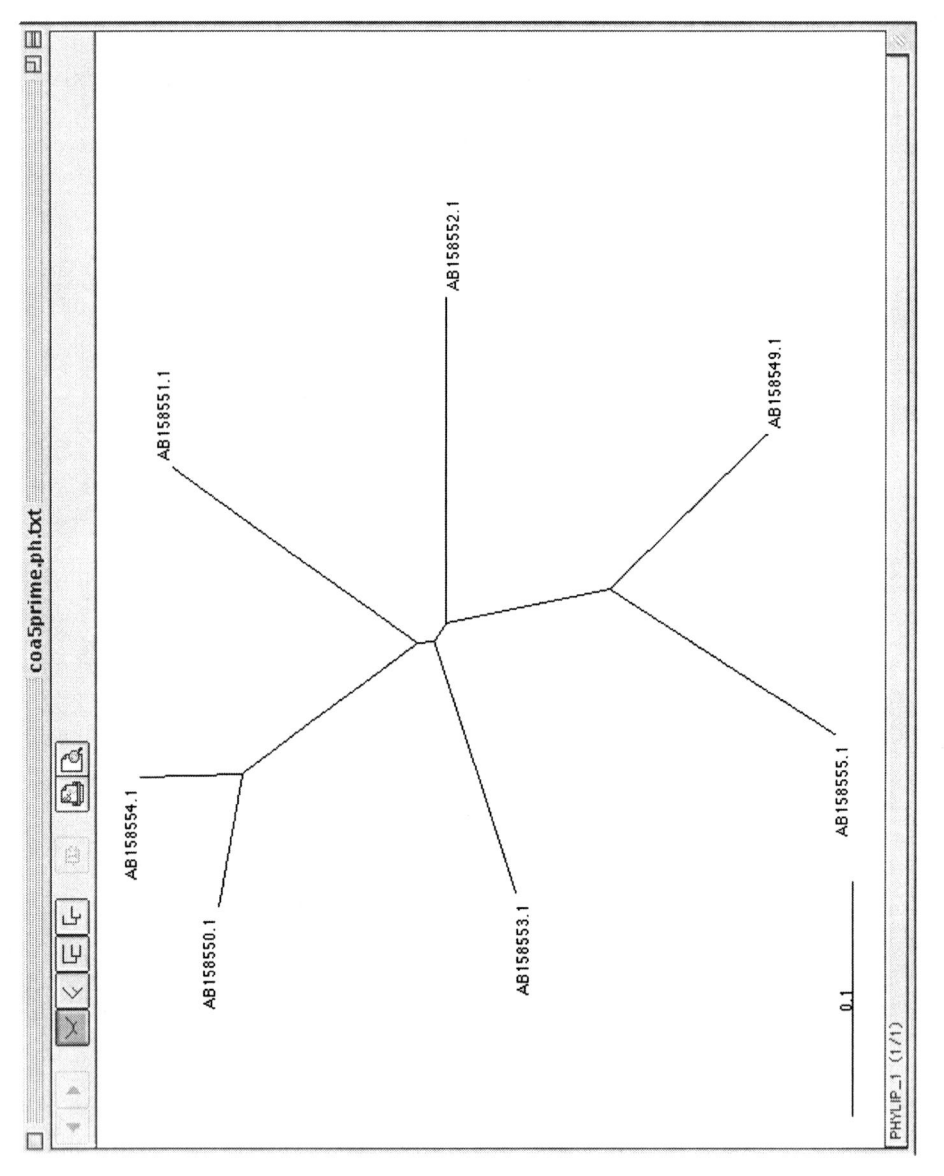

Fig. 5. Phylogenetic tree plotted by TreeView.

3.6. Pathways—KEGG

1. Access the home page of the URL *(20)*.
2. Find a link to "PATHWAY". On the linked page, there is a list of biochemical pathways categorized by reaction names.
3. Click any subcategorized reaction names of interest. At first, a "reference pathway" is shown, which is a collection of biochemical pathways from any known organisms. To view the basic features of enzymes, click the EC numbers in the boxes. Then select specific organisms and/or strains from the pulldown menu on the top of the pathway map. One is able to know, e.g., if an organism requires specific amino acids, sugars, or vitamins by viewing combinations of biochemical reactions that the organism can perform (*see* **Notes 13–15**).

4. Notes

1. These tasks for annotation themselves are not comparative genomics but are necessary for analysis of the genes within a DNA sequence that have been determined. They are not necessary when simply comparing the nucleotide sequences.
2. Further biological information for this example appears in Chapter 7 and in **ref.** *21*.
3. MUMmer is similar to bl2seq-GATA but exclusively computes matches that appear only once in each sequence. Thus, it cannot find repetitive matches but is better for grabbing the rough similarity among sequences, and the computation is much faster.
4. Programs such as bl2seq-GATA and MUMmer extract chromosomal rearrangements such as translocations or inversions. The unmatched regions between chromosome sequences would correspond to transposons, prophages, pathogenicity islands, or other genomic islands, which have been acquired by horizontal transfer.
5. Whole-genome alignment of megabase-long sequences still is a computationally hard task, in the sense that it is difficult to automatically obtain "the alignment that you intended." Current programs give correct alignment for the conserved regions. However, for less conserved regions, one might need to manually align the local regions by more accurate alignment tools (which in turn cannot handle long alignments).
6. The analysis in this example lists differences between the two strains N315 and Mu50, which can suggest the genes resulting in their difference in resistance to the antibiotic vancomycin (*see* **ref.** *22* for further discussion).
7. If one refers to the data from the KEGG services, cite the publication listed in the home page. One of them is by Kanehisa and Goto *(20)*.
8. Very importantly, the KEGG databases assign their own annotation results to genes identified from whole-genome sequences, and if necessary, update the assignment, no matter what annotations were originally assigned. This keeps the databases highly reliable.
9. The inverted regions between two sequences appear as segments connected by red lines in **Fig. 2**. Repetitive sequences, which here correspond to the insertion

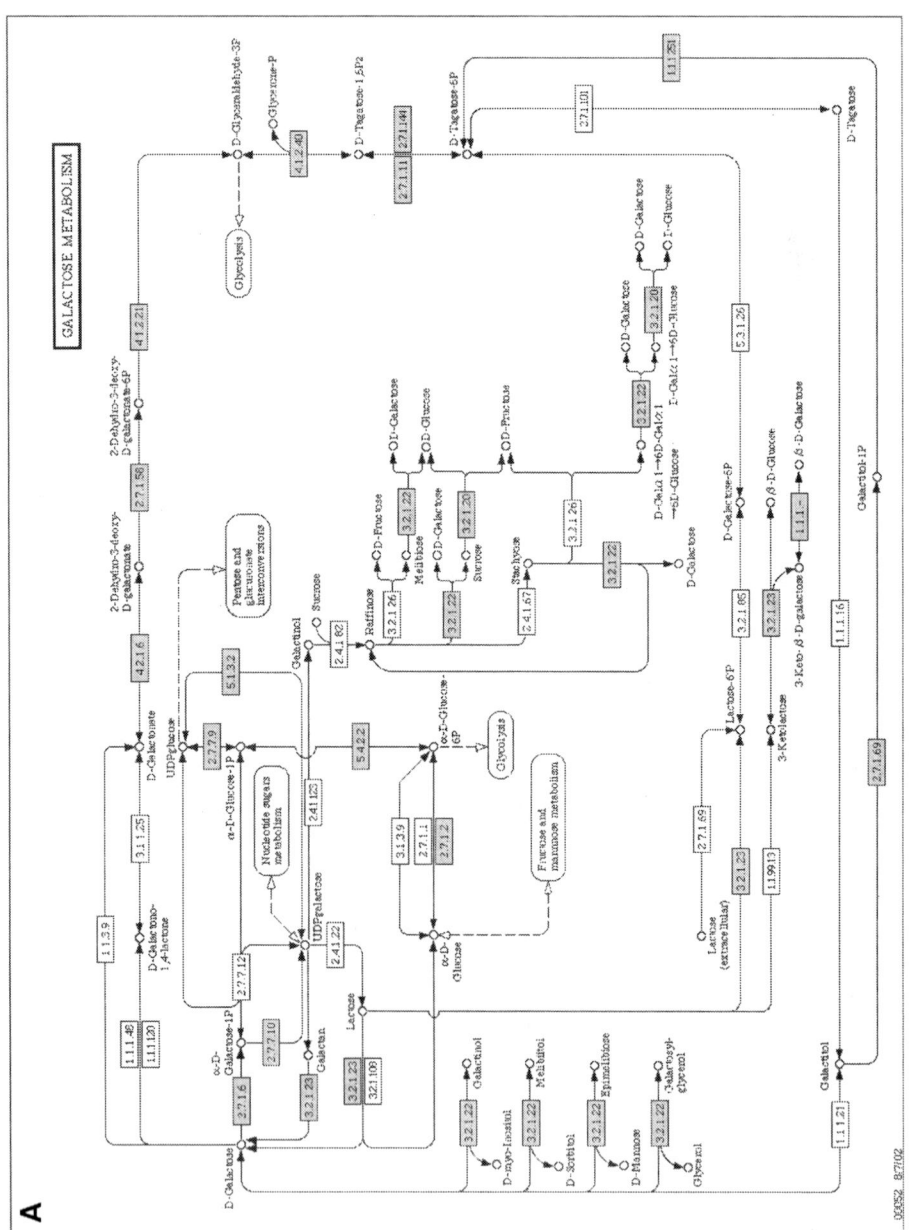

Fig. 6. (*Continued*)

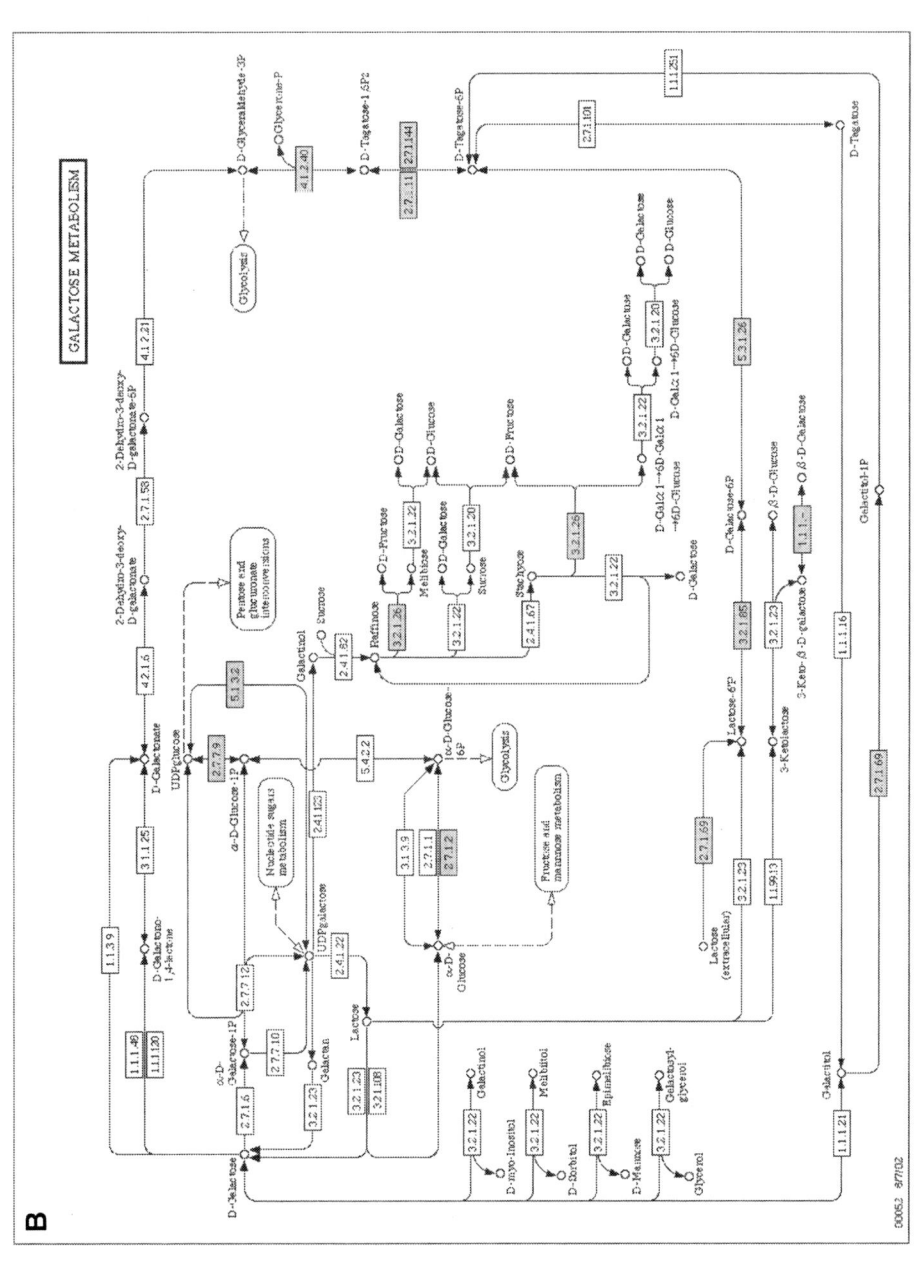

Fig. 6. Example of KEGG Pathway Database showing differences between *E. coli* (**A**) and *S. aureus* (**B**) in their lactose metabolism pathways.

sequence IS*431*, appear in two copies in type II and four copies in type III (marked by arrows in the figure), which are connected by black lines, indicating their similarity.
10. The red short diagonal line of accumulating points in the left bottom corner of **Fig. 3** indicates forward match, which corresponds to the conserved region around the origin of replication. The blue long line from the top left to bottom right indicate inverted match, which shows that most of the remaining chromosome is collinear but inverted. The gaps interrupting the line correspond to genomic islands. Further explanation of this example can be found in **ref. 23**.
11. We computed the phylogenetic tree based on the alignment of the 5′ half of *coa* gene, to avoid the effect of the repetitive sequence existing in the rest of the gene. Further biological information for this example can be found in **ref. 24**.
12. Many other variations for computing phylogenetic trees are implemented in the PHYLIP *(25)* package.
13. An example, **Fig. 6** shows the differences in lactose metabolism pathways between *S. aureus* strain N315 and *Escherichia coli* K-12. Each organism has enzymes that EC numbers are drawn with green background but lacks the enzymes with white background numbers. This shows that *S. aureus* can efficiently utilize lactose via the "tagatose pathway" that *E. coli* lacks.
14. Quite reasonably, most *S. aureus* strains whose whole genome has been determined share a common biochemical pathway; however, strain-specific pathways may be found if whole genomes of other *S. aureus* strains are determined in the future.
15. The existence of a gene for an enzyme does not always mean the gene is expressed, for example because of promoter malfunction by mutation. Indeed, the KEGG pathway databases explain that *S. aureus* strain N315 possesses biosynthetic pathways for all of the amino acids; however, the strain in fact requires alanine, glycine, isoleucine, proline, arginine, and valine. Unfortunately, there is no practical procedure for predicting each *S. aureus* promoter activity only from its sequence and, therefore, further knowledge on the *S. aureus* gene expression system is important.

Acknowledgments

This study was partially supported by a Grant-in-Aid for Young Scientists (B) and a Grant-in-Aid for 21st Century COE from the Ministry of Education, Culture, Sports, Science and Technology (MEXT).

References

1. Brown, T. A. (2006) *Genomes*, Garland Science, New York.
2. Mount, D. W. (2001) *Bioinformatics*, Cold Spring Harbor Laboratory Press, Cold Spring Harbor, NY.
3. Henry-Stocker, S. and Bartlett, K. (2003) *Unix for Mac: Your Visual Blueprint to Maximizing the Foundation of Mac OS X*, John Wiley & Sons, New York.
4. Gibas, C., Jambeck, P., and Fenton, J. M. (2002) *Developing Bioinformatics Computer Skills*, O'Reilly, Sebastopol, Calif.

5. Benson, D. A., Karsch-Mizrachi, I., Lipman, D. J., Ostell, J., and Wheeler, D. L. (2005) GenBank. *Nucleic Acids Res.* **33**, D34–D38 (http://www.ncbi.nlm.nih.gov/Genbank/index.html).
6. Delcher, A. L., Harmon, D., Kasif, S., White, O., and Salzberg, S. L. (1999) Improved microbial gene identification with GLIMMER. *Nucleic Acids Res.* **27**, 4636–4641 (http://cbcb.umd.edu/software/glimmer/).
7. Suzek, B. E., Ermolaeva, M. D., Schreiber, M., and Salzberg, S. L. (2001) A probabilistic method for identifying start codons in bacterial genomes. *Bioinformatics* **17**, 1123–1130 (ftp://ftp.tigr.org/pub/software/ RBSfinder/).
8. Madden, T. L., Tatusov, R. L., and Zhang, J. (1996) Applications of network BLAST server. *Methods Enzymol.* **266**, 131–141 (http://www.ncbi.nlm.nih.gov/blast/).
9. Xanagen *GenomeGambler* (http://www.xanagen.com/).
10. Rutherford, K., Parkhill, J., Crook, J., et al. (2000) Artemis: sequence visualization and annotation. *Bioinformatics* **16**, 944, 945 (http://www.sanger.ac.uk/Software/Artemis/).
11. Nano+Bio-Center. Services in computational biology. (http://nbc11.biologie.uni-kl.de/annotation_suite_tutorial/).
12. Altschul, S. F., Gish, W., Miller, W., Myers, E. W., and Lipman, D. J. (1990) Basic local alignment search tool. *J. Mol. Biol.* **215**, 403–410.
13. Nix, D. *GATA (Graphic Alignment Tool for Comparative Sequence Analysis)* (http://gata.sourceforge.net/).
14. Tanaka, N., Abe, T., Miyazaki, S., and Sugawara, H. (2006) G-InforBIO: Integrated system for microbial genomics BMC Bioinformatics 7, 368 (www.wdcm.org/inforbio/G-InforBIO/download.html).
15. Kurtz, S., Phillippy, A., Delcher, A. L., et al. (2004) Versatile and open software for comparing large genomes. *Genome Biol.* **5**, R12 (http://mummer.sourceforge.net/).
16. Thompson, J. D., Higgins, D. G., and Gibson, T. J. (1994) CLUSTAL W: improving the sensitivity of progressive multiple sequence alignment through sequence weighting, position-specific gap penalties and weight matrix choice. *Nucleic Acids Res.* **22**, 4673–4680.
17. Thompson, J. D., Gibson, T. J., Plewniak, F., Jeanmougin, F., and Higgins, D. G. (1997) The CLUSTAL_X windows interface: flexible strategies for multiple sequence alignment aided by quality analysis tools. *Nucleic Acids Res.* **25**, 4876–4882.
18. Page, R. D. (1996) TreeView: an application to display phylogenetic trees on personal computers. *Comput. Appl. Biosci.* **12**, 357–358 (http://taxonomy.zoology.gla.ac.uk/rod/treeview.html).
19. Brudno, M., Do, C. B., Cooper, G. M., et al. (2003) LAGAN and Multi-LAGAN: efficient tools for large-scale multiple alignment of genomic DNA. *Genome Res.* **13**, 721–731 (http://lagan.stanford.edu/lagan_web/index.shtml).
20. Kanehisa, M. and Goto, S. (2000) KEGG: kyoto encyclopedia of genes and genomes. *Nucleic Acids Res.* **28**, 27–30 (http://www.genome.jp/kegg/).
21. Ito, T., Katayama, Y., Asada, K., et al. (2001) Structural comparison of three types of staphylococcal cassette chromosome mec integrated in the chromosome in

methicillin-resistant Staphylococcus aureus. *Antimicrob. Agents Chemother.* **45,** 1323–1336.
22. Ohta, T., Hirakawa, H., Morikawa, K., et al. (2004) Nucleotide substitutions in *Staphylococcus aureus* strains, Mu50, Mu3, and N315. *DNA Res.* **11,** 51–56.
23. Takeuchi, F., Watanabe, S., Baba, T., et al. (2005) Whole-genome sequencing of *Staphylococcus haemolyticus* uncovers the extreme plasticity of its genome and the evolution of human-colonizing staphylococcal species. *J. Bacteriol.* **187,** 7292–7308.
24. Watanabe, S., Ito, T., Takeuchi, F., Endo, M., Okuno, E., and Hiramatsu, K. (2005) Structural comparison of ten serotypes of staphylocoagulases in *Staphylococcus aureus. J. Bacteriol.* **187,** 3698–3707.
25. Felsenstein, J. (2005) *PHYLIP (Phylogeny Inference Package)* (http://evolution.genetics.washington.edu/phylip.html).

13

Genomic Analysis of Gene Expression of *Staphylococcus aureus*

Chuanxin Yu, Junsong Sun, Li Zheng, and Yinduo Ji

Summary

The microarray has shown tremendous potential for investigating gene expression profiles and expression levels in comparative biology; exploring the regulation mechanisms of gene expression; and evaluating target gene for developing new chemotherapeutic agents, vaccine, and diagnostic methods. In this chapter, we provide a detailed protocol for scientists who wish to investigate gene expression profiles by performing a microarray analysis, including different methods of RNA purification, decontamination, cDNA synthesis, fragmentation, and biotin labeling for hybridization using Affymetrix *Staphylococcus aureus* chips.

Key Words: Microarray analysis; real-time reverse transcription polymerase chain reaction (RT-PCR); gene expression.

1. Introduction

There are thousands of genes and their products (i.e., RNA and proteins) in a given living organism that function in a complicated and orchestrated way that creates the mystery of life. Traditional methods in molecular biology generally work on a "one gene in one experiment" basis, which means that the throughput is very limited and the whole picture of gene function is hard to obtain. Elucidating the global interaction of all of the genes in an organism and the expression dynamics of certain pivotal genes during the growth, development, and pathogenic processes of some pathogens can provide better clues for developing antimicrobial agents and vaccines *(1,2)*. To achieve these aims, microarray and real-time RT-PCR provide researchers with powerful tools.

A microarray, also known as a biochip, is an orderly arrangement of samples that provides a medium for matching known and unknown DNA samples based on base-pairing rules and automation of the process of identifying the unknowns *(1)*.

With the microarray assay, investigators are able to measure changes of the whole genome on a single chip and see the big picture of the interactions among thousands of genes simultaneously. Regulatory systems are part of important networks modulating the expression of *Staphylococcus aureus* genes, including genes that control antibiotic resistance. Therefore, elucidation of the regulons of these regulatory systems is important for researchers to better understand the molecular mechanisms of pathogenesis. Using a microarray-based approach, different *S. aureus* regulons including Agr, ArlRS, Sar, SigB, Rot, and Mgr have been revealed *(3–7)*. Recently, DNA microarray techniques have increasingly been used to characterize the mode of action for drugs against different bacterial pathogens *(8–11)*.

Real-time RT-PCR has also been widely applied in investigating gene expression levels, owing to its high specificity and sensitivity, the ease of calculating the exact amount of original template DNA molecules, and the fact that it is easier to manipulate compared to Northern blotting *(12,13)*. We have been using microarrays and real-time RT-PCR to explore the functions of certain genes important for survival and/or pathogenesis of *S. aureus*. In this chapter, we introduce the protocols of RNA preparation, microarray, and real-time RT-PCR assays.

2. Material

2.1. Cell Culture and Lysis

1. Trypic soy broth (TSB) medium (BD Biosciences, San Jose, CA).
2. Bacterium strain *S. aureus*.
3. Lysostaphin (Sigma, St. Louis, MO).
4. Lysozyme (Sigma).

2.2. RNA Purification (see Note 1)

1. Wizard® SV Total RNA Isolation System (Promega, Madison, WI).
2. Nuclease-free tubes (15 and 1.5 mL) and tips.
3. Diethylpyrocarbonate (DEPC)-treated water.
4. Isopropanol, 70% ethanol, and dehydrated ethanol.
5. Phenol:chloroform, pH 4.5 (Ambion; phenol:chloroform:isoamyl alcohol is 25:24:1).
6. DNA-free™ Kit (Ambion, Austin, TX).
7. Mini-Bead Beater-8 (BioSpec, Bartlesville, OK).
8. Lytic buffer.

2.3. cDNA Synthesis and Fragmentation

1. Random hexamers (250 ng/µL).
2. 10 mM dNTP (2.5 mM each of dATP, dCTP, dTTP, and dGTP).
3. SuperScript™ II Reverse Transcriptase (200 U/µL), 5X first-strand cDNA synthesis buffer and 0.1 M dithiothreitol [DTT] (Invitrogen, Carlsbad, CA).
4. RNaseH (5 U/µL) (Biolabs).
5. QIAquick PCR purification Kit (Qiagen, Valencia, CA).

2.4. cDNA Labeling

1. Terminal transferase with 5X reaction buffer and $CoCl_2$ (25 mM), Biotin-ddUTP (Roche, Indianapolis, IN).
2. Wizard® SV Gel and PCR Clean-up System (Promega).
3. NeutrAvidin.

2.5. Real-Time PCR

1. Control DNA template, usually titrated vector DNA or PCR products (10–10^7 copies).
2. Primers used for real-time PCR.
3. Spectrofluorometric Thermal Cycler (optimal cycling conditions will vary on different real-time instruments). In this protocol, Mx3000PTM from Strategene is employed.
4. Real-time PCR Kit, including 2X Brilliant® SYBR® Green qPCR Master Mix (Strategene, La Jolla, CA).
5. 96-Well plate for RT-PCR.
6. 96-Well plate–sealing film for RT-PCR.

3. Methods
3.1. RNA Extraction and Purification of S. aureus
3.1.1. Method 1: Isolation of Total RNA from S. aureus Using Wizard® SV Total RNA Isolation System

1. Incubate *S. aureus* isolates overnight at 37°C in TSB with appropriate antibiotics and with shaking (225 rpm). The following day, inoculate the overnight cultures in 1% dilution in fresh TSB and incubated until the OD_{600nm} reaches 0.4–0.5. This should take only a few hours. If growth is too slow, reduce the dilution factor.
2. Transfer 3 mL of culture to a 10-mL tube. Centrifuge for 2 min at 14,000g.
3. Carefully remove the supernatant leaving the pellet as dry as possible.
4. Resuspend the pellet in 100 µL of freshly prepared TE, add 6.5 µL of lysostaphin (2 mg/mL) and 6 µL of lysozyme (50 mg/mL), mix, and incubate the tube at 37°C for 6–8 min (not longer than 10 min). Mix one time during the incubation period.
5. Add 75 µL of SV RNA lysis buffer.
6. Add 350 µL of RNA dilution buffer. Gently mix by inversion until the content becomes clear. Do not centrifuge.
7. Add 200 µL of 95% ethanol to the cleared lysate, and mix by pipetting to cut the genomic DNA until the content becomes clear.
8. Transfer the transparent content into a spin column that has been put in a collection tube, and centrifuge the spin column assembly at 14,000g for 1 min.
9. Discard the follow-through solution, and add 600 µL of SV RNA wash solution to the spin column. Centrifuge the spin column assembly at 14,000g for 1 min.
10. Empty the collection tube as before and prepare the DNaseI incubation mix by combining (in this order) 40 µL of yellow core buffer, 5 µL of 0.09 M $MnCl_2$,

and 5 μL of DNaseI enzyme per sample in a sterile tube. Apply 50 μL of this freshly prepared DNase1 incubation mix directly to the membrane inside the spin basket making sure that the solution is in contact with and thoroughly covering the membrane.

11. Incubate for 15 min at room temperature, and then add 200 μL of SV DNase stop solution to the spin column and centrifuge at 14,000g for 1 min.
12. Add 600 μL of SV RNA wash solution and centrifuge at 14,000g for 1 min.
13. Empty the collection tube and add 250 μL of SV RNA wash solution. Centrifuge at high speed for 2 min.
14. Transfer the spin column to a sterilized 1.5-mL centrifuge tube, remove the cap of the tube, and apply 100 μL of nuclease-free water to the column membrane. Be sure to completely cover the surface of the membrane with the water. Centrifuge at 14,000g for 1 min. Remove the spin column and discard. Cap the elution tube containing the purified RNA and store at −70°C.
15. Add 11 μL of 10X DNaseI buffer and 2 μL of DNaseI (2 U/μL) to the RNA solution to digest the contaminated genomic DNA. Mix and spin briefly to collect the solution at the bottom of the tube, and keep the tube at 37°C for 30 min.
16. Add 1/10 vol of DNaseI inactivator (Ambion) to the RNA solution to remove the DNaseI. Mix and keep it at room temperature for 2 min (mixing once during this period).
17. Centrifuge the reaction at 12,000g for 2 min. Carefully transfer the supernatant containing the RNA to a sterilized clean tube.
18. Measure the quality and quantity of purified RNA using a photometer (1 OD_{260} = 40 μg; the ratio of OD_{260}/OD_{280} should be between 1.8 and 2.1). Store the purified RNA at −70°C.

3.1.2. Method 2: Extraction of Total RNA from S. aureus Using Phenol (14)

1. Culture the bacterium at 37°C to the early stationary phase (OD_{600nm} = 1.20–1.30). Then dilute the cells by 1:100 with fresh TSB and continue to culture at 37°C to the mid-exponential phase (OD_{600} = 0.4–0.5).
2. Collect the bacteria of 30-mL cultures by centrifuging at 8000g for 10 min and discard the supernatants fully.
3. Resuspend the pellets in 1 mL of TE containing glucose (25%).
4. Add 10–15 μL of lysostaphin (2 mg/mL), mix gently, and then incubate at 37°C for 10 min. Alternatively, use this method to disrupt the bacterial cell wall: Suspend the bacteria pellets with 1.5 mL of DEPC water, and then transfer the cell suspension to a 2-mL Eppendorf tube containing approximately equal volumes (about 1 mL) of 0.1-mm silica/zirconia beads. Insert the tube into the wells of the shaker of a Mini-Bead Beater-8, and shake the tube vigorously at maximum speed for 3 min. After the shaker machine stops, transfer 1 mL of supernatant to a sterilized Eppendorf tube (*see* **Note 2**).
5. Add 1 mL of lytic buffer to the bacteria solution and gently mix by inversion several times until the mixture becomes clear.

Gene Expression of S. aureus

6. Add 2 mL of acidic phenol/chloroform; vigorously mix the mixture by inversion for 3–5 min.
7. Centrifuge the mixture at 12,000g for 20 min at 4°C, carefully transfer the upper aqueous phase to a new tube, and extract the supernatant again with acidic phenol/chloroform and then chloroform separately.
8. Carefully transfer the supernatant to a new tube without disturbing the interface.
9. Add an equal volume of cold isopropyl alcohol to the solution and mix it by inversion. Keep the tube at −20°C for 3 h or overnight to deposit the RNA.
10. Centrifuge the RNA at 12,000g for 15 min at 4°C. Remove the solution, wash the RNA pellets with 70% alcohol, and dry the RNA pellets at room temperature for 5–10 min.
11. Redissolve the RNA pellets with 100 μL of nuclease-free water.
12. Remove the genomic DNA contamination, and check the quality and quantity of the RNA following **steps 15–18** in **Subheading 3.1.1**.

3.2. Synthesis of First-Strand cDNA

1. Put the reagents in a sterilized 0.2-mL tube as follows: total RNA × μL (12 μg), 3 μL (750 ng) of random primer.
2. Add DEPC water to a total volume of 30 μL, mix, and centrifuge briefly.
3. Incubate at 70°C for 10 min and keep at 25°C for 10 min. Finally, put the reaction on ice for at least 2 min.
4. Add the following components to the RNA/primer mixture in the indicated order:
   a. 12 μL of 5X first-strand cDNA synthesis buffer.
   b. 6 μL of 0.1 *M* DTT.
   c. 3 μL of 10 m*M* dNTP.
   d. 1.5 μL of Superase In.
   e. 5 μL of SuperScriptII.
   f. 2.5 μL of nuclease-free H$_2$O.

   The total volume of the reaction is 60 μL. Mix by pipetting up and down several times and centrifuge briefly. Incubate at 25°C for 10 min, at 37°C for 60 min, then at 42°C for 60 min.
5. Terminate the reaction by keeping the tube at 70°C for 10 min; then hold at 4°C.
6. Centrifuge the reaction tube briefly to collect the products, add 20 μL of 1 *N* NaOH, mix, and keep at 65°C for 30 min to remove the RNA. Then add 10 μL of 1 *N* HCl to neutralize the solution. The synthesizing cDNA products can be stored at −20°C.

3.3. Purification and Quantization of cDNA Synthesis Products

Use the QIAquick PCR Purification Kit to clean up the cDNA synthesis product following the protocol provided by the supplier (Qiagen) (*see* **Note 3**).

3.4. cDNA Fragmentation

1. Prepare the following reaction mixture on ice (*see* **Note 4**): 5 μL of 10X DNase I buffer, 40 μL (6 μg) of cDNA, 0.18 μL (the concentration is 0.6 U/μg of cDNA) of DNase I (2 U/μL), and nuclease-free H$_2$O up to 50 μL.

2. Mix gently and centrifuge briefly to collect the reaction mixture. Incubate the reaction at 37°C for 10 min. Inactivate the enzyme at 98–100°C for 10 min.

3.5. Quality Control/Agarose Gel Electrophoresis

For quality control, pre- and postfragmented cDNA samples are analyzed by agarose gel electrophoresis (**Fig. 1**).

1. Cast a 2% agarose gel containing 0.5 µg/mL of ethidium bromide using TAE.
2. Analyze the fragmented cDNA (~500 ng) and the total cDNA (~500 ng) as a control by electrophoresis. Use a 100-bp DNA ladder (Promega) as a marker for size determination. The desired result should yield a majority of the DNA fragments within a distribution of 50–200 bp.

3.6. Terminal Labeling

The 3N termini of the fragmentized cDNA are labeled using Biotin-ddUTP by terminal transferase (Roche).

1. Prepare the following reaction mix: 14 µL of 5X reaction buffer, 14 µL of 5X $CoCl_2$ (25 mM), 1 µL of Biotin ddUTP, 2 µL of terminal deoxyribonucleotide transferase, and 39 µL (4–5 µg) of fragmentized cDNA, for a total volume of 70 µL.
2. Incubate the reaction for 1 h at 37°C. Add 1.5 µL of 0.5 M EDTA to terminate the reaction. (This labeled fragmented cDNA can be used for microarray directly.)
3. Remove unincorporated biotin label either by using Qiagen RNA/DNA Mini Columns, or by ethanol precipitation. If using ethanol, add 50 µg of glycogen as carrier, 1/10 vol of 3 M sodium acetate, and 2.5 vol of ethanol to samples to precipitate the labeled fragmentized cDNA. Follow this by washing the pellets twice with 750 µL of 70% ethanol. Then, for either method, dissolve the RNA in 20–30 µL of nuclease-free water.
4. Quantify the RNA product by 260-nm absorbance. Typical yields for the procedure are 3 to 4 µg of RNA.

3.7. Analysis of Labeling Efficiency by Gel Shift Assay

1. Prepare a NeutrAvidin solution (2 mg/mL in 50 mM Tris-HCl, pH 7.0).
2. Prepare 150- to 200-ng aliquots of fragmented and biotinylated sample in a fresh tube, add 5 µL of 2 mg/mL NeutrAvidin to each sample, mix, and incubate at room temperature for 5 min.
3. Run the RNA sample in a 4–20% TBE polyacrylamide gel at 150 V until the front dye (red) almost reaches the bottom.
4. Stain the gel with a 1X solution of SYBR Green II or 0.5 µg/mL of Ethidium Bromide (EB) solution.
5. Place the gel on an ultraviolet light box and produce an image.

3.8. Hybridization, Signal Detection, Image Scanning, and Data Collection

The Genechips of *S. aureus* are supplied by Affymetrix. Follow the supplier's protocol for hybridization, signal detection, image scanning, and data collection.

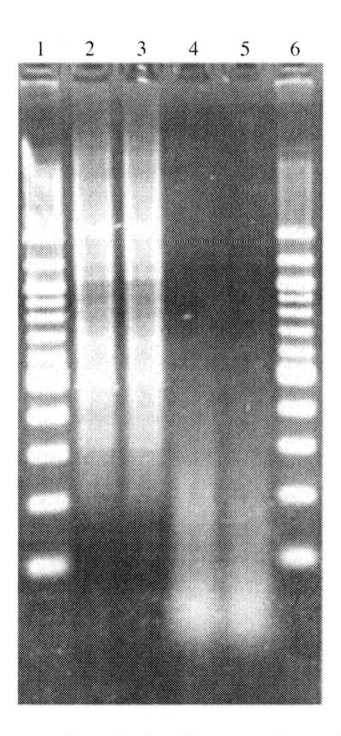

Fig. 1. Two percent agarose gel analysis of pre- and postfragmented cDNA. Lanes 1 and 6, 100-bp DNA marker; lanes 2 and 3, total cDNA; lanes 4 and 5 fragmented cDNA.

3.9. Real-Time PCR

3.9.1. cDNA Synthesis

1. Set up a 20-µL reaction by adding the following components: 1 to 2 µL of random primers (100 ng, included in the cDNA synthesis kit), 1 µL of dNTP mix (10 mM each), and 1 µg of total RNA.
2. Add nuclease-free water to a 13-µL total volume.
3. Heat the mixture at 65°C for 10 min, and then chill on ice immediately.
4. Collect the contents of the tube by brief centrifugation. Add the following reagents to the tube: 4 µL of 5X first-strand buffer, 2 µL of 0.1 M DTT.
5. Incubate the reaction at 25°C for 2 min, add 1 µL of SuperScript II Reverse transcriptase (200 U), and incubate the reaction at 25°C for 10 min followed by 42°C for 50 min.
6. Inactivate the enzymes by heating the reaction to 70°C for 15 min.

*3.9.2. Preparation of Real-Time PCR (see **Note 5**)*

1. Thaw the Brilliant SYBR Green qPCR master mix at room temperature, store it on ice, and keep the unused portion at 4°C in a dark container (*see* **Note 6**). Dilute

the reference dye 1:500 in nuclease-free water if it will be used in the reaction. Dilute the primers to a concentration of 3 μM. Dilute the template cDNA in 20 ng/μL of final concentration.
2. Set up the reaction mixture. For a single reaction mixture of 15 μL, add the following components to each well of 96-well plate: 7.5 μL of 2X master mix; 1.0 μL of upstream and downstream primer (3.0 μM each), respectively; 0.225 μL of diluted preference dye (optional, final concentration: 30 nM), and 1.0 μL of diluted cDNA template. (The amount of cDNA can vary from 1 to 1000 ng depending on the abundance of the specific mRNA in the cells or the diluted genomic DNA in the control experiment.) Adjust the final volume to 15 μL with nuclease-free water. Gently mix the reactions without creating bubbles, and centrifuge the reactions briefly (*see* **Note 7**).
3. Seal the cover of the 96-well plate with thermostable film. Put the plate with the reaction mixtures onto the real-time PCR instrument (*see* **Note 8**).
4. Run the PCR program according to the following procedure being applied to the amplification of a 100- to 300-bp DNA fragment: 1 denature cycle at 95°C for 5 min; 40 amplification cycles at 95°C for 30 s, 55–60°C for 1 min, and 72°C for 30 s; and 1 dissociation cycle at 95°C for 1 min and 53–58°C for 1 min. After the program is finished (about 2.5 h), save and then analyze the data on the computer using the software provided by the manufacturer.

4. Notes

1. Ribonucleases are extremely difficult to inactivate. Avoid inadvertently introducing RNase activity into the RNA sample during or after the isolation procedure. Use sterile technique when handling all the reagents and wear gloves at all times. Treat nondisposable glassware and plastic ware before use to ensure that it is free of RNase. Bake glassware at 200°C overnight, and thoroughly rinse plastic ware with 0.1 N NaOH, 1 mM EDTA followed by RNase-free water. Treat solutions supplied by the user by adding DEPC to 0.1% and then incubating overnight at room temperature and autoclaving for 30 min to remove any trace of DEPC.
2. To utilize this method, we can isolate large amounts of total RNA from *S. aureus*. Therefore, it is dependent on how much RNA one wishes to prepare for the assay.
3. Removal of all residual ethanol from the spin column is critical for the success and reproducibility of the following step. To ensure ethanol removal, centrifuge the column for at least 10 min at high speed in a clean Eppendorf tube following the wash step.
4. The amount of DNase1 required for the cDNA fragmentation can vary from supplier to supplier. A titration experiment should be performed with each new batch of enzyme. The high active enzyme should be diluted with 1X DNase1 buffer first, and the DNase1 should be added to the reaction last.
5. When designing the real-time PCR experiment, standard and blank control, and/or no reverse transcriptase control must be considered. Good quantitative PCR (qPCR) results must also contain those data.

6. The SYBR Green dye and the reference dye are light sensitive. They should be stored away from light.
7. The bubbles in the reaction mixture may interfere with fluorescence detection. Prepare the reaction mixture carefully to avoid the formation of bubbles.
8. The 96-well plate with reaction mixtures should be sealed tightly with thermostable film, because evaporation of the reaction mixture may lead to the loss of the reaction volume and cause the polymerase to malfunction.

Acknowledgment

This work was supported by USPHS research grant AI057451 from the National Institute of Allergy and Infectious Disease.

References

1. Ramsay, G. (1998) DNA chips: state-of-the art. *Nat. Biotechnol.* **16,** 40–44.
2. Chin, K. V. and Kong, A. (2002) Application of DNA microarrays in pharmacogenomics and toxicogenomics. *Pharm. Res.* **19,** 1173–1178.
3. Dunman, P., Murphy, E., Haney, S., et al. (2001) Transcription profiling-based identification of *Staphylococcus aureus* genes regulated by the *agr* and/or *sarA* loci. *J. Bacteriol.* **183,** 7341–7353.
4. Bischoff, M., Dunman, P., Kormanec, J., et al. (2004) Microarray-based analysis of the *Staphylococcus aureus* σB regulon. *J. Bacteriol.* **186,** 4085–4099.
5. Liang, X., Zheng, L., Landwehr, C., Lunsford, D., Holmes, D., and Ji, Y. (2005) Global regulation of gene expression by ArlRS, a two-component signal transduction regulatory system of *Staphylococcus aureus. J. Bacteriol.* **187,** 5486–5492.
6. Saïd-Salim, B., Dunman, P., McAleese, F., et al. (2003) Global regulation of *Staphylococcus aureus* genes by Rot. *J. Bacteriol.* **185,** 610–619.
7. Luong, T., Dunman, P., Murphy, E., Projan, S., and Lee, C. (2006) Transcription profiling of the *mgrA* regulon in *Staphylococcus aureus. J. Bacteriol.* **188,** 1899–1910.
8. Bammert, G. and Fostel, J. (2000) Genome-wide expression patterns in *Saccharomyces cerevisiae*: comparison of drug treatments and genetic alterations affecting biosynthesis of ergosterol. *Antimicrob. Agents Chemother.* **44,** 1255–1265.
9. Gmuender, H., Kuratli, K., Di Padova, K., Gray, C., Keck, W., and Evers, S. (2001) Gene expression changes triggered by exposure of *Haemophilus influenzae* to novobiocin or ciprofloxacin: combined transcription and translation analysis. *Genome Res.* **11,** 28–42.
10. Khodursky, A., Peter, B., Schmid, M., DeRisi, J., Botstein, D., Brown, P., and Cozzarelli, N. (2000) Analysis of topoisomerase function in bacterial replication fork movement: use of DNA microarrays. *Proc. Natl. Acad. Sci. USA* **97,** 9419–9424.
11. Wilson, M., DeRisi, J., Kristensen, H., Imboden, P., Rane, S., Brown, P., and Schoolnik, G. (1999) Exploring drug-induced alterations in gene expression in *Mycobacterium tuberculosis* by microarray hybridization. *Proc. Natl. Acad. Sci. USA* **96,** 12,833–12,838.
12. Wilhelm, J. and Pingoud, A. (2003) Real-time polymerase chain reaction. *Chembiochem.* **4,** 1120–1128.

13. Arya, M., Shergill, I. S., Williamson, M., Gommersall, L., Arya, N., and Patel, H. R. (2005) Basic principles of real-time quantitative PCR. *Expert Rev. Mol. Diagn.* **5,** 209–219.
14. Cheung, A. (1994) A method to isolate RNA from Gram-positive bacteria and Mycobacteria. *Anal. Biochem.* **222,** 511–514.

14

Proteomic Approach to Investigate MRSA

Patrice Francois, Alexander Scherl, Denis Hochstrasser, and Jacques Schrenzel

Summary

Over the past decade numerous genomes of pathogenic bacteria were fully sequenced and annotated, while others are continuously being sequenced and published. To date, the sequences of >440 bacterial genomes are publicly available for research purposes. These efforts in high-throughput sequencing parallel major improvements in methods permitting the study of whole transcriptome and proteome of bacteria. This provides a basis for a comprehensive understanding of the bacterial metabolism, adaptability to the environment, regulation, resistance pathways, and pathogenicity mechanisms of pathogens. *Staphylococcus aureus* is a Gram-positive human pathogen causing a wide variety of infections ranging from benign skin infections to life-threatening diseases. Furthermore, the spreading of multiresistance strains requiring the use of last-barrier drugs has resulted in the medical and scientific community focusing particularly on this pathogen. We describe here proteomic methods to prepare, identify, and analyze protein fractions, allowing the study of *S. aureus* on the organism level. Coupled with methods analyzing the whole bacterial transcriptome, this approach might contribute to the development of rapid diagnostic tests and to the identification of new drug targets.

Key Words: *Staphylococcus aureus*; methicillin-resistant *S. aureus*; proteomics; protein fractionation; separation; identification; quantification; expression.

1. Introduction

Staphylococcus aureus is a Gram-positive bacteria member of the Micrococcaceae family. The bacterium is able to grow under aerobic and anaerobic conditions and causes various infections, ranging from mild skin infections and food poisoning to life-threatening diseases, such as pneumonia, sepsis, osteomyelitis, and infectious endocarditis *(1)*. *S. aureus* showed a peculiar ability to rapidly develop multiple resistances to antimicrobial agents currently used in human medicine.

The pathogenicity of *S. aureus* is particularly complex, involving numerous bacterial products as well as elaborated regulation pathways *(2)*. *S. aureus* is able to produce a wide range of toxins showing a deleterious effect on cell integrity and functions. Most of these factors (e.g., toxic shock syndrome toxin-1, exfoliatin toxins A and B, Panton-Valentine leukocidin, enterotoxins, and hemolysins) contribute to the virulence of clinical isolates in the context of acute infections *(1,2)*. In addition to these excreted compounds, *S. aureus* is able to produce several cell wall–associated proteins allowing interactions with host plasma or extracellular proteins, such as fibronectin, fibrinogen, collagen, vitronectin, laminin, and bone sialoprotein *(3)*. The contribution of these bacterial compounds to the virulence of *S. aureus* and the regulation of their expression is partly known. For example, the contribution of fibrinogen- or fibronectin-binding proteins to infective endocarditis *(4,5)*, the role of the collagen-binding protein in septic arthritis *(6)*, the role of hemolysins in corneal infections *(7,8)*, and the ability of *S. aureus* to persist intracellularly have been documented *(9)*, a crucial aspect in the context of chronic infection *(10–12)*. Most of these epidemiological or clinical behaviors are not totally understood because these important studies relied on the utilization of single mutant strains of *S. aureus* genetically engineered and showed the impact of a single loss of function, thus limiting the evaluation of compensation mechanisms using such strategies.

To date, seven *S. aureus* genomes are already publicly available, allowing the utilization of massively parallel techniques such as microarrays and proteomics to study this bacterium as a biological system. The postgenomic era of *S. aureus* began in 2001 with the release of the two first whole-genome sequences of hospital-acquired strains, published by Kuroda et al. *(13)*. More recently, the whole-genome sequence of the community-acquired strain (community-acquired methicillin-resistant *S. aureus* [MRSA]) MW2 has been released by the same group *(14)*. These sequenced strains contain approx 2,800,000 bp forming circular genomes and coding for approx 2600 proteins. Using appropriate bioinformatics tools, the availability of these genome sequences allowed prediction of the vast majority of open reading frames (ORFs), then deducing of the amino acid sequence of the whole proteome. An average 40% of these putative ORFs encode for proteins, showing a high degree of homology with proteins having known functions in other organisms (orthologues). However, the majority of putative ORFs correspond to hypothetical proteins, never isolated or encoding for hypothetical proteins of unknown function. More recently, a whole sequence of strains COL (an MRSA laboratory strain) *(15)*, MRSA252, and MSSA476 *(16)* was published. This considerable amount of information allows researchers to develop strategies to study in more detail the genetic backgrounds of clinical isolates *(17)*. Progress in bacterial genomics and proteomics will benefit the medical community by providing an opportunity to evaluate virulence on a more global view using techniques allowing the study of whole-organism contents in RNA and proteins during a single experiment.

The proteomic approach provides new information on metabolism, regulation pathways, as well as regulation under stress conditions including the presence of antimicrobial drugs. Proteomics also reveals instrumental for confirming gene annotations by identifying hypothetical or unpredicted molecules. Finally, the detection of specific proteomic signatures corresponding to peculiar resistance profiles or particular metabolic states has a major impact on the comprehension of global regulatory networks involved in the virulence or resistance of strains. Ultimately, these signatures will allow the identification of new drug targets and/or the development of new diagnostic tools. These advances will likely contribute to the understanding of *S. aureus* pathogenicity.

In this chapter, we describe proteomic procedures allowing (1) the preparation of total or membrane *S. aureus* protein extracts, (2) the separation of proteins using one-dimensional (1D)- or two-dimensional (2D) electrophoresis, and (3) the identification of proteins using matrix-assisted laser desorption ionization time of flight (MALDI-TOF) or mass spectrometry (MS). Together, these methods were used to improve knowledge about global mechanisms involved in virulence and to characterize specific properties of the bacteria.

2. Materials
2.1. S. aureus *Culture and Lysis*

1. *S. aureus* suspensions: Prepare from overnight culture in Mueller-Hinton broth (Difco, Detroit, MI).
2. Lysis medium: phosphate-buffered saline (PBS) containing 1 mM CaCl$_2$ and 1 mM MgCl$_2$ enriched with one tablet of protease inhibitors (Complete®; Boehringer) for 50 mL of PBS.
3. Lysis buffer containing 100 µg/mL of lysostaphin (Ambicin): This is used to incubate bacterial pellets for 15 min at 37°C with agitation.
4. DNase (100 U) (Sigma, St. Louis, MO), for bacterial cell wall hydrolysis (*see* **Note 1**).
5. After removal of bacterial debris, protein concentration is assessed from a 10 µl aliquot (Pierce). Generally, a total amount of 15–25 mg of proteins is obtained using this procedure from 100-mL suspensions. For a large volume of suspension, *see* **Note 2**.

2.2. *Preparation of Membrane Protein Extracts*

1. Isotonic buffer: 10 mM Tris-HCl, pH 7.4, 1.5 mM MgCl$_2$, 10 mM KCl, 0.5 mM DTE, 1.1 M saccharose, and protease inhibitors.
2. Lysostaphin (100 µg/mL).
3. MilliQ water containing protease inhibitors and 100 U of DNase I, for incubation of protoplasts (*see* **Note 1**).

2.3. Protein Separation Using 2D Electrophoresis (see Note 3)

1. Nonlinear immobilized pH gradients (3.5–10.0, 180-mm length): These are used as the first dimension (Pharmacia).
2. Immobilized pH gradient (IPG) strips: Rehydrate overnight in the dedicated cassette with 25 mL of a buffer containing 8 M urea, 2% (w/v) cholamidopropyl dimethylammonio propanesulfonate (CHAPS), 10 mM dithioerythritol (DTE), Resolyte, pH 3.5–10.0 (2% [v/v]), and bromophenol blue.
3. Strip tray (Pharmacia), low-viscosity paraffin oil (Merck): Remove the strips from the rehydratation cassette and transfer to the strip tray. Cover the electrodes and loading cups with low-viscosity paraffin oil.
4. Dedicated power supply (Pharmacia): Perform isoelectric focusing (IEF) at linearly increasing voltage (300–3500 V) during the first 3 h, followed by another 3 h at 3500 V, and finally to 5000 V.
5. Second dimension buffer: 50 mM Tris-HCl, pH 8.4, 6 M urea, 30% (v/v) glycerol, 2% (w/v) sodium dodecyl sulfate (SOS), and 2%(w/v) DTE: This is used to equilibrate the strips after the first dimension.
6. Blocking solution: 100 mL of a solution containing Tris-HCl (50 mM), pH 6.8, urea (6 M), glycerol (30% [v/v]), SDS (2% [w/v]), iodoacetamide (2.5% [w/v]), and traces of bromophenol blue (see **Note 4**): This is used to block –SH groups.

2.4. Sodium Dodecyl Sulfate-Polyacrylamide Gel Electrophoresis

This procedure is performed essentially as described by Laemmli *(18)* with some minor modifications.

1. Separating buffer (4X): 1.5 M Tris-HCl, pH 8.7, 0.4% SDS. Store at room temperature.
2. Stacking buffer (4X): 0.5 M Tris-HCl, pH 6.8, 0.4% SDS. Store at room temperature.
3. 30% Acrylamide/bisacrylamide solution (Protogel; National Diagnostics) (see **Note 5**) and TEMED (Bio-Rad, Hercules, CA) (see **Notes 6–8**).
4. Agarose (0.5%), molecular biology quality in Tris-glycine-SDS (25–198 mM, 0.1% SDS), pH 8.3: This is required to cover the IPG strip, which is then carefully dipped into the agarose until making contact with the running gel. Migration is then initiated for 5 h in a cold room (8–12°C) at a constant current of 40 mA/gel. Generally, a voltage of 200–400 V is currently observed.

2.5. Protein Detection

Application of the 2D polyacrylamide gel electrophoresis (PAGE) technology to separate, analyze, and characterize proteins contained in biological samples would not have been possible without the development of complementary detection methods. Depending on the type of analysis (either analytic or quantitative), the most currently used protein coloration is silver staining, which is 100-fold more sensitive than Coomassie Brilliant Blue. The silver-staining protocol is performed under constant rotary agitation. For silver staining, large volumes of bi-distilled water and several solvents are required:

1. Ethanol:acetic acid:water (40:10:50).
2. Ethanol:acetic acid:water (5:5:90).
3. Glutaraldehyde (1%) and sodium acetate (0.5 M).
4. 2,7-Naphtalene-disulfonic acid solution (0.05% [w/v]) (see **Note 9**).
5. Ammoniacal silver nitrate solution: To prepare 750 mL of this solution, dissolve 6 g of silver nitrate in 30 mL of deionized water; then slowly mix into a solution containing 160 mL of water, 10 mL of concentrated ammonia (25%), and 1.5 mL of NaOH (10 N). A transient brown precipitate might form, after spontaneous removal; add water to give the final volume.
6. Solution containing citric acid (0.01% [w/v]) and formaldehyde (0.1% [v/v]): This is used to develop images, which takes 5–10 min (see **Note 10**).
7. Solution of Tris (5% [w/v]) and acetic acid (2% [v/v]), to stop staining development.

2.6. Protein Trypsinization-In-Gel Digestion

For moderately complex protein extracts, SDS gel allows a first separation of proteins; further analysis of protein composition by MALDI-TOF is performed after protein digestion. For such analysis, 1D gel is stained using Coomassie staining.

1. Perform staining in Coomassie blue R-250 solution (0.1% in 50% methanol).
2. Destaining solution: 10% acetic acid, 40% methanol in MilliQ water.
3. Use 10 mM DTE solution (1.54 mg in 1 mL of 50 mM bicarbonate [BA]) to reduce proteins.
4. Prepare fresh 55 mM iodoacetamide solution (10.2 mg in 1 mL of 50 mM BA) to alkylate proteins in the presence of 2 M urea and 0.05% SDS.
5. 50 mM Ammonium bicarbonate, pH 8.0, to dehydrate.
6. Put the dried gel pieces (see **Note 11**) on ice and prepare trypsin solution (6.25 ng/µL).
7. Concentrate the peptides and desalt using an Oasis HLB 1-cc 10-mg solid-phase extraction cartridge (Waters, Milford, MA).
8. Analytical steps will require the utilization of 0.1% trifluoroacetic acid (TFA) in 50% CH$_3$CN.
9. Perform biphasic multidimensional liquid chromatography in a 100-µm-id column first packed with a C$_{18}$ reverse phase over 10 cm and then with a 5-cm strong cation-exchange phase (polysulfoethyl A BMSE05-03, PolyLC Inc., Columbia, MD).
10. Spot fractions obtained with 0, 10, 20, 40, 80, 150, 300, and 500 mM KCl on MALDI target plates and analyze by MALDI-TOF/TOF.
11. Perform database search and validation with MASCOT 1.8 software using the same search parameters and validation criteria as previously described *(19)*.

2.7. Peptide Analysis Using Mass Spectrometry Techniques

Mass spectrometry is currently used for protein analysis and is useful in the identification of proteins separated by 2D gel electrophoresis. The most common mass spectrometry protein identification technique (peptide mass fingerprinting) relies on the determination of peptide masses by spectrometric techniques after protein digestion with residue-specific proteases as described in **Subheading 2.6**. Peptide masses are then analyzed against theoretical peptide libraries generated from protein sequence databases, generated from nucleotide sequence information.

Developments of soft ionization methods such as MALDI and electrospray ionization (ESI) permitted the analysis of large biopolymers. Mass spectrometers measure the mass to charge ratio (m/z) of gas-phase ions. The instrumentation consists of three basic components: (1) the ion source, (2) the mass analyzer, and (3) the detector. The sample is ionized in the ion source; the ions are separated according to their m/z ratio in the mass analyzer before striking the detector. Protein identification is performed by comparison between experimental ion masses and theoretical ion masses deduced from protein sequence databases. An excellent first-reading introduction is given in **ref. 20**.

In MALDI *(21,22)*, the analyte is cocrystallized with a solid matrix, usually an aromatic acid. This matrix absorbs the photonic energy from a laser, which results in an explosive phase transition from solid to gas of matrix neutrals, entraining analyte molecules. During this gas jet, energy is transferred from the matrix to the analyte molecules, resulting mostly in the formation of singly charged gas-phase analyte ions, following an extensively studied, but not entirely understood, mechanism *(23)*.

Compared to MALDI, ESI is performed in liquid solution *(23,24)*. By increasing the potential applied to the liquid, a critical value is reached; the liquid is no longer able to hold charges and blows apart into a thin cloud of small charged droplets. The resulting ions have therefore relatively low m/z ratios compared to MALDI ions. Typically, ESI spectra are recorded from 400 to 1600 Th, where most macromolecules appear as multicharged species. Since ESI is performed in liquid, these ion sources are particularly adapted for direct (or on-line) coupling with high-performance liquid chromatography systems. Quadrupoles, linear ion traps, three-dimensional ion traps, and TOFs are widely used for protein and peptide analysis. More recently, commercially available instruments with Fourier-transform ion-cyclotron-resonance mass analyzers have been introduced, providing ultrahigh mass accuracy and resolution. In addition, the recently introduced orbitrap mass analyzer promises great performance for applications in the area of proteomics *(25)*.

3. Methods
3.1. Preparation of Total Protein from MRSA

1. Perform *S. aureus* culture with agitation at 37°C in Mueller-Hinton broth (50 mL in a 500-mL flask). At postexponential phase (OD_{540nm} = 5 to 6 Absorbance Unit (AU), corresponding to 2 to 3 × 10^9 cells/mL), chill on ice and harvest by centrifuging at 8000g for 5 min at 4°C.
2. Resuspend the pellet in 5 mL of lysis buffer, and add 100 µg/mL of lysostaphin for 10 min at 37°C with constant shaking.
3. After initiation of lysis add 10 µg/mL of DNase I.
4. Recover the total protein extract after centrifuging at 8000g for 15 min.
5. Assay the protein concentration.

3.2. Analysis of MRSA Proteome: The Challenge of Membrane Protein Characterization in the Study of S. aureus

Membrane proteins play an important role in signal transduction, transport, endocytosis, cell adhesions, drug resistance, and many other cellular functions. The majority of all drugs targets are membrane-associated proteins *(26)*. These proteins belong to two main classes: (1) proteins that span the lipid bilayer, with a cytoplasmic and an extracellular domain; and (2) proteins only partially embedded in the membrane, or attached to the membrane by a linker. The first category is called "integral membrane proteins" or "intrinsic membrane proteins." The transmembrane (TM) domain (or segment) spans the lipid bilayer. Integral membrane proteins show one or numerous TM domains composed of hydrophobic amino acids yielding highly hydrophobic moiety in the protein sequence, as schematized in **Fig. 1**. To create a catalog of proteins from *S. aureus* containing membrane proteins, we used the following strategy:

1. Perform *S. aureus* culture with agitation at 37°C in Mueller-Hinton broth (200 mL in a 1000-mL flask). At postexponential phase (OD_{540nm} = 5 to 6 AU, corresponding to 2 to 3 H 10^9 cells/mL), chill on ice and harvest by centrifuging at 8000g for 5 min at 4°C.
2. For preparation of membrane extracts, prepare 20-mL culture aliquots and wash in 1.1 M saccharose-containing buffer. Then suspend in 2-mL aliquots of the same buffer containing 100 µg/mL of lysostaphin for 10 min at 37°C.
3. Recover protoplasts by centrifuging for 30 min at 8000g, and perform hypoosmotic shock in the presence of 10 µg/mL of DNase I.
4. Obtain the membrane pellets after ultracentrifuging at 50,000g for 50 min in a Beckman Optima TLX.
5. Solubilize one of the fractions in 2 mL of 0.1% SDS solution for determination of protein concentration.
6. For 300 to 1000 µg of membrane-enriched protein fractions, solubilize in 150 µL of 50 mM ammonium bicarbonate.

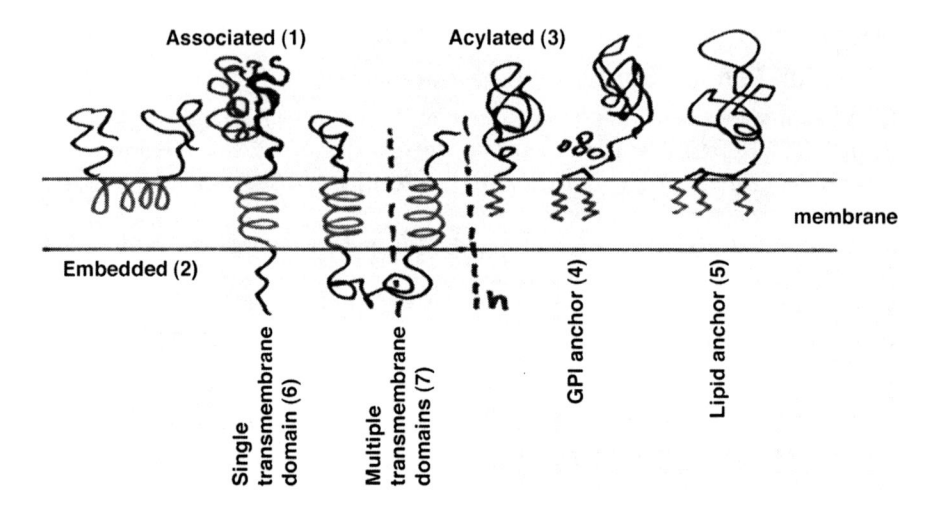

Fig. 1. Schematic representation of different categories of membrane proteins. Associated membrane proteins can be either bound to the membrane via interactions to another membrane protein (1), embedded into the membrane (2), or attached to it through a posttranslationally added group such as a GPI (4) or a lipid anchor (5). Integral membrane proteins have one (6) or multiple (7) TM domains. (Kindly provided by J. Deshusses.)

7. Add 1 mL of a 2,2,2-trifluoroethanol (TFE)/CHCl$_3$ (1:1 [v/v]) mixture, shake vigorously, and maintain at 0°C for 1 h with periodic vortexing (see **Note 12**).
8. Centrifuge at 10,000g for 4 min to allow separation of the mixture into three phases *(27)*. The lower chloroformic and the upper aqueous phases are separated from the insoluble interphase.
9. Carefully collect each phase and concentrate in a vacuum evaporator.
10. Perform 1D electrophoresis according to Laemmli *(18)* using 55-mm-long gels to check the quality of the fractionation process.
11. Perform preparative 2D electrophoresis and follow the analytic work flow.

3.3. In-Gel Digestion of MRSA Proteins and Identification

Total protein and enriched membrane fractions subjected to electrophoresis separation, either 1D or 2D, are digested in-gel by trypsin.

1. Excise spots or slices (see **Note 13**) of acrylamide gel and perform in-gel trypsin digestion.
2. Analyze using a 4700 MALDI-TOF/TOF tandem mass spectrometer (Applied Biosystems, Framingham, MA). Load extracted peptides onto a 10-cm-long homemade column with a 75-μm id packed with C$_{18}$ reverse phase (YMS-ODS-AQ200; Michrom BioResource, Auburn, CA), and elute on a MALDI target.

The analysis of hydrophobic proteins, such as integral membrane proteins, is a major limiting factor in 2-dimensional electrophoresis (2DE). Continuous efforts were made to improve buffers for hydrophobic proteins compatible with IEF. Despite these efforts, an abundance of membrane proteins is clearly underrepresented after 2DE separation of proteins *(28–30)*. Owing to these limitations, alternative protein and peptide separation techniques, such as multidimensional liquid chromatography *(31)*, IEF of digested protein mixtures *(32)* and other electrophoretic-based methods were developed, providing powerful alternatives to 2DE-based analytical techniques.

Differences in solubility are observed between peripheral membrane proteins and integral membrane proteins. A majority of peripheral membrane proteins requires only mild treatments, such as increased ionic strength, to dissociate them from the membrane. In such conditions, these proteins dissociate free from lipids and are relatively soluble in neutral aqueous buffers *(33,34)*. Many different detergents from different classes (ionic, nonionic, and zwitterionic) have been extensively studied for the solubilization of membrane proteins (for a review, *see* **ref. 35**).

Because of their hydrophobicity, it is possible to solubilize integral membrane proteins with hydrophobic, nonpolar solvents such as chloroform. This approach has been used successfully to extract proteins containing multiple TM segments from crude membrane preparations *(36–38)*. Other organic solvents or cosolvents can be used for the extraction of hydrophobic proteins. A buffered 60% methanol solution was used with success for the solubilization of diverse membrane protein preparations *(39)*. The use of 50% TFE as cosolvent in aqueous solutions was applied with success in our laboratory to improve the solubility of membrane protein extracts *(27)* from membrane fractions of *S. aureus* (*see* **Note 12** and **Fig. 2**).

However, despite continual improvements for the solubilization of hydrophobic proteins, membrane proteins are generally not recovered from 2DE gels. This is illustrated by an experiment performed in our laboratory: We applied water/trifluoroethanol/chloroform extractions to a crude *S. aureus* membrane preparation. Hydrophobic proteins recovered from the interphase (between the aqueous and the organic phase) were never retrieved on 2DE gels (**Fig. 3**; unpublished data). However, when the same protein fractions were separated by SDS-PAGE *(18)*, the hydrophobic proteins were recovered (**Fig. 4**).

According to the thickness of bacterial cytoplasmic membrane, the TM domains of proteins should be constituted of 10–25 hydrophobic amino acids, forming generally α-helixes. In addition, TM pores formed by β-sheet structures are also observed, such as in the pore-forming toxins of *S. aureus* *(40–42)*. By analyzing amino acid sequences of proteins, bioinformatic tools permit the

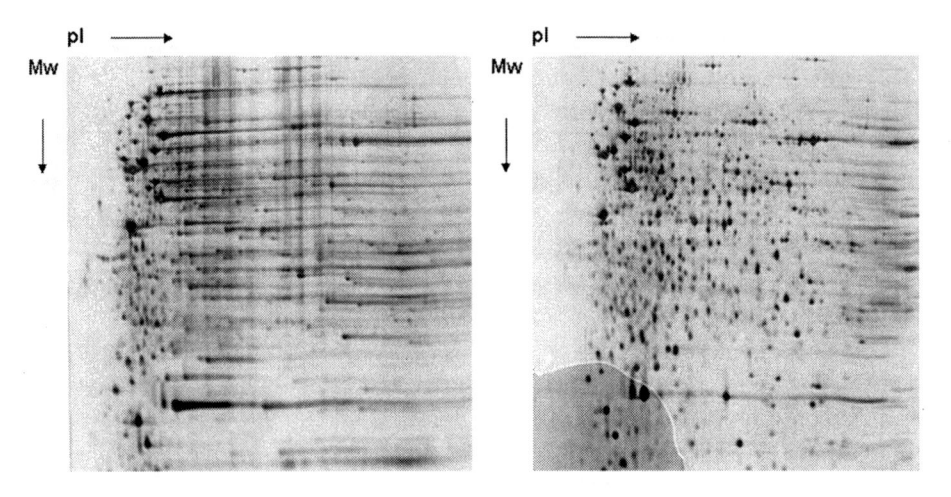

Fig. 2. Silver-stained 2DE gels of membrane protein fractions from N315 using standard procedure (*left*) or in presence of 50% TFE during IEF (*right*).

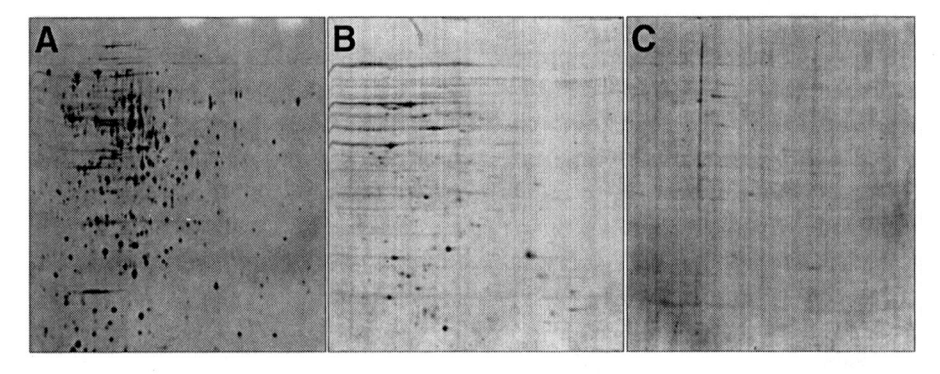

Fig. 3. Silver-stained 2D gels of protein fractions after chloroform/water/trifluoroethanol extractions of a crude *S. aureus* N315 membrane extract: (**A**) soluble proteins; (**B**) total membrane extracts showing limited number of unfocused proteins; (**C**) absence of proteins from hydrophobic fraction at interface between chloroformic and aqueous phase. (A detailed procedure is described in **ref. 27**.)

prediction of hydrophobic segments corresponding to TM sequence (see http://expasy.org/tools/). Within the deduced proteome of *S. aureus*, the proportion of predicted integral membrane proteins is estimated to be 20–30% of all proteins, a value in accordance with what is generally found in all living organisms *(43)*. This evaluation was performed on *S. aureus* strain N315 *(44)*. Based on the sequence-deduced 2575 ORFs, a total of 668 proteins are predicted to contain one or more TM domains after cleavage of the signal peptide, representing 26%

Proteomic Approach to Investigate MRSA

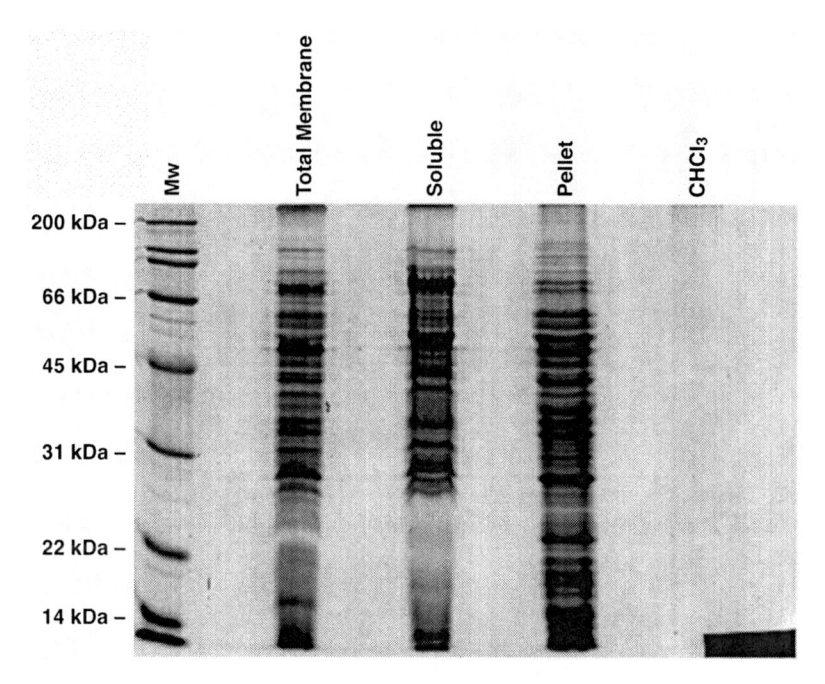

Fig. 4. Silver-stained SDS-PAGE gel of protein fractions after chloroform/water/ trifluoroethanol extractions of a crude *S. aureus* membrane extract. No proteins were detected from the chloroformic phase. Mw, molecular weight standards. Tot. Memb, crude membrane extracts; Soluble, protein from the aqueous phase; Pellet, proteins from the interphase; CHCl$_3$, content of the chloroformic phase.

of all ORFs. The number of proteins as a function of the number of predicted TM domains is shown in **Fig. 5**.

3.4. Application of Proteomic Methods to Study of MRSA

After several tests relying on medium composition during all analytical steps, we used the following preparation scheme for the isolation of total or subcellular fractions of N315 proteins, in order to build an extensive catalog of characterized proteins (**Fig. 6**). For sample preparation, *S. aureus* N315 was grown in Mueller-Hinton broth (Difco) as previously described. For total protein extracts, cells were lysed with 20 µg/mL of lysostaphin (Ambicin; Applied Microbiology, Tarrytown, NY) for 15 min at 37°C in Tris-EDTA buffer. Insoluble material was removed by centrifuging at 5000g for 10 min. Protoplasts were prepared using the same amount of lysostaphin in the presence of 1.1 *M* sucrose. Protoplasts were then lysed in hypotonic buffer and membrane extracts were collected by centrifuging at 33,000g for 1 h. 2DE was performed

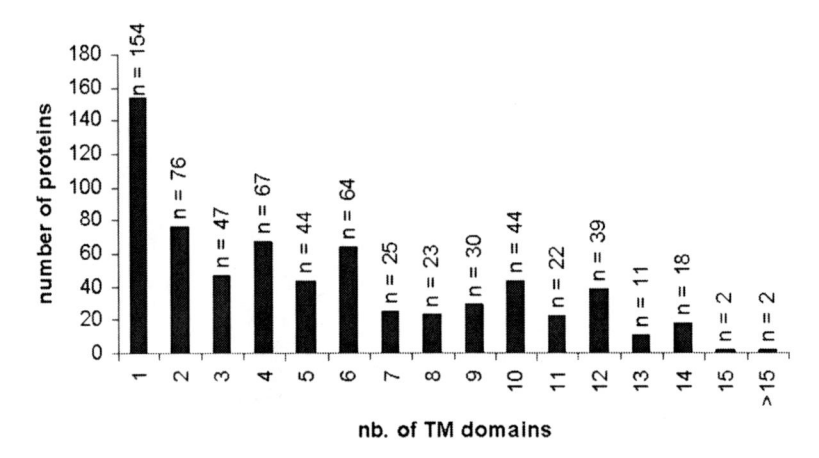

Fig. 5. Number of the 2575 predicted ORFs from *S. aureus* strain N315 displaying one or more TM segments. In total, 668 (26%) of the ORFs are integral membrane proteins.

in the presence of trifluoroethanol during IEF as described previously *(27)*. SDS-PAGE was then performed using a standard procedure with 12.5% acrylamide. Finally, 2DE gels were analyzed using Melanie 3 software (Genebio, Geneva, Switzerland).

Publicly available genome sequences allowed our group to develop software for the design of oligonucleotides showing optimal physicochemical properties, usable for the manufacture of microarrays covering the whole genome of *S. aureus* *(45)*. In a recent study, a combined proteomic and transcriptomic analysis of *S. aureus* strain N315 was performed to analyze a sequenced strain at the system level *(44)*. Total protein and membrane protein extracts were prepared and analyzed using various proteomic work flows (**Fig. 6**) including 2DE, SDS-PAGE combined with microcapillary liquid chromatography (LC)-MALDI-MS/MS, and multidimensional LC. The results obtained for SDS-PAGE work flow are presented in **Fig. 7**. The presence of a protein was then correlated with its respective transcript levels from *S. aureus* cells grown under the same conditions. Gene-expression data revealed that 97% of the 2596 ORFs were detected during the postexponential phase. At the protein level, 23% of these ORFs (591 proteins) were identified. Correlation of the two data sets revealed that 42% of the identified proteins (248 proteins) were among the top 25% of genes with the highest mRNA signal intensities, and 69% of the identified proteins (406 proteins) were among the top 50% with the highest mRNA signal intensities. The fact that the remaining 31% of proteins were not strongly expressed at the RNA level indicates either that some low-abundance proteins

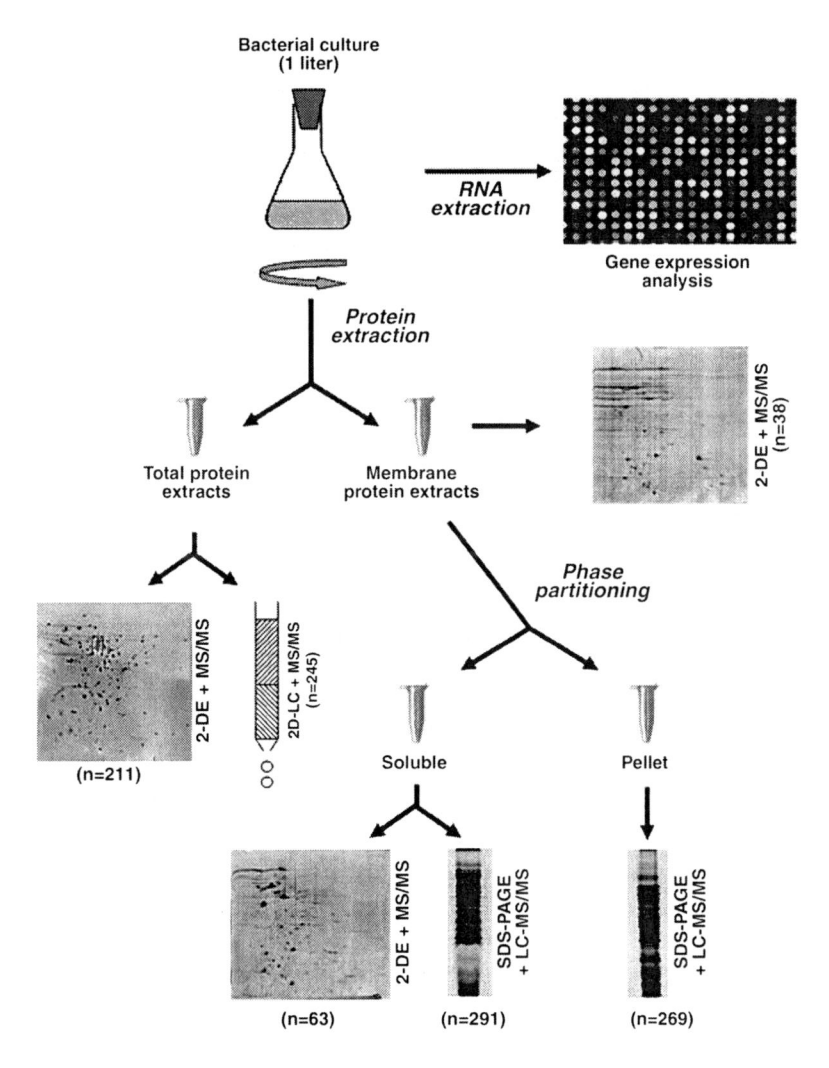

Fig. 6. Protein and RNA sample preparation with analysis steps. Bacteria were grown until postexponential phase (OD_{540nm} = 5 to 6 AU, corresponding to 2 to 3 × 10^9 cells/mL). Total bacterial RNA was labeled, hybridized, and analyzed on a custom genomewide oligoarray. Total protein extracts were analyzed by 2DE LC-MS/MS or multidimensional LC-MS/MS. Total membrane extracts were either analyzed by 2DE/LC-MS/MS or subjected to a phase-partitioning procedure. Soluble membrane extracts were analyzed by 2DE/LC-MS/MS or 1DE/LC-MS/MS. Insoluble membrane extracts were analyzed by 1DE/LC-MS/MS. Numbers in parentheses refer to the number of unique proteins identified by each specific work flow.

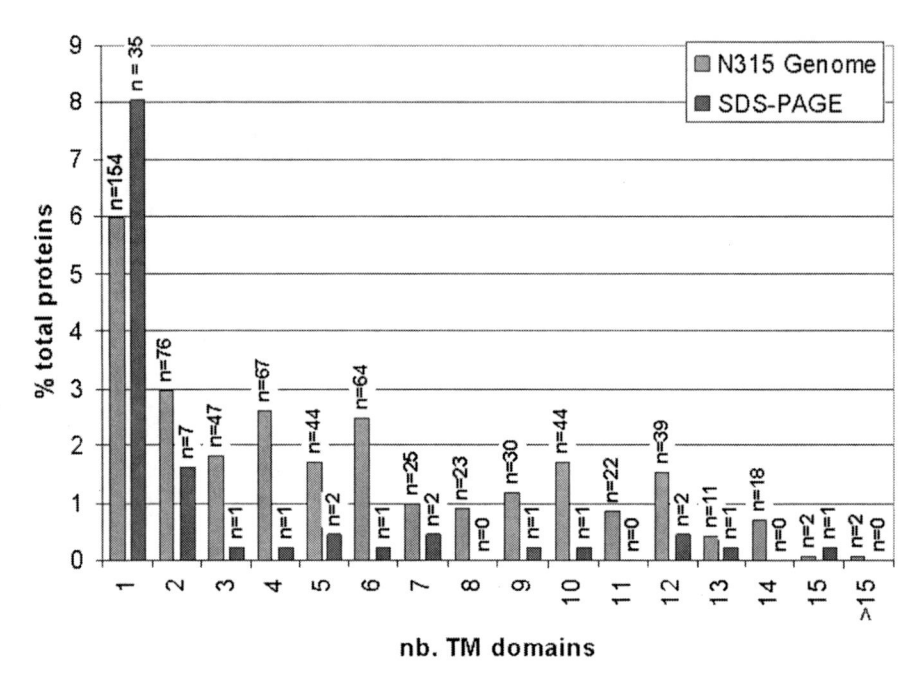

Fig. 7. Proteins with TM domain(s) identified from membrane fractions isolated as shown in **Fig. 6** and compared to N315 genome.

were identified or that some transcripts or proteins showed extended half-lives. The most abundant classes identified with the combined proteomic and transcriptomic approach involved energy production, translational activities, and nucleotide transport, reflecting an active metabolism *(45a)*. The simultaneous large-scale analysis of transcriptomes and proteomes enabled a global and holistic view of the *S. aureus* biology, allowing the parallel study of multiple active events in an organism.

S. aureus is an important human pathogen, and numerous groups of researchers have contributed to important advances in this regard. Cordwell et al. *(46)* initiated a protein map of the bacterium, then compared the response of methicillin-susceptible *S. aureus* and MRSA strains *(47)* using strain COL, the prototype of MRSA sequenced recently *(15)*, in the presence of Triton X-100. They were able to identify a short list of proteins probably involved in the resistance against methicillin using refined proteomic analysis *(47)*.

Recent efforts in the field of proteomics allow quantitative information about protein abundance to be obtained *(48–50)*. These considerable technological advances now allow the study of organisms as biological systems considering that the analysis of their transcriptome and proteome can possibly be done simultaneously.

Such combined studies have already been successfully used to elucidate the vast majority of genomes and proteomes of *S. aureus* bacteriophages. In a recent study, Kwan et al. *(51)* discovered that *S. aureus* bacteriophages belong to three major categories and that variation processes included not only recombination between phages but also between phage and the host *S. aureus*. Such a study provides important information including complete genome sequences and putative phage protein content and also allows the study of phage evolution and diversity origins *(51)*, an important issue in the epidemiology of clinical isolates of MRSA. The biology of *S. aureus* has also benefited from this type of approach. Kohler et al. *(52)* elucidated potential mechanisms underlying a particular phenotype of *S. aureus* known as small colony variant. In addition to the reduced amount of exo-proteins, a characteristic already described, Kohler et al. *(52)* identified possible adenosine triphosphate generation pathways in these slow-growing organisms presumably clinically relevant in the context of chronic infection *(11,53)*. In the clinical context, the expression of virulence factors is also of major importance. Ziebandt et al. *(55)* identified at the protein level the impact of global regulators on the extracellular proteome of *S. aureus*. This study allowed these investigators to precisely identify different stages of regulation during the growth of *S. aureus* and to understand how the bacterium tightly controlled temporal expression of these virulence factors. This important study suggests that pathogenicity is strictly temporally controlled through a network of interacting regulons, controlling the virulence of the strain *(54)*.

3.5. Applications of Proteomics Dedicated to MRSA

S. aureus is one of the leading causes of infections in immunocompromised patients and is the major cause of nosocomial infections in developed countries. Simultaneously, its exceptional adaptability to environmental changes and its particular ability to rapidly develop resistances against antimicrobials justify important measures to control its spread. The recent apparition of multiresistant strains even against last barrier drugs is also a major concern. The emergence and the potential spread of such dangerous clones urgently demand new antimicrobial molecules. Massively parallel techniques such as whole-organism proteomics and/or genomics might be instrumental in the development of such compounds. Despite the fact that one-third of the bacterial genome still consists of genes with unknown function, their utilization as possible drug targets is not restricted because part of these genes is required for bacterium viability. However, despite the huge promise of the genomics era, new classes of antibiotics targeting novel enzymes resulting exclusively in genomics studies are still not commercialized. Of course, the processes before commercialization are time-consuming and the steps required upstream, such as identification of new target, screening of inhibitors, synthesis, evaluation of the spectrum of activity, membrane permeability, determination of optimal concentration, and pharmacokinetics, are costly and difficult to ensure (for reviews *see* **refs. 55–57**). In such a context, the contribution of proteomics will permit restriction of the number of potential targets of interest, thus restricting the risk of developing new drugs for nontranslated genes or rapidly identifying genes whose product is really required for bacterial survival.

Consequently, quantitative proteomics will be crucial for the development of rapid diagnostic tests or the development of antibacterial vaccines. For example, by rapid screening of infected patient sera, Vytvytska et al. *(58)* identified a list of *S. aureus* antigens of potential interest for vaccine development. A similar strategy has been used for another bacteria of interest, *Helicobacter pylori (59)*. To date, this type of strategy has been successfully used to identify circulating cancer markers, allowing the rapid diagnosis of the disease by testing patient sera *(60,61)*. In the field of microbial diagnosis, future development of such tests targeting markers of bacterial identification, resistance, epidemiology, and virulence predictors will be instrumental to researchers.

The sequencing of multiple genomes of *S. aureus* corresponds to a new era in the study of this important human pathogen. Simultaneously to this advance, proteomic tools have evolved considerably and both genomic and proteomic information is now assessable using adapted strategies. These highly parallel methods will help researchers to elucidate the complex molecular mechanisms involved in virulence and the resistance to environmental stresses and antimicrobials. In addition to molecular genetics, these technological advances will bring new information about the complex "lifestyle" of *S. aureus*.

4. Notes

1. Depending on the density of bacterial suspensions, lysis medium will become very viscous. For digestion of dense suspension and preparation of membrane protein extracts that require a reduction in the lysis volume before ultracentrifugation, the addition of 100–200 U of DNase drastically reduces the viscosity of the digestion mixture.
2. Most *S. aureus* strains are digested by a high concentration of the murolytic enzyme lysostaphin, as indicated in **Subheading 2.1**. However, some clinical isolates appear particularly resistant when suspensions are very dense. We fractionate the original suspension into 10-mL aliquots for the digestion step. Fractions are then pooled for the following steps.
3. If very high concentrations of cytosolic or cell wall–associated proteins are required for preparative analysis, precipitation can be performed using trichloroacetic acid (TCA). By adjusting to 10% TCA using a concentrated stock solution, soluble proteins precipitate efficiently after a few hours at 4°C (overnight

incubation allows excellent recovery). After centrifuging for 30 min at 8000g, the pellets are washed with ethanol:acetone (1:1) to remove traces of TCA. Pellets are then solubilized in the desired buffer.
4. This step is required to reduce S-S bonds and to solubilize proteins after IEF in a buffer compatible with SDS-PAGE.
5. Acrylamide/bisacrylamide monomer (before polymerization) is a recognized neurotoxin. Be careful to avoid direct exposure.
6. TEMED in powder is stored at ambient condition in desiccators. Solutions are commercially available, but we have observed that the time for polymerization increases with the age of the solution.
7. SDS acrylamide gels are not polymerized in the presence of SDS. This seems to prevent the formation of micelles that contain acrylamide monomer, thus increasing the homogeneity of pore size. The SDS used in the gel running buffer is sufficient to maintain the necessary negative charge on the proteins.
8. Piperazine diacrylyl can be used as crosslinker of acrylamide gel. It results probably in the reduction of N-terminal protein blockage, gives better resolution, and reduces diamine silver-staining background.
9. A solution of 2,7 naphtalene-disulfonic acid (0.05% [w/v]) allows homogeneous dark brown staining of proteins to be obtained.
10. For silver staining of acrylamide gel, we have observed that the addition of sodium thiosulfate drastically reduces the background of staining of the gels.
11. If gel pieces are still blue after destaining and reducing steps, repeat steps 1–5 until destaining is complete.
12. TFE acts as a cosolvent, aggregating around the hydrophobic portions of the proteins. This excludes water and therefore makes impossible the formation of hydrogen bonds. In addition, the cosolvent provides a low dielectric environment favoring the formation of intraprotein hydrogen bonds, promoting the stability of the secondary structure. The secondary structure is stabilized at the expense of the tertiary structure. The unfolded protein in its cosolvent shell is therefore maintained in solution.
13. Typically, a 1D electrophoresis gel is sliced into 30–50 pieces.

References

1. Lowy, F. D. (1998) *Staphylococcus aureus* infections. *N. Engl. J. Med.* **339,** 520–532.
2. Bronner, S., Monteil, H., and Prevost, G. (2004) Regulation of virulence determinants in *Staphylococcus aureus*: complexity and applications. *FEMS Microbiol. Rev.* **28,** 183–200.
3. Foster, T. J. and Hook, M. (1998) Surface protein adhesins of *Staphylococcus aureus*. *Trends Microbiol.* **6,** 484–488.
4. Que, Y. A., Haefliger, J. A., Piroth, L., et al. (2005) Fibrinogen and fibronectin binding cooperate for valve infection and invasion in *Staphylococcus aureus* experimental endocarditis. *J. Exp. Med.* **201,** 1627–1635.
5. Moreillon, P., Entenza, J. M., Francioli, P., McDevitt, D., Foster, T. J., Francois, P., and Vaudaux, P. (1995) Role of *Staphylococcus aureus* coagulase and

clumping factor in pathogenesis of experimental endocarditis. *Infect. Immun.* **63,** 4738–4743.
6. Switalski, L. M., Patti, J. M., Butcher, W., Gristina, A. G., Speziale, P., and Hook, M. (1993) A collagen receptor on *Staphylococcus aureus* strains isolated from patients with septic arthritis mediates adhesion to cartilage. *Mol. Microbiol.* **7,** 99–107.
7. Supersac, G., Piemont, Y., Kubina, M., Prevost, G., and Foster, T. J. (1998) Assessment of the role of gamma-toxin in experimental endophthalmitis using a *hlg*-deficient mutant of *Staphylococcus aureus. Microb. Pathog.* **24,** 241–251.
8. O'Callaghan, R. J., Callegan, M. C., Moreau, J. M., et al. (1997) Specific roles of alpha-toxin and beta-toxin during *Staphylococcus aureus* corneal infection. *Infect. Immun.* **65,** 1571–1578.
9. Sinha, B., Francois, P. P., Nusse, O., et al. (1999) Fibronectin-binding protein acts as *Staphylococcus aureus* invasin via fibronectin bridging to integrin a5b1. *Cell Microbiol.* **1,** 101–117.
10. Clement, S., Vaudaux, P., Francois, P., et al. (2005) Evidence of an intracellular reservoir in the nasal mucosa of patients with recurrent *Staphylococcus aureus* rhinosinusitis. *J. Infect. Dis.* **192,** 1023–1028.
11. Proctor, R. A., van Langevelde, P., Kristjansson, M., Maslow, J. N., and Arbeit, R. D. (1995) Persistent and relapsing infections associated with small-colony variants of *Staphylococcus aureus. Clin. Infect. Dis.* **20,** 95–102.
12. Vaudaux, P., Francois, P., Bisognano, C., et al. (2002) Increased expression of clumping factor and fibronectin-binding proteins by *hemB* mutants of *Staphylococcus aureus* expressing small colony variant phenotypes. *Infect. Immun.* **70,** 5428–5437.
13. Kuroda, M., Ohta, T., Uchiyama, I., et al. (2001) Whole genome sequencing of methicillin-resistant *Staphylococcus aureus. Lancet* **357,** 1225–1240.
14. Baba, T., Takeuchi, F., Kuroda, M., et al. (2002) Genome and virulence determinants of high virulence community-acquired MRSA. *Lancet* **359,** 1819–1827.
15. Gill, S. R. (2004) *Staphylococcus aureus* COL genome. www.tigr.org/tigr-scripts/CMR2/GenomePage3.spl?database=gsa.
16. Holden, M. T., Feil, E. J., Lindsay, J. A., et al. (2004) Complete genomes of two clinical *Staphylococcus aureus* strains: evidence for the rapid evolution of virulence and drug resistance. *Proc. Natl. Acad. Sci. USA* **101,** 9786–9791.
17. Fitzgerald, J. R., Sturdevant, D. E., Mackie, S. M., Gill, S. R., and Musser, J. M. (2001) Evolutionary genomics of *Staphylococcus aureus*: insights into the origin of methicillin-resistant strains and the toxic shock syndrome epidemic. *Proc. Natl. Acad. Sci. USA* **98,** 8821–8826.
18. Laemmli, U. K. (1970) Cleavage of structural proteins during the assembly of the head of bacteriophage T4. *Nature* **227,** 680–685.
19. Scherl, A., Francois, P., Converset, V., et al. (2004) Nonredundant mass spectrometry: a strategy to integrate mass spectrometry acquisition and analysis. *Proteomics* **4,** 917–927.
20. Steen, H. and Mann, M. (2004) The ABC's (and XYZ's) of peptide sequencing. *Nat. Rev. Mol. Cell Biol.* **5,** 699–711.
21. Nakazawa, T., Yamaguchi, M., Nishida, K., et al. (2004) Enhanced responses in matrix-assisted laser desorption/ionization mass spectrometry of peptides derivatized

with arginine via a C-terminal oxazolone. *Rapid Commun. Mass Spectrom.* **18,** 799–807.
22. Karas, M. and Hillenkamp, F. (1988) Laser desorption ionization of proteins with molecular masses exceeding 10,000 daltons. *Anal. Chem.* **60,** 2299–2301.
23. Zenobi, R. and Knochenmuss, R. (1998) Ion formation in MALDI mass spectrometry. *Mass Spectrom. Rev.* **17,** 337–366.
24. Zeleny, J. (1914) The electrical discharge from liquid points, and a hydrostatic method measuring the electric intensity at their surfaces. *Phys. Rev.* **3,** 69–91.
25. Hu, Q., Noll, R. J., Li, H., Makarov, A., Hardman, M., and Graham, C. R. (2005) The Orbitrap: a new mass spectrometer. *J. Mass Spectrom.* **40,** 430–443.
26. Russell, R. B. and Eggleston, D. S. (2000) New roles for structure in biology and drug discovery. *Nat. Struct. Biol.* **7(Suppl.),** 928–930.
27. Deshusses, J. M., Burgess, J. A., Scherl, A., et al. (2003) Exploitation of specific properties of trifluoroethanol for extraction and separation of membrane proteins. *Proteomics* **3,** 1418–1424.
28. Wilkins, M. R., Gasteiger, E., Sanchez, J. C., Bairoch, A., and Hochstrasser, D. F. (1998) Two-dimensional gel electrophoresis for proteome projects: the effects of protein hydrophobicity and copy number. *Electrophoresis* **19,** 1501–1505.
29. Rabilloud, T., Adessi, C., Giraudel, A., and Lunardi, J. (1997) Improvement of the solubilization of proteins in two-dimensional electrophoresis with immobilized pH gradients. *Electrophoresis* **18,** 307–316.
30. Adessi, C., Miege, C., Albrieux, C., and Rabilloud, T. (1997) Two-dimensional electrophoresis of membrane proteins: a current challenge for immobilized pH gradients. *Electrophoresis* **18,** 127–135.
31. Washburn, M. P., Wolters, D., and Yates, J. R. III (2001) Large-scale analysis of the yeast proteome by multidimensional protein identification technology. *Nat. Biotechnol.* **19,** 242–247.
32. Cargile, B. J., Talley, D. L., and Stephenson, J. L. Jr. (2004) Immobilized pH gradients as a first dimension in shotgun proteomics and analysis of the accuracy of pI predictability of peptides. *Electrophoresis* **25,** 936–945.
33. Singer, S. J. and Nicolson, G. L. (1972) The fluid mosaic model of the structure of cell membranes. *Science* **175,** 720–731.
34. Richardson, S. H., Hultin, H. O., and Green, D. E. (1963) Structural proteins of membrane systems. *Proc. Natl. Acad. Sci. USA* **50,** 821–827.
35. Helenius, A. and Simons, K. (1975) Solubilization of membranes by detergents. *Biochim. Biophys. Acta* **415,** 29–79.
36. Brugiere, S., Kowalski, S., Ferro, M., et al. (2004) The hydrophobic proteome of mitochondrial membranes from *Arabidopsis* cell suspensions. *Phytochemistry* **65,** 1693–1707.
37. Ferro, M., Salvi, D., Riviere-Rolland, H., et al. (2002) Integral membrane proteins of the chloroplast envelope: identification and subcellular localization of new transporters. *Proc. Natl. Acad. Sci. USA* **99,** 11,487–11,492.
38. Ferro, M., Seigneurin-Berny, D., Rolland, N., et al. (2000) Organic solvent extraction as a versatile procedure to identify hydrophobic chloroplast membrane proteins. *Electrophoresis* **21,** 3517–3526.

39. Blonder, J., Conrads, T. P., Yu, L. R., et al. (2004) A detergent- and cyanogen bromide-free method for integral membrane proteomics: application to *Halobacterium purple* membranes and the human epidermal membrane proteome. *Proteomics* **4**, 31–45.
40. Guillet, V., Roblin, P., Werner, S., et al. (2004) Crystal structure of leucotoxin S component: new insight into the Staphylococcal beta-barrel pore-forming toxins. *J. Biol. Chem.* **279**, 41,028–41,037.
41. Guillet, V., Keller, D., Prevost, G., and Mourey, L. (2004) Crystallization and preliminary crystallographic data of a leucotoxin S component from *Staphylococcus aureus*. *Acta Crystallogr. D Biol. Crystallogr.* **60**, 310–313.
42. Menestrina, G., Dalla, S. M., Comai, M., et al. (2003) Ion channels and bacterial infection: the case of beta-barrel pore-forming protein toxins of *Staphylococcus aureus*. *FEBS Lett.* **552**, 54–60.
43. Wallin, E. and Von Heijne, G. (1998) Genome-wide analysis of integral membrane proteins from eubacterial, archaean, and eukaryotic organisms. *Protein Sci.* **7**, 1029–1038.
44. Scherl, A., Francois, P., Bento, M., et al. (2005) Correlation of proteomic and transcriptomic profiles of *Staphylococcus aureus* during the post-exponential phase of growth. *J. Microbiol. Methods* **60**, 247–257.
45. Charbonnier, Y., Gettler, B. M., Francois, P., et al. (2005) A generic approach for the design of whole-genome oligoarrays, validated for genomotyping, deletion mapping and gene expression analysis on *Staphylococcus aureus*. *BMC Genomics* **6**, 95.
45a. Scherl, A., Francois, P., Charbonnier, Y., et al. (2006) Exploring glycopeptide resistance in *Staphylococcus aureus*: a combined proteomics and transcriptomics approach for the identification of resistance-related markers. *BMC Genomics* **7**, 296.
46. Cordwell, S. J., Nouwens, A. S., Verrills, N. M., McPherson, J. C., Hains, P. G., Van Dyk, D. D., and Walsh, B. J. (1999) The microbial proteome database—an automated laboratory catalogue for monitoring protein expression in bacteria. *Electrophoresis* **20**, 3580–3588.
47. Cordwell, S. J., Larsen, M. R., Cole, R. T., and Walsh, B. J. (2002) Comparative proteomics of *Staphylococcus aureus* and the response of methicillin-resistant and methicillin-sensitive strains to Triton X-100. *Microbiology* **148**, 2765–2781.
48. DeSouza, L., Diehl, G., Rodrigues, M. J., Guo, J., Romaschin, A. D., Colgan, T. J., and Siu, K. W. (2005) Search for cancer markers from endometrial tissues using differentially labeled tags iTRAQ and cICAT with multidimensional liquid chromatography and tandem mass spectrometry. *J. Proteome Res.* **4**, 377–386.
49. Sechi, S. and Oda, Y. (2003) Quantitative proteomics using mass spectrometry. *Curr. Opin. Chem. Biol.* **7**, 70–77.
50. Pasquarello, C., Sanchez, J. C., Hochstrasser, D. F., and Corthals, G. L. (2004) N-t-butyliodoacetamide and iodoacetanilide: two new cysteine alkylating reagents for relative quantitation of proteins. *Rapid Commun. Mass Spectrom.* **18**, 117–127.
51. Kwan, T., Liu, J., DuBow, M., Gros, P., and Pelletier, J. (2005) The complete genomes and proteomes of 27 *Staphylococcus aureus* bacteriophages. *Proc. Natl. Acad. Sci. USA* **102**, 5174–5179.

52. Kohler, C., von Eiff, C., Peters, G., Proctor, R. A., Hecker, M., and Engelmann, S. (2003) Physiological characterization of a heme-deficient mutant of *Staphylococcus aureus* by a proteomic approach. *J. Bacteriol.* **185,** 6928–6937.
53. Kahl, B., Herrmann, M., Everding, A. S., et al. (1998) Persistent infection with small colony variant strains of *Staphylococcus aureus* in patients with cystic fibrosis. *J. Infect. Dis.* **177,** 1023–1029.
54. Ziebandt, A. K., Weber, H., Rudolph, J., Schmid, R., Hoper, D., Engelmann, S., and Hecker, M. (2001) Extracellular proteins of *Staphylococcus aureus* and the role of SarA and *sigma B*. *Proteomics* **1,** 480–493.
55. Stoughton, R. B. and Friend, S. H. (2005) How molecular profiling could revolutionize drug discovery. *Nat. Rev. Drug Discov.* **4,** 345–350.
56. Berger, A. B., Vitorino, P. M., and Bogyo, M. (2004) Activity-based protein profiling: applications to biomarker discovery, in vivo imaging and drug discovery. *Am. J. Pharmacogenomics* **4,** 371–381.
57. Lau, A. T., He, Q. Y., and Chiu, J. F. (2003) Proteomic technology and its biomedical applications. *Sheng Wu Hua Xue Yu Sheng Wu Wu Li Xue Bao (Shanghai)* **35,** 965–975.
58. Vytvytska, O., Nagy, E., Bluggel, M., Meyer, H. E., Kurzbauer, R., Huber, L. A., and Klade, C. S. (2002) Identification of vaccine candidate antigens of *Staphylococcus aureus* by serological proteome analysis. *Proteomics* **2,** 580–590.
59. Haas, G., Karaali, G., Ebermayer, K., et al. (2002) Immunoproteomics of *Helicobacter pylori* infection and relation to gastric disease. *Proteomics* **2,** 313–324.
60. Shoshan, S. H. and Admon, A. (2005) Proteomics in cancer vaccine development. *Expert Rev. Proteomics* **2,** 229–241.
61. Le Naour, F. (2001) Contribution of proteomics to tumor immunology. *Proteomics* **1,** 1295–1302.

15

Environmental Surveillance for MRSA

J. Scott Weese

Summary

The role of the inanimate environment, including the air, in the transmission of methicillin-resistant *Staphylococcus aureus* (MRSA) infection is unclear; however, there are certain situations when evaluation of MRSA contamination of the environment is indicated. At this point, conventional culture methods are predominantly used, with molecular methods reserved for characterization of recovered isolates. A variety of methods are available for environmental sampling, and the objectives of sampling must be considered when choosing the appropriate technique.

Key Words: Methicillin-resistant *Staphylococcus aureus*; environment; disinfection; contamination; screening.

1. Introduction

The close contacts that we have with the inanimate environment and the potential for widespread bacterial contamination of this environment inevitably raise concerns about the environment as a source of infection with a variety of microorganisms. The environment is clearly a source of infection for some pathogens (e.g., *Legionella*), whereas its role is unclear for many others. *Staphylococcus aureus* is in the latter category, but concerns exist because of its prevalence in the human population and its survival characteristics. *S. aureus* is tolerant to many environmental conditions and can remain viable in the environment for a variable period of time depending on environmental conditions such as temperature, humidity, and exposure to ultraviolet light and disinfectants. One study reported the survival of *S. aureus* in the environment for up to 20 d *(1)*. Additionally, it has been shown that outbreak strains of methicillin-resistant *S. aureus* (MRSA) can live longer in the environment than other strains *(2)*.

The epidemiology of nosocomial MRSA has been extensively studied and numerous studies evaluating MRSA contamination of the environment have been performed, but the role of the environment in MRSA transmission remains unclear. Although the environment is generally considered not to be a major source of MRSA infection, it has been suggested that the environment may play a role in certain situations *(3–6)*.

MRSA can be found throughout the environment in human and veterinary hospitals, on floor surfaces, in personnel contact areas, and in patient contact sites *(7–10)*. Boyce et al. *(11)* reported isolation of MRSA from the environment of 73% of hospital rooms housing infected patients, and 69% of rooms housing colonized patients. In particular, frequent hand contact sites such as door handles, bed linens, bedside tables and chairs, and window ledges are often contaminated *(7–10)*. MRSA can also be found in the households of colonized individuals *(12)*.

Although the hands of patients and health-care personnel are accepted as the most common source of MRSA infection, there is considerable debate about the role of airborne transmission of MRSA. Bacteria can become airborne attached to skin squames, lint, and dust and spread via air currents *(13,14)*, and outbreaks of MRSA postoperative infection have been linked to airborne transmission *(15)*. Other sources of airborne *S. aureus* include dispersal during bed making of colonized patients *(16)* and contamination of ventilation systems *(17)*, but the clinical relevance of these is unclear.

The main problem in evaluating the role of environmental pathogens such as MRSA is not identification of MRSA in the environment. Rather, the problem involves interpreting the results and drawing objective conclusions. The environment cannot be evaluated in a simple dichotomous view of either sterile or not *(3)*. The inanimate environment is not sterile, nor is it required to be so in virtually any circumstance. Indeed, it would be unrealistic to assume that the environment of a person infected or colonized with MRSA would not become passively contaminated through normal activities. Isolation of an organism after inoculation onto an inanimate surface is considered the weakest form of evidence that an environmental surface or fomite is a source of infection *(3)*. Therefore, although culture and evaluation of isolates can be important tools for evaluation of environmental contamination with bacterial pathogens, they can provide only limited information. Although providing further evidence, molecular analysis of isolates is not as useful in this context compared to other areas of epidemiology *(3)*. For example, identification of indistinguishable MRSA isolates from a patient developing nosocomial MRSA infection does not prove a causal relationship (or indeed any relationship) between the patient and the environment. Both sites could have been infected by a common source (i.e., health-care workers' hands); the patient could have been infected first, then contaminated the environment; or the environment could have been contaminated with a common strain prior to the patient's admission.

Evaluation of MRSA contamination of the environment or air has typically involved outbreak investigations or research studies, and results must be interpreted with care. Environmental screening might be useful in a few situations. These include using the environment as an indicator of whether MRSA has recently been present in the population in a specific area (in areas where MRSA is rarely encountered or has not been previously identified), assessing disinfection if performed in a controlled manner immediately after disinfection, conducting outbreak investigation in the hospital or in the community, detecting disinfectant-resistant strains, monitoring quality control of sterilization, educating individuals of the need for proper hand hygiene, and determining whether an environmental site is a potential source of infection with a properly designed study. Routine sampling in normal (endemic) situations should be minimal *(18)*.

A variety of sampling methods are available, and there is little objective information regarding the relative effectiveness of different methods for recovery of MRSA. The choice of method will depend on the surface type being sampled, the desired sensitivity, the cost, and laboratory convenience. In general, broth enrichment methods are likely to be superior to direct culture in terms of MRSA recovery. Further, sampling methods that sample a large surface area and that are able to get into cracks and other potential surface defects may be more effective in many environments. Volumetric air sampling can provide more specified quantitative data, and is more commonly used; however, passive air sampling with settle plates (plates open left on surfaces) might be more representative in situations in which pathogens in the air can settle on sites such as wounds *(3)*.

Similarly, a variety of culture methods are available. Among the factors to be considered when choosing the appropriate media are the expected environmental MRSA burden, level of expected presence of other organisms, growth characteristics of other anticipated organisms, presence of inhibitors, desired sensitivity, cost, availability, and laboratory convenience. The main choices that must be made involve the use of selective vs nonselective culture media, and the use of direct culture vs broth enrichment. Each option may be preferable in certain situations, and a combination of methods may be useful in many circumstances.

The role of molecular methods for the detection of environmental MRSA is unclear. Polymerase chain reaction (PCR), particularly real-time PCR, might be useful because of the potential for high sensitivity. PCR can be performed directly off environmental samples, or following enrichment procedures. PCR has been used to detect a variety of surface and airborne pathogens *(19,20)*, but there has been minimal evaluation of the usefulness of PCR in the detection of environmental

S. aureus. The main concern about using molecular methods such as real-time PCR for environmental screening is the inability to differentiate between viable and dead bacteria. This may be important in some circumstances because the clinical significance of detecting MRSA that was killed via appropriate cleaning and disinfection may be minimal. If quantification is desired, real-time or competitive PCR would be required. Currently, there is inadequate information to recommend routine use of molecular means for primary evaluation of environmental MRSA. This chapter focuses on the use of conventional culture techniques to isolate MRSA that can be further evaluated using molecular methods.

2. Materials
2.1. General Culture
2.1.1. Broth Enrichment

1. Selective enrichment broth with or without antimicrobials: 10 g/L Tryptone T, 75 g/L NaCl, 10 g/L mannitol, and 2.5 g/L yeast extract; Todd-Hewitt broth; Mueller-Hinton broth with NaCl and oxacillin.
2. Nonselective enrichment broth: brain heart infusion broth or typticase soy broth.

2.1.2. Direct Culture

1. Nonselective agar: blood agar, trypticase soy agar.
2. Selective agar: mannitol salt agar with 2 µg/mL oxacillin, mannitol salt agar with 4–10 µg/mL of cefoxitin, Mueller-Hinton agar with 2–4 µg/mL of oxacillin, oxacillin resistance screening agar media.

2.2. General Identification

1. Gram stain supplies.
2. Hydrogen peroxide.
3. Rabbit plasma for coagulase testing.
4. *S. aureus* latex agglutination test (LAT).
5. Penicillin binding protein 2a (PBP2a) LAT.

2.3. Collection with Contact Plates

1. Contact plates with selective or nonselective agar.

2.4. Collection with Swabs

1. Sterile swabs with transport medium (i.e., Stuart's or Amies).

2.5. Collection with Electrostatic Cloth

1. Commercial electrostatic cloth (i.e., Swiffer™).
2. Sterile sealable bags of adequate size and strength to hold a cloth ± culture broth.

Environmental Surveillance for MRSA

2.6. Passive Air Sampling

1. Nonselective or selective agar.

2.7. Active Air Sampling

1. Active air impacter with the ability to select a specific volume of air to be sampled.
2. Nonselective or selective culture plates.

2.8. Molecular Analysis of Isolates

See Chapters 5–8 for specific tests and requirements.

3. Methods

3.1. General Culture

3.1.1. Broth Enrichment

1. Place collection device (swab, cloth) in enrichment broth. For swabs, break the swab into 2 mL of broth. For cloths, add 90 mL of broth to the sterile bag containing the cloth.
2. Incubate inoculated broth at 35°C for 24 h.
3. Subculture onto selective or nonselective agar as described in **Subheading 3.1.2.** (*see* **Notes 1** and **2**).

3.1.2. Direct Culture

1. Incubate culture plates aerobically at 35°C for 24–48 h.
2. Subculture colonies with morphology consistent with *S. aureus*. This will vary with different culture media (*see* **Note 2**).

3.2. General Identification

1. Identify as *S. aureus* based on Gram stain morphology (Gram-positive cocci), catalase reaction (positive), coagulase reaction (positive), and biochemical tests or *S. aureus* LAT.
2. Confirm as MRSA via PBP2a LAT or detection of *mecA* via PCR.

3.3. Collection with Contact Plates

1. Press the contact plate onto the desired surface and remove. For some surfaces, a wider area can be sampled by sweeping the plate over the particular surface (*see* **Note 3**).
2. Replace the lid. Incubate and identify MRSA as described in **Subheadings 3.1.2.** and **3.2**.

3.4. Collection with Swabs

1. Moisten a sterile cotton-tipped swab in sterile saline.
2. Rub the swab over a predetermined size of the area to be sampled.

3. Transport to the laboratory in standard aerobic transportation medium (i.e., Stuart's or Amies medium), or place the swab directly in a tube containing enrichment broth.
4. Break the shaft of the swab into the tube containing enrichment broth. Incubate at 35°C for 24 h.
5. Subculture onto selective agar or blood agar and incubate aerobically at 35°C for 24 h.
6. Identify MRSA *S. aureus* as described in **Subheading 3.2**.

3.5. Collection with Electrostatic Cloth (Note 3)

1. Hold the cloth in a gloved hand or on a disinfected holder.
2. Wipe the cloth over the surface to be sampled.
3. Place the cloth in a sterile bag or container and transport to the laboratory. If samples must be shipped or stored prior to delivery to the laboratory, store them at 4°C.
4. Add 90 mL of broth and incubate aerobically at 35°C for 24 h.
5. Inoculate approx 100 µL of broth onto selective agar or nonselective agar and incubate at 35°C for 24 h.
6. Identify MRSA as described in **Subheading 3.2**.

3.6. Passive Air Sampling

1. Place the culture plates on the desired surface with the lid removed.
2. Leave the plates for a predetermined period of time (i.e., 2 h).
3. Incubate and identify MRSA as described in **Subheadings 3.1.2. and 3.2**.

3.7. Active Air Sampling

1. Place a nonspecific or selective agar plate in an air sampler.
2. Place the air sampler in a predetermined location.
3. Sample a predetermined volume of air (i.e., 100 L).
4. Incubate the plate and identify MRSA as described in in **Subheadings 3.1.2. and 3.2**.
5. If applicable, use the conversion factor provided with the air sampler to adjust colony counts to account for the potential for double impaction of bacteria through the same sampling portal.

3.8. Molecular Analysis of Isolates

In some cases, further analysis of isolates may be required. The methods chosen will depend on the desired objective. For comparison with local clinical isolates, a method with good discriminatory power and repeatability such as pulsed-field gel electrophoresis would be indicated. For broader environmental study, multilocus sequence typing would be useful. In either case, adjunctive techniques such as identification of SCC*mec* type and demonstration of different virulence genes could also be beneficial. These methods are described in Chapters 5–7.

4. Notes

1. If quantification of MRSA is required, enrichment methods are not appropriate. Sometimes a combination of direct and enrichment methods is optimal; direct methods allow quantification and decrease the chance of contaminant overgrowth with enrichment, allowing for recovery of lower numbers of less viable MRSA in samples.
2. Refinement or modification of culture techniques may be required depending on the original objectives and preliminary results. If excessive growth of other organisms is required, more selective media may be needed. If poor overall recovery is encountered with direct culture methods, enrichment should be considered.
3. There are a variety of other potential methods for environmental sampling that were not discussed. The methods described in **Subheadings 3.6.** and **3.7.** were included because of the frequency of use or the author's subjective assessment of relative ease and usefulness. Some alternative methods might be useful in certain situations. Because there is no true "gold standard" and there can be marked differences sampling different sites, intralaboratory comparison of different methods for a specific objective may be required.

References

1. Clarke, P. and Humphreys, H. (2001) Persistence of vancomycin-resistant enterococci (VRE) and other bacteria in the environment. *Ir. Med. J.* **94,** 277, 278.
2. Wagenvoort, J. H., Sluijsmans, W., and Penders, R. J. (2000) Better environmental survival of outbreak vs. sporadic MRSA isolates. *J. Hosp. Infect.* **45,** 231–234.
3. Rhame, F. S. (1998) The inanimate environment, in *Hospital Infections,* 4th ed. (Bennett, J. V. and Brachman, P. S., eds.), Lippincott-Raven, New York, pp. 299–324.
4. Embil, J. M., McLeod, J. A., Al-Barrak, A. M., et al. (2001) An outbreak of methicillin resistant *Staphylococcus aureus* on a burn unit: potential role of contaminated hydrotherapy equipment. *Burns* **27,** 681–688.
5. French, G. L., Otter, J. A., Shannon, K. P., Adams, N. M., Watling, D., and Parks, M. J. (2004) Tackling contamination of the hospital environment by methicillin-resistant *Staphylococcus aureus* (MRSA): a comparison between conventional terminal cleaning and hydrogen peroxide vapour decontamination. *J. Hosp. Infect.* **57,** 31–37.
6. Edmiston, C. E. Jr., Seabrook, G. R., Cambria, R. A., et al. (2005) Molecular epidemiology of microbial contamination in the operating room environment: is there a risk for infection? *Surgery* **138,** 573–579.
7. Oie, S., Hosokawa, I., and Kamiya, A. (2002) Contamination of room door handles by methicillin-sensitive/methicillin-resistant *Staphylococcus aureus. J. Hosp. Infect.* **51,** 140–143.
8. Weese, J. S., Goth, K., Ethier, M., and Boehnke, K. (2004) Isolation of methicillin-resistant *Staphylococcus aureus* from the environment in a veterinary teaching hospital. *J. Vet. Int. Med.* **18,** 468–470.
9. Loeffler, A., Boag, A. K., Sung, J., et al. (2005). Prevalence of methicillin-resistant *Staphylococcus aureus* among staff and pets in a small animal referral hospital in the UK. *J. Antimicrob. Chemother.* **56,** 592–597.

10. Sexton, T., Clarke, P., O'Neill, E., Dillane, T., and Humphreys, H. (2005) Environmental reservoirs of methicillin-resistant *Staphylococcus aureus* in isolation rooms: correlation with patient isolates and implications for hospital hygiene. *J. Hosp. Infect.* **62,** 187–194.
11. Boyce, J. M., Potter-Bynoe, G., Chenevert, C., and King, T. (1997) Environmental contamination due to methicillin-resistant *Staphylococcus aureus*: possible infection control implications. *Infect. Control Hosp. Epidemiol.* **18,** 622–627.
12. Allen, K. D., Anson, J. J., Parsons, L. A., and Frost, N. G. (1997) Staff carriage of methicillin-resistant *Staphylococcus aureus* (EMRSA 15) and the home environment: a case report. *J. Hosp. Infect.* **35,** 307–311.
13. Rampling, A., Wiseman, S., Davis, L., Hyett, A. P., Walbridge, A. N., Payne, G. C., and Cornaby, A. J. (2001) Evidence that hospital hygiene is important in the control of methicillin-resistant *Staphylococcus aureus*. *J. Hosp. Infect.* **49,** 109–116.
14. Dharan, S. and Pittet, D. (2002) Environmental controls in operating theatres. *J. Hosp. Infect.* **51,** 79–84.
15. Mortimer, E. A. Jr., Wolinsky, E., Gonzaga, A. J., and Rammelkamp, C. H. Jr. (1966) Role of airborne transmission in staphylococcal infections. *Br. Med. J.* **5483,** 319–322.
16. Shiomori, T., Miyamoto, H., Makishima, K., et al. (2002) Evaluation of bedmaking-related airborne and surface methicillin-resistant *Staphylococcus aureus* contamination. *J. Hosp. Infect.* **50,** 30–35.
17. Kumari, D. N., Haji, T. C., Keer, V., Hawkey, P. M., Duncanson, V., and Flower, E. (1998) Ventilation grilles as a potential source of methicillin-resistant *Staphylococcus aureus* causing an outbreak in an orthopaedic ward at a district general hospital. *J. Hosp. Infect.* **39,** 127–133.
18. McGowan, J. E. Jr. and Weinstein, R. A. (1998) The role of the laboratory in control of nosocomial infection, in *Hospital Infections*, 4th ed. (Bennett, J. V. and Brachman, P. S., eds.), Lippincott-Raven, New York, pp. 143–164.
19. Makino, S. and Cheun, H. I. (2003) Application of the real-time PCR for the detection of airborne microbial pathogens in reference to the anthrax spores. *J. Microbiol. Methods* **53,** 141–147.
20. Stetzenbach, L. D., Buttner, M. P., and Cruz, P. (2004) Detection and enumeration of airborne biocontaminants. *Curr. Opin. Biotechnol.* **115,** 170–174.

16

Control and Prevention of MRSA Infections

Liangsu Wang and John F. Barrett

Summary

Methicillin-resistant *Staphylococcus aureus* (MRSA) has posed an immense problem for clinicians in the hospital setting for years, emerging as the most frequent nosocomial infection. To deal with this problem pathogen and others, infectious disease specialists have developed a variety of procedures for their control and prevention, involving options from preventative measures such as decolonization and isolation of MRSA-confirmed patients, to the more simple procedures of hand washing, expanding glove use, and reducing time in the hospital. With the realization that MRSA is now a community problem, there are expanded efforts toward more direct intervention, such as the use of anti-MRSA antibacterials and vaccines, in an attempt to reduce the overall burden of MRSA.

Key Words: Methicillin-resistant *Staphylococcus aureus*; infection; control; colonization; decolonization; resistance; prevention; treatment.

1. Introduction

The treatment of infectious diseases remains an unmet medical need, covering a variety of infectious disease conditions including viral, fungal, bacterial, and parasitic-mediated. This continuous need for prevention, control, and treatment stems from both the dissemination of infectious diseases throughout the world and the emergence of resistance among many of the once-thought-controlled microbes and viruses. One of the most intriguing of these infectious disease pathogens is methicillin-resistant *Staphylococcus aureus* (MRSA), and the epidemiology and emergence of resistance of MRSA has drawn great attention over the past decade as it continues to increase in both the hospital and community infection settings *(1–9)*.

The pathogen *S. aureus*, initially viewed as a minor opportunistic pathogen, began its climb to being one of the most feared human pathogens when it first acquired resistance to penicillin within a year after its introduction in the 1940s

(1,4,5). Its resistance to the semisynthetic penicillin, methicillin, which ultimately became known as the drug susceptibility marker for β-lactam resistance, has since emerged as being synonymous with multidrug-resistant *S. aureus* *(4)*. First presenting as a problem in the United States in the mid-1980s, MRSA is now the number one priority for prevention and control in the hospital setting *(1,6,10,11)*.

2. MRSA as a Contagious Entity

MRSA is contagious. It is spread by direct contact with MRSA-infected people or objects. It is almost ubiquitous in the hospital setting and, as such, has become the leading nosocomial pathogen in hospitals, and with its more recent identification as a community pathogen, it has the potential for explosive growth outside the hospital *(5,12,14–17)*. MRSA is both an opportunistic pathogen and an overt pathogen and can normally be found in a large percentage (~25–30%) of the population as an opportunistic pathogen as part of the normal colonized flora on the skin and nasal passages of humans *(7,18–20)*.

MRSA is a hardy microbial life form, able to grow in virtually any environment, with minimal nutrient requirements, and with the ability to adapt to a hostile environment (i.e., human host) by selection of a spontaneous mutant variant under stress or antibiotic pressure *(7–9,12,22)*. There are numerous publications on both the ancestral nature of clonal dissemination of MRSA *(8,12,18,23–25)* and more recent typing of the evolving community-acquired MRSA (CA-MRSA), which are now presenting as serious outbreaks.

3. Why Control MRSA?

MRSA is a recognized, overt bacterial pathogen and a contaminant of many environmental surfaces, both in and out of the hospital setting *(3,6,20,23,26)*. MRSA is also found as a communal, opportunistic pathogen on the human skin and nasal passages *(6,7,16,18,19)*. The result of this broad-based exposure of this opportunistic pathogen is the risk of infection ranging from minor skin infection to bacteremia.

MRSA is a prolific pathogen and is frequently multidrug resistant *(3–5,9)*. The resistance to all β-lactams is by the acquisition of an exogenous, redundant penicillin-binding protein, PBP2a *(8,25,27–29)*, and the production of β-lactamases *(30)*. MRSA has sweeping β-lactam resistance with minimum inhibitory concentrations (MICs) of >16–128 μg/mL *(2,31–33)*. In addition, MRSA generally has the susceptibility phenotype of multidrug resistance (MDR) against other antibiotics *(2,31–34)*.

MRSA is currently a problem in hospitals *(2,3,13)* and is continuing to increase in magnitude. MRSA is frequently associated with high rates of morbidity and mortality, with the highest in surgery and burn patients *(33,35,36)*. Moreover, even in CA-MRSA, where examined, susceptibility is evolving to the MDR phenotype *(7,9,12)*. Buckingham et al. *(14)* examined MRSA isolates as part of an epidemiology study within an MRSA outbreak in pediatric patients at a Memphis, TN, hospital. Thirty-eight percent (46 of 122 MRSA isolates) were confirmed as CA-MRSA during just the first 18 mo of the study. In the last year of the study, researchers recorded 63% (106 of 167 isolates) as CA-MRSA *(14)*. Pulse-field gel electrophoresis analyses revealed that 15 of 16 community-associated isolates examined shared a common pulsed-field type commonly observed in CA-MRSA infections in the United States, and that these strains were therefore not likely to be sourced from nosocomial infections. The fresh isolates were characterized as having the smaller staphylococcal cassette chromosome *mec* (SCC*mec*) type IV, which is characterized as clindamycin susceptible, which is typical of CA-MRSA *(14)*. With the mobility of the SCC*mec* genetic element, the options for MDR may be unlimited where a mobile resistance exists in the microbial ecology.

4. Control and Prevention of MRSA

The control and prevention of MRSA may begin with the prevention of spread, both before and after hospital admission, but it is a complex infectious disease with multiple variables that risks infection by MRSA in virtually any environmental setting *(5,37–40)*. Factors associated with the spread of MRSA have included the well-described risks of exposure to the hospital environment as well as the more recent characterization of community-originated MRSA *(11–13,16,23,24,42)*. Found to be more than just different origins of infections, the clones that constitute both nosocomial and community MRSA are quite distinct *(13,18,22–25,27)*.

4.1. Control of MRSA

The control of MRSA is a multipronged approach, combining prevention, containment, decolonization, and treatment efforts. Ultimately, only a vaccine against *S. aureus*, regardless of methicillin-susceptibility/resistance phenotypes, will eliminate this problem pathogen.

Among the options toward mitigation of the problem of MRSA are the following:

1. Improved infection control *(7,26,41,43–46)*; the spread of MRSA can be linked to suboptimal containment procedures in the hospital setting (spread clonally by poor hand washing, inadequate disinfection, and so on).
2. Decolonization of MRSA carriers *(20,47–49)*.
3. More prudent use of antibiotics in both the intensive care unit (ICU) and the non-ICU setting *(50–52)*.
4. *Appropriate* use of antibiotics in the ICU and non-ICU setting *(50,53–56)*. This is not a matter of semantics but, rather, a matter of choosing the "right" antibacterial agent up front in the treatment of confirmed or *presumed* MRSA, since recent

Table 1
Antibacterials in Clinical Development with Anti-MRSA Activity

Drug name/ designation (company)	Class	Target	Status
Dalbavancin (Vicuron/Aventis)	Glycopeptide	Cell wall	Phase III/NDA filed (approvable letter)[a,b]
Oritavancin (LY333328) (Intermune/Lilly)	Glycopeptide	Cell wall	Phase III[a,b]
BAL5788 (Basilea/Roche)	Cephalosporin	Cell wall	Phase III[a,b]
TD-6424 (Theravance)	Lipoglycopeptide	Cell wall	Phase III[a,b]
CS-023 (R-1558) (Sankyo/Roche)	Carbapenem	Cell wall	Phase I[a]
MC-02479/ RWJ-54428/442831 (Trine; J&J)[c]	Cephalosporin	Cell wall	Phase I[a]
MC-04546/ RWJ-333441/333442 (Trine; J&J)[c]	Cephalosporin	Cell wall	Phase I[a]
VRC-4887 (LBM415) (Vicuron/Novartis)	Hydroxamate	Peptide deformylase	Phase I[a]

[a]IDdB database.
[b]Company Web site, press release, or analyst meeting.
[c]Prodrug of active component.

reports have suggested that certain classes of antibacterials drive MRSA selection *(50,53)*.

5. Successful completion of development/approval/launch of late-stage anti-MRSA compounds (**Table 1**) *(5,57)*.
6. Ramped-up drug discovery efforts in big pharmaceutical companies toward the discovery of new anti-MRSA agents *(5,57–60)*.
7. Development of anti-MRSA target drugs *(61–64)* and MRSA vaccines *(65–67)*.

4.2. Prevention of MRSA Colonization and Infection

Infected and colonized patients constitute the major reservoir of MRSA *(6,21,43,44,49,68)*. Colonization of MRSA can occur on virtually any skin surface, including in the nose, throat, and groin, and as such, MRSA colonization is a major risk factor in becoming infected with MRSA *(21,47,49)*. The spread of MRSA occurs primarily by contact with live or inanimate reservoirs of MRSA *(6,7,37,41,44,46,47,69)*. It is now common knowledge that the best defense against the spread of MRSA is maintaining a clean environment

(37,38,43,45,48,69,70) in the hospital through good infection control procedures, but it is impossible to completely sterilize the environment or maintain complete sterility in the hospital environment, owing to the "traffic" of people contact in the hospital.

It is believed that colonization usually precedes infection, and the infecting organism is usually the same subtype as the colonizing strain *(47,72)*. This offers an opportunity to prevent MRSA infection by prophylactic measures. There are several pathways for MRSA colonization. Proper interventions to these pathways can potentially prevent the transmission and spread of MRSA.

4.2.1. Hand Hygiene

Hand hygiene has been considered one of the cornerstones of infection control. The predominant mode of MRSA transmission is believed to be from patient to patient through the hands of health-care workers *(37,48,69,71,73)*. It is reported that the number of microorganisms steadily increases with the duration of patient care, on average 16 microorganisms/min on ungloved hands, with more microorganisms on the hands after direct contact with patients, respiratory care, and handling of body fluid secretions *(73)*. Although gloves can reduce contamination of the hands, they do not provide complete protection owing to leaks after use *(74,75)*. Hand cleansing is thus required regardless of whether gloves are used or changed.

Many studies have reported that compliance with recommendations on hand hygiene among health-care workers is extremely low, thus providing a high-risk environment for the transfer of MRSA. Several factors have been identified for the low compliance, including time constraints, skin irritation from hand hygiene agents, and poor accessibility to sinks *(76)*.

Pittet and Boyce *(77)*, researchers belonging to an international group, have revisited the guidelines for hand hygiene in health-care settings. The most revolutionary recommendation concerns the use of alcohol-based hand rub formulations as the new standard of care. These alcohol-based hand rubs often have an application time in clinical practice shorter than 30 s, yet this practice show antimicrobial activity equal to that of the routine liquid preparations. Currently, several alcohol-based hand rubs are available commercially. There are studies on tolerance and user acceptability of these hand rubs, and the harshness of alcohol-based preps clearly plays a major factor in compliance *(78)*. Thus, whether these agents will improve hand hygiene compliance remains to be seen.

4.2.2. Environmental Disinfection and Personal Protective Equipment

Although transmission of MRSA is most likely to be directly through hands from one person to another, the role of a contaminated environment as a possible reservoir of MRSA in hospitals should not be underestimated. It has been a concern that cross-infection with MRSA could occur if patients were admitted to a contaminated area, especially where the compliance of hand hygiene remains poor.

There are observational studies on the occurrences of environmental contamination, which may serve as a theoretical basis for the need to disinfect inanimate objects to prevent the transmission of MRSA. In a recent study, Bhalla et al. *(69)* reported the high frequency of acquisition of bacterial pathogens on investigators' hands after contacting environmental surfaces near patients; the rates varied to a high of up to 53%. In a detailed study conducted by Boyce et al. *(46)*, environmental contamination of MRSA occurred in the rooms of 73% of all infected patients and 69% of all colonized patients. Furthermore, about 65% of nurses contaminated their gowns with MRSA after their morning patient-care activities, and 42% of hospital personnel contaminated their gloves with MRSA after touching contaminated surfaces without direct contact with MRSA patients. Changing gloves after contact with patients and/or contaminated surfaces in patients' rooms and frequently changing gowns are beneficial in preventing the spread of MRSA.

Although MRSA guidelines recommend decontamination of patients' environment following discharge from a ward, environmental contamination of MRSA after terminal cleaning still persists. In one study, 46% of rooms that were previously occupied by MRSA patients showed evidence of MRSA in the environment after terminal cleaning *(79)*. Frequently contaminated objects include mattresses, pillows, chairs, floors, and electrical equipment. Some surface areas may be easily overlooked. Implementation of a good decontamination process and environmental screening is most important in preventing MRSA cross-contamination and transmission.

4.2.3. Active Screening and Isolation of Source

Some investigators consider isolation of colonized or infected MRSA patients in combination with active screening an important strategy in preventing cross-infections. Numerous studies have been performed to evaluate the effectiveness of isolation and screening practices in reducing transmission of epidemic or endemic MRSA. Results often vary from study to study and are sometimes conflicting. This leads to the debate over whether or not isolation policies would benefit in reducing the incidence of MRSA.

In a systematic review of literature published from 1966 to 2000 on the effect of isolation in the hospital management of MRSA, Cooper et al. *(68)* found that there are major methodological weaknesses in most of these studies, such as lack of adjustment for potential confounders and lack of or inappropriate statistical analysis. They reported that no conclusions could be drawn about the effect of isolation in one-third of the studies analyzed. Nonetheless, they concluded that most of the other studies provided either weak or strong evidence consistent with a reduction in MRSA. Stochastic and deterministic models showed that improving either the detection rate or isolation capacity would decrease endemic levels with substantial savings *(68)*. Additional well-designed studies are required for isolation to be an evidence-based measure in preventing the transmission of MRSA.

4.2.4. Prudent Use of Antibiotics

Antibiotic-selective pressure might play a big role in the genesis of endemic MRSA. Some theoretical evidence and observational studies have shown that rates of MRSA are directly associated with inappropriate use of antibiotics, including both suboptimal and excessive use. Most staphylococci in bacterial cultures are heterogeneous; even an individual colony of methicillin-susceptible *S. aureus* (MSSA) is heterogeneous with a small number of individual organisms having high MICs. Administration of antibiotics to bacteria in suboptimal concentrations confers selective advantage for the survival of those harboring resistant mechanisms by turning on salvage survival mechanisms in the absence of incomplete bacterial infection. It has been reported that antibiotics used for the treatment of infections or used as presurgical prophylaxis can exert selective pressure and lead to changeover of colonizing flora from MSSA to MRSA by selecting and amplifying MRSA *(51)*. Sufficient exposure and avoiding prolonged use of antibiotics during antibiotic therapy are crucial to reduce MRSA colonization and infection.

Antibiotics are currently administrated almost empirically owing to the lack of rapid and accurate diagnostic tests for point-of-care therapy. This leads to excessive or inappropriate use of antibiotics. Heavy antibiotic use could disturb the indigenous natural anaerobic flora in a human body and lead to overgrowth with pathogenic and/or resistance bacteria *(80)*. Lines of evidence have been found to support dose-effect relationships between MRSA and antimicrobial use and plausible biological models to explain this relationship *(81)*. Prior exposure of antibiotics such as quinolones has been linked to infection with MRSA *(50)*. Prudent antibiotic use is thus important in preventing selection and amplification of MRSA.

4.2.5. Education of Staff

It has been recommended that the education and training of health-care workers on infection control and prevention practices be included as a vital component of MRSA control and prevention programs. Education is an important facilitator for other preventative measures. It can raise awareness of guidelines and improve compliance as well as remind staff of the importance of good hand hygiene techniques and the principals of cross-infections. Knowledge of potential risks of the transmission of MRSA to patients as well as potential risks of staff colonization or infection acquired from patients may facilitate contact precautions and active surveillance and promote the disinfection process. Education efforts may also help clinicians to connect inappropriate use of antibiotics to the selection and spread of MRSA.

4.2.6. Health Habits in the Community

With increasing numbers of patients being treated at home as outpatients to reduce medical costs and with CA-MRSA strains being isolated and trending toward epidemic levels throughout the United States, preventative measures are becoming important in preventing the transmission of MRSA in the community.

Health habits in community medicine are the most important tools to limit the spread of MRSA. Just as in the hospital, hand hygiene is the most important health habit in preventing the transmission of MRSA in the community from both colonized and infected populations. Colonized or infected patients, healthy people, and medical staff such as visiting nurses should all establish good hand hygiene habits. Room separation for outpatients and disinfection of the environment around the patients should be considered as well in the outpatient or home treatment setting.

Education is another tool for community medicine in preventing the transmission of MRSA. Such education includes helping outpatients to understand their infections and how these infections might spread to other people, as well as helping the community to understand the importance of hand hygiene.

4.3. Treatment of MRSA

The nosocomial MRSA frequency of ~30% began its climb in 1994–1995 *(2,3,33,35,36,82)* and by 1999 had accounted for approx 50% of all *S. aureus* isolates found in US ICUs *(2,3,7,33,35,36,82)*. The incidence of MRSA has continued to rise, reaching an infection rate of >60% in ICU patients in 2003–2004 *(2,3,7,33,36)* and ~40% in non-ICU patients in 2003 *(7)*. Because the newer-generation antibacterial follow-on agents, including the quinolones, have failed to expand coverage to MRSA, it has emerged as the primary nosocomial pathogen in the hospital over the past decade *(12,70)*.

There are several therapeutic options for the treatment of MRSA (**Table 2**), but they all have limitations. Beyond the commonly used anti-MRSA agents for serious infections (**Table 2**), there are a number of other agents for which CA-MRSA may still be used (clindamycin, ciprofloxacin, levofloxacin, Trimethoprim-sulfamethoxazole, azithromycin, or clarithromycin) *(7,12,13,33,83)*; the changing susceptibility patterns of these drugs facilitated by the transfer of multiple, independent SCC*mec* genetic elements, however, will increasingly threaten the empirical use of many of these agents.

Table 2
Anti-MRSA Drugs in Clinical Use

Drug	Class	Target
Linezolid (Zyvox™)	Oxazolidinone	Protein synthesis
Daptomycin (Cubicin™)	Lipopeptide	Cell membrane
Quinupristin/dalfopristin (Synercid™)	Streptogramins	Protein synthesis
Vancomycin[a]	Glycopeptide	Cell wall
Clindamycin[b]	Aminoglycoside	Protein synthesis
Trimethoprim-sulfamethoxazole	Sulfa	Folate biosynthesis
Tigacil™ (Tigecycline; GAR-936)	Glycylcycline	Protein synthesis

[a]For use in MRSA infections with confirmed vancomycin susceptibility.
[b]Not for use in iMLS$_B$ resistance (inducible macrolide-lincosamide-streptogramin).

A number of drugs that could represent a significant improvement in the anti-MRSA arsenal of antibacterial agents are also in development *(5,6,57,63)*. Among these new agents are multiple drug classes targeting different anti-MRSA-essential genes (**Table 1**), but because virtually all of these later-stage compounds in development are only hospital-based iv products, the solution for emerging CA-MRSA treatment in the community is not imminent.

5. Risk Factors

There are numerous documented risk factors for contracting MRSA, among them hospital stay, length of hospital stay, underlying illness, nasal carriage *(38)*, extranasal colonization *(38)*, iv drug use, invasive lines or tubes such as catheters, and prior antibiotic exposure *(26)*. Multiple reports have described these risk factors and others elsewhere *(37,43,45–48,68,71,72)*.

6. Detection and Diagnostics of MRSA

Increased surveillance, including screening of high-risk patients, has been recognized as an important component for initiating effective control and preventing further spread of the MRSA pathogen. Rapid and accurate detection of MRSA is thus crucial. The traditional methods for detecting MRSA have been culture and susceptibility tests such as disk diffusion or broth microdilution, which require 2 to 3 d to perform. In addition to the lag of detection time, they may fail to detect resistance when phenotypic expression of an isolate is heterogeneous. Nevertheless, these conventional methods are still being widely used by clinical laboratories.

Several rapid detection methods have been developed and are available as commercial kits, such as BBL Crystal™ MRSA ID system (Becton-Dickinson, Cockeysville, MD), Velogene™ Rapid MRSA identification assay (ID

Biomedical, Vancouver, BC, Canada), and latex agglutination MRSA-Screen (Denka Seiken, Tokyo, Japan). The BBL Crystal MRSA ID system has been designed to allow detection of oxacillin resistance in *S. aureus* within 4 to 5 h when organisms are cultured in oxacillin-containing broth or agar after an initial culture isolation. The method uses an oxygen-sensitive fluorescent indicator, which is quenched and colorless if organisms do not consume available oxygen in the presence of oxacillin. Fluorescence is easily detected when oxygen is removed from the broth or agar by respiration of the MRSA strains. Various studies have reported a varied sensitivity range (86.5–100%) and high specificity (about 97.6–100%) for the method *(84–86)*. The Velogene is a 90-min assay that utilizes colorimetric cycling probe technology to detect the *mecA* gene. It uses a DNA-RNA-DNA *mecA* probe labeled with fluorescein at the 5′ terminus and biotin at the 3′ terminus, which is sensitive to RNase H when bound to the *mecA* gene. The colorimetric enzyme immunoassay produces a blue color from the uncut probe whereas *mecA*-positive strains result in no color. The latex agglutination MRSA-Screen method is a 15-min latex agglutination test that uses latex particles sensitized with anti-PBP2a monoclonal antibody. Both methods have been reported to have a high sensitivity and high specificity of >98% *(85,86)*.

Although these rapid methods have the advantage of short assay time in comparison with 24- to 48-h conventional susceptibility tests, they still require the initial culture isolation and subculture, which takes at least 1 to 2 d. More rapid methods are sought to detect MRSA within the same day as specimens are collected. To date, several products have been approved by the Food and Drug Administration in the United States and Europe as in vitro diagnostics. The IDI-MRSA™ test from GeneOhm Sciences is a test for the direct detection of MRSA sequence from nasal swabs. The test uses real-time PCR on SmartCycler (Cepheid) to amplify the sequence in SCC*mec* followed by fluorogenic detection using molecular beacons, hairpin-forming single-stranded oligonucleotides labeled with a quencher at one end and fluorophore at the other end. The hairpin structure opens on beacon/target hybridization, resulting in emission of fluorescence. The method takes <2 h from nasal swabs to results, with about 92.5% sensitivity and 96.4% specificity (IDI-MRSA package insert).

The MRSA Evigene™ from AdvanDx utilizes fluorescent *in situ* hybridization using peptide nucleic acid probes to detect species-specific 16S-rRNA from *S. aureus*, the *mecA* gene, and the *nuc* gene to differentiate MRSA from coagulase-negative staphylococci. The abundance of 16S-rRNA molecules in each microorganism allows detection of individual cells without PCR. MRSA can be detected directly from positive blood cultures or clinical isolates *(87)*. The MRSA Evigene is claimed to have about 100% sensitivity and specificity (package insert). The time required is the time for blood culture (~6 h) plus ~3 h for assay; results are therefore obtained on the same day as specimen collection.

Control and Prevention of MRSA Infections

BacLite® Rapid MRSA screening (Acolyte Biomedica, UK) is another product for rapid detection of MRSA. It is based around conventional microbiological techniques and utilizes adenylate kinase (AK) bioluminescence for end-point detection. AK converts adenosine 5′-diphosphate (ADP) into adenosine triphosphate (ATP) in the presence of excess ADP, and ATP is measured using bioluminescence. The method takes a clinical nasal swab through selective culture with oxacillin, ciprofloxacin, and colistin, followed by ultrasensitive AK bioluminescence detection. Results are produced directly from the clinical swab in 5 h. The method has 93.4% sensitivity and 95.7% specificity (Acolyte Biomedica Ltd. Salisbury, Wiltshire, UK, Spring 2005 Report. http://www.acolytebiomedica.com).

The described rapid MRSA detection and diagnostic methods can enable hospitals and reference centers to improve dramatically the identification of MRSA and, thus, enhance MRSA prevention and control programs. Although results are generated within the same day, these methods are still steps away from point-of-care diagnostics for various practical reasons. In addition, it is hard to differentiate MRSA infection from MRSA colonization using these methods. Additional superior methods are awaited for the diagnosis of MRSA infections and for monitoring treatment of MRSA infections.

7. Development of Vaccine

The development of vaccine against *S. aureus* may provide an alternative prophylactic approach for the prevention of MRSA infection or even MRSA colonization. The challenges of developing such agents are identifying appropriate antigenic targets for vaccine and defining proper immune effectors mediating protective immunity against MRSA infection.

The current most effective antigenic targets are believed to be bacterial surface polysaccharides, especially capsular polysaccharides. Of the 12 capsular serotypes identified to date, type 5 and 8 are predominant and expressed in the vast majority of *S. aureus* clinical isolates. Efforts have been made to conjugate these two types of capsular polysaccharides to recombinant genetically detoxified exoprotein A from *Pseudomonas aeruginosa* (rEPA). The vaccine has been shown to be safe and immunogenic in healthy volunteers and confers statistically significant protective immunity to systemic infection in patients with end-stage renal disease undergoing hemodialysis *(65–67)*. An immunogenicity study in cardiovascular patients with implanted devices is under way. A second surface polysaccharide, poly-*N*-acetyl glucosamine, is associated with a number of important biological and pathological properties of *S. aureus* and other clinically important strains of staphylococci and has been proposed to have broadly protective vaccine potential *(88)*. Other antigenic targets under investigation include *S. aureus* virulence factors such as surface-expressed adhesins that influence bacterial adhesions and toxins associated with septic shock or toxic shock syndrome. Such immunogenic targets, like surface polysaccharides, may be used as targets for vaccines and/or immunotherapy. The availability of a wide selection of antigens would provide opportunities to design targeted vaccines against various pathogenic properties of *S. aureus*.

Although the current efforts in developing *S. aureus* vaccines do not differentiate between MSSA and MRSA, if successful vaccines are developed, they should reduce the incidence of *S. aureus* and the severity of MRSA infections in general. On the other hand, it remains to be seen whether vaccination with *S. aureus* vaccines could also lead to prevention of colonization, such as by reducing nasal carriage among health-care workers and high-risk patients.

8. Summary

With the prevalence of MRSA and its treatment challenges owing to MDR, the use of multitude approaches is critical to the successful control and prevention of this "superbug." Measures such as practicing good hand hygiene, taking precautions against personal contact, cleaning the environment, utilizing active surveillance/screening, ensuring the prudent use of antibiotics, and educating staff are important in preventing the spread of MRSA. Additional treatment options and the development of effective *S. aureus* vaccines will eventually control this pathogen.

Acknowledgment

We thank Christine Jenkins for technical and editorial assistance with preparation of the manuscript.

References

1. Enright, M. C., Robinson, D. A., Randle, G., Feil, E. J., Grundmann, H., and Spratt, B. G. (2002) The evolutionary history of methicillin-resistant *Staphylococcus aureus* (MRSA). *Proc. Natl. Acad. Sci. USA* **99**, 7687–7692.
2. Solomon, S., Horan, T., Andrus, M., et al. (2003) National Nosocomial Infections Surveillance (NNIS) System Report, data summary from January 1992 through June 2003, issued 2003. *Am. J. Infect. Control.* **31**, 481–498.
3. Chopra, I. (2003) Antibiotic resistance in *Staphylococcus aureus*: concerns, causes and cures. *Expert Rev. Anti-infect. Ther.* **1**, 45–55.
4. Foster, T. J. (2004) The *Staphylococcus aureus* "superbug." *J. Clin. Invest.* **114**, 1693–1696.
5. Barrett, J. F. (2005) MRSA—what is it, and how do we deal with the problem? *Expert Opin. Ther. Targets* **9**, 253–265.
6. Cooper, B. S., Medley, G. F., Stone, S. P., et al. (2004) Methicillin-resistant *Staphylococcus aureus* in hospitals and the community: stealth dynamics and control catastrophes. *Proc. Natl. Acad. Sci. USA* **101**, 10,223–10,228.
7. Boyce, J. M. (2003) Update on resistant *Staphylococcus aureus* infections. *Clin. Updates Infect. Dis.* **VI**, 1–4.

8. Hiramatsku, K., Cui, L., Kuroda, M., and Ito, T. (2001) The emergence and evolution of methicillin-resistant *Staphylococcus aureus*. *Trends Microbiol.* **9**, 486–493.
9. Lindsay, J. A. and Holden, M. T. G. (2004) *Staphylococcus aureus*: superbug, super genome? *Trends Microbiol.* **12**, 378–385.
10. Wisplinghoff, H., Bischoff, T., Tallent, S. M., Seifert, H., Wenzel, R. P., and Edmond, M. B. (2004) Nosocomial bloodstream infections in US hospitals: analysis of 24,179 cases from a prospective nationwide surveillance study. *Clin. Infect. Dis.* **39**, 309–317.
11. Verhoff, J. (2001) Stopping short the spread of methicillin-resistant *Staphylococcus aureus*. *Can. Med. Assoc. J.* **165**, 31, 32.
12. Chambers, H. F. (2001) The changing epidemiology of *Staphylococcus aureus*? *Emerg. Infect. Dis.* **7**, 178–182.
13. Marshall, C., Kossmann, T., Wesselingh, S., and Spelman, D. (2004) Methicillin-resistant *Staphylococcus aureus* and beyond: what's new in the world of the 'golden staph'? *ANZ J. Surg.* **74**, 465–469.
14. Buckingham, S. C., McDougal, L. K., Cathey, L. D., et al. (2004) Emergence of community-associated methicillin-resistant *Staphylococcus aureus* at a Memphis, Tennessee Children's Hospital. *Pediatr. Infect. Dis. J.* **23**, 619–624.
15. Dufour, P., Gillet, Y., Bes, M., et al. (2002) Community-acquired methicillin-resistant *Staphylococcus aureus* infections in France: emergence of a single clone that produces Panton-Valentine leucocidin. *Clin. Infect. Dis.* **35**, 819–824.
16. Weber, J. T. (2005) Community-associated methicillin-resistant *Staphylococcus aureus*. *Clin. Infect. Dis.* **41**, S269–S272.
17. Eady, E. A. and Cove, J. H. (2003) Staphylococcal resistance revisited: community acquired methicillin resistant *Staphylococcus aureus*—an emerging problem of the management of skin and soft tissue infections. *Curr. Opin. Infect. Dis.* **16**, 103–124.
18. Melles, D. C., Gorkink, F. J., Boelens, A. M., et al. Natural population dynamics and expansion of pathogenic clones of *Staphylococcus aureus*. *J. Clin. Invest.* **114**, 1732–1740.
19. Lee, Y.-L., Cesario, T., Pax, A., Tran, C., Ghouri, A., and Thrupp, L. D. (1999) Nasal colonization by *Staphylococcus aureus* in active, independent, community seniors. *Age Ageing* **28**, 229–232.
20. Lucet, J.-C., Chevret, S., Durand-Zaleski, I., Chastang, C., and Regnier, B. (2003) Prevalence and risk factors for carriage of methicillin-resistant *Staphylococcus aureus* at admission to the intensive care unit. *Arch. Intern. Med.* **163**, 181–188.
21. Kluytmans, J., Van Belkum, A., and Verbrugh, H. (1997) Nasal carriage of *Staphylococcus aureus*: epidemiology, underlying mechanisms, and associated risks. *Clin. Microbiol. Rev.* **10**, 505–520.
22. Fluit, A. D. and Schmitz, F.-J., eds. (2003) *MRSA: Current Perspectives*, Csiter Academic Press, Norfolk, UK.
23. Fey, P. D., Said-Salim, B., Rupp, M. E., et al. (2003) Comparative molecular analysis of community- or hospital-acquired methicillin-resistant *Staphylococcus aureus*. *Antimicrob. Agents Chemother.* **47**, 196–203.

24. Robinson, D. A. and Enright, M. C. (2003) Evolutionary models of the emergence of methicillin-resistant *Staphylococcus aureus*. *Antimicrob. Agents Chemother.* **47(12),** 3926–3934.
25. Robinson, D. A. and Enright, M. C. (2004) Multilocus sequence typing and the evolution of methicillin-resistant *Staphylococcus aureus*. *Clin. Microbiol. Infect.* **10,** 92–97.
26. Graffunder, E. M. and Venezia, R. A. (2002) Risk factors associated with nosocomial methicillin-resistant *Staphylococcus aureus* (MRSA) infection including previous use of antimicrobials. *J. Antimicrob. Chemother.* **49,** 999–1005.
27. Crisostomo, M. I., Westh, H., Tomasz, A., Chung, M., Oliveira, D. C., and De Lencastre, H. (2001) The evolution of methicillin resistance in *Staphylococcus aureus*: similarity of genetic backgrounds in historically early methicillin-susceptible and -resistant isolates and contemporary epidemic clones. *PNAS* **98,** 9965–9870.
28. Giesbrecht, P., Kersten T., Maidof, H., and Wecke, J. (1998) Staphylococcal cell wall: morphogenesis and fatal variations in the presence of penicillin. *Microbiol. Mol. Rev.* **62,** 1371–1414.
29. Pinho, M. G., De Lencastre, H., and Tomasz, A. (2001) An acquired and a native penicillin-binding protein cooperate in building the cell wall of drug-resistant staphylococci. *PNAS* **98,** 10,886–10,891.
30. Johnson, A. P., Mushtaq, S., Warner, M., and Livermore, D. M. (2004) Activity of daptomycin against multi-resistant Gram-positive bacteria including enterococci and *Staphylococcus aureus* resistant to linezolid. *Int. J. Antimicrob. Agents* **24,** 315–319.
31. Anonymous. (2004) Proportion of *S. aureus* nosocomial infections resistant to oxacillin (MRSA) among intensive care unit patients, 1989–2003. *NNIS System.* Available at www.cdc.gov. last accessed 2004.
32. Jones, M. E., Draghi, D. C., Thornsberry, C., Karlowsky, J. A., Sahm, D. C., and Wenzel, R. P. (2004) Emerging resistance among bacterial pathogens in the intensive care unit—a European and North American surveillance study (2000–2002). *Ann. Clin. Microbiol. Antimicrob.* **3,** 14–24.
33. Bouchillon, S. K., Hoban, D. J., Johnson, J. L., et al. (2004) In vitro activity of gemifloxacin and contemporary oral antimicrobial agents against 27,247 Gram-positive and Gram-negative aerobic isolates: a global surveillance study. *Int. J. Antimicrob. Agents* **23,** 181–196.
34. Speller, D. C. E., Johnson, A. P., James, D., Marples, R. R., Charlett, A., and George, R. C. (1997) Resistance to methicillin and other antibiotics in isolates of *Staphylococcus aureus* from blood and cerebrospinal fluid, England and Wales, 1989–95. *Lancet* **350,** 323–325.
35. Reynolds, R., Potz, N., Colman, M., Williams, A., Livermore, D., MacGowan, A., and the BSAC Extended Working Party on Bacteraemia Resistance Surveillance. (2004) *J. Antimicrob. Chemother.* **53(6),** 1018–1032.
36. Oteo, J., Baquero, F., Vindel, N. A., Campos, J., and the European Antimicrobial Resistance Surveillance System (EARSS) (2004) *J. Antimicrob. Chemother.* **53(6),** 1033–1038.

37. Farr, B. M. (2004) Prevention and control of methicillin-resistant *Staphylococcus aureus* infections. *Curr. Opin. Infect. Dis.* **17**, 317–322.
38. Marshall, C., Wesselingh, S., McDonald, M., and Spelman, D. (2004) Control of endemic MRSA—what is the evidence? *J. Hosp. Infect.* **56**, 253–268.
39. Sheagren, J. N. (1984) *Staphylococcus aureus*: the persistent pathogen (part 1). *N. Engl. J. Med.* **310**, 1368–1373.
40. Sheagren, J. N. (1984) *Staphylococcus aureus*: the persistent pathogen (part 2). *N. Engl. J. Med.* **310**, 1437–1442.
41. Platt, A. T. (2001) MRSA in intensive care. *Nurs. Stand.* **15(31)**, 27–32.
42. Karchmer, A. W. (2000) Nosocomial bloodstream infections: organisms, risk factors and implications. *Clin. Infect. Dis.* **31**, S139–S143.
43. Bissett, L. (2005) Controlling the risk of MRSA infection: screening and isolating patients. *Br. J. Nurs.* **14**, 386–390.
44. Boyce, J. M., Havill, N. L., Kohan, C., Dumigan, D. G., and Ligi, C. E. (2004) Do infection control measures work for methicillin-resistant *Staphylococcus aureus*? *Infect. Contr. Hosp. Epidemiol.* **25**, 395–401.
45. Dettenkofer, M. and Block, C. (2005) Hospital disinfection: efficacy and safety issues. *Curr. Opin. Infect. Dis.* **18**, 320–325.
46. Boyce, J. M., Potter-Bynoe, G., Chenevert, C., and King, T. (1997) Environmental contamination due to methicillin-resistant *Staphylococcus aureus*: possible infection control implications. *Infect. Control Hosp. Epidemiol.* **18**, 622–627.
47. Keene, A., Vavagiakis, P., Lee, M. H., et al. (2005) *Staphylococcus aureus* colonization and the risk of infection in critically ill patients. *Infect. Control Hosp. Epidemiol.* **26**, 622–628.
48. Myatt, R. and Langley, S. (2003) Changes in infection control practice to reduce MRSA infection. *Br. J. Nurs.* **12**, 675–681.
49. Von Eiff, C., Becker, K., Machka, K., Stammer, H., and Peters, G. (2001) Nasal carriage as a source of *Staphylococcus aureus* bacteria. *N. Engl. J. Med.* **344**, 11–16.
50. Paterson, D. L. (2004) "Collateral damage" from cephalosporin or quinolone antibiotic therapy. *Clin. Infect. Dis.* **15**, S341–S345.
51. Schentag, J. J., Hyatt, J. M., Carr, J. R., et al. (1998) Genesis of methicillin-resistant *Staphylococcus aureus* (MRSA), how treatment of MRSA infections has selected for vancomycin-resistant Enterococcus faecium, and the importance of antibiotic management and infection control. *Clin. Infect. Dis.* **26**, 1204–1214.
52. Tenover, F. C. and McDonald, L. C. (2005) Vancomycin-resistant staphylococci and enterococci: epidemiology and control. *Curr. Opin. Infect. Dis.* **18**, 300–305.
53. Hosein, I. K., Hill, D. W., Jenkins, L. E., and Magee, J. T. (2002) Clinical significance of the emergence of bacterial resistance in the hospital environment. *Symp. Ser. Soc. Appl. Microbiol.* **31**, 90S–97S.
54. Courvalin, P. and Trieu-Cuot, P. (2001) Minimizing potential resistance: the molecular view. *Clin. Infect. Dis.* **33**, S138–S146.
55. Drlica, K. and Zhao, X. L. (2004) Is 'dosing-to-cure' appropriate in the face of antimicrobial resistance? *Rev. Med. Microbiol.* **15**, 73–80.

56. Blondeau, J. M., Hansen, G., Metzler, K., and Hedlin, P. (2004) The role of PK/PD parameters to avoid selection and increase of resistance: mutant prevention concentration. *J. Chemother.* **16(3),** 1–19.
57. Bryskier, A. (2005) Anti-MRSA agents: under investigation, in the exploratory phase and clinically available. *Expert Rev. Anti. Infect. Ther.* **3,** 505–553.
58. Walsh, C. (2003) Where will new antibiotics come from? *Nat. Rev. Microbiol.* **1,** 65–70.
59. Labischinski, H., Ehlert, K., and Wieland, B. (1998) Novel anti-staphylococcal targets and compounds. *Expert Opin. Invest. Drugs* **7,** 1245–1256.
60. Shlaes, D. (2003) The abandonment of antibacterials: why and wherefore? *Curr. Opin. Pharmacol.* **3,** 470–473.
61. Long, T. E. (2003) Recent progress towards the clinical development of new anti-MRSA antibiotics. *IDrugs* **6,** 351–359.
62. Khare, M. and Keady, D. (2003) Antimicrobial therapy of methicillin-resistant *Staphylococcus aureus* infections. *Expert Opin. Pharmacother.* **4,** 165–177.
63. Wagenlehner, F. M. E. and Naber, K. G. (2004) New drugs for Gram-positive uropathogens. *Int. J. Antimicrob. Agents* **24S,** S39–S43.
64. Abbanat, D., Macielag, M., and Bush, K. (2003) Novel antibacterial agents for the treatment of serious Gram-positive infections. *Expert Opin. Invest. Drugs* **12,** 379–399.
65. Shinefield, H., Black, S., Fattom, A., et al. (2002) Use of a *Staphylococcus aureus* conjugate vaccine in patients receiving hemodialysis. *N. Engl. J. Med.* **346,** 491–496.
66. Welch, P. G., Fattom, A., Moore, J. Jr., et al. (1996) Safety and immunogenicity of *Staphylococcus aureus* type 5 capsular polysaccharide-*Pseudomonas aeruginosa* recombinant exoprotein A conjugate vaccine in patients on hemodialysis. *J. Am. Soc. Nephrol.* **7,** 247–253.
67. Fattom, A., Fuller, S., Propst, M., et al. (2004) Safety and immunogenicity of a booster dose of *Staphylococcus aureus* types 5 and 8 capsular polysaccharide conjugate vaccine (StaphVAX) in hemodialysis patients. *Vaccine* **16,** 656–663.
68. Cooper, B. S., Stone, S. P., Kibbler, C. C., et al. (2003) Systematic review of isolation policies in the hospital management of methicillin-resistant *Staphylococcus aureus*: a review of the literature with epidemiological and economic modelling. *Health Technol. Assess.* **7,** 1–194.
69. Bhalla, A., Pultz, N. J., Gries, D. M., et al. (2004) Acquisition of nosocomial pathogens on hands after contact with environmental surfaces near hospitalized patients. *Infect. Control Hosp. Epidemiol.* **25,** 164–167.
70. Rice, L. B. (2003) Controlling antibiotic resistance in the ICU: different bacteria, different strategies. *Cleveland Clinic J. Med.* **70,** 793–800.
71. Kampf, G. (2003) State-of-the-art hand hygiene in community medicine. *Int. J. Hyg. Environ. Health* **206,** 465–472.
72. Pujol, M., Pena, C., Pallares, R., et al. (1996) Nosocomial *Staphylococcus aureus* bacteremia among nasal carriers of methicillin-resistant and methicillin-susceptible strains. *Am. J. Med.* **100,** 509–516.

73. Pittet, D., Mourougam, P., and Perneger, T. V. (1999) Compliance with handwashing in a teaching hospital. Infection Control Program. *Ann. Intern. Med.* **130,** 126–130.
74. Dirschka, T., Winter, K., Kralj, N., and Hofmann, F. (2004) Glove perforation in outpatient dermatologic surgery. *Dermatol. Surg.* **30,** 1210–1213.
75. Yangco, B. G. and Yangco, N. F. (1989) What is leaky can be risky: a study of the integrity of hospital gloves. *Infect. Control Hosp. Epidemiol.* **10,** 553–556.
76. Pittet, D., Dharan, S., Touveneau, S., Sauvan, V., and Perneger, T. V. (1999) Bacterial contamination of the hands of hospital staff during routine care. *Arch. Intern. Med.* **159,** 821–826.
77. Pittet, D. and Boyce, J. M. (2003) Revolutionising hand hygiene in health-care settings: guidelines revisited. *Lancet Infect. Dis.* **3,** 269, 270.
78. Kramer, A., Bernig, T., and Kampf, G. (2002) Clinical double-blind trial on the dermal tolerance and user acceptability of six alcohol-based hand disinfectants for hygienic hand disinfection. *J. Hosp. Infect.* **51,** 114–120.
79. Blythe, D., Keenlyside, D., Dawson, S. J., and Galloway, A. (1998) Environmental contamination due to methicillin-resistant *Staphylococcus aureus* (MRSA). *J. Hosp. Infect.* **38,** 67–69.
80. Donskey, C. J. (2004) The role of the intestinal tract as a reservoir and source for transmission of nosocomial pathogens. *Clin. Infect. Dis.* **39,** 219–226; Epub 2004 Jun 25.
81. Monnet, D. L. (1998) Methicillin-resistant *Staphylococcus aureus* and its relationship to antimicrobial use: possible implications for control. *Infect. Control Hosp. Epidemiol.* **19,** 552–559.
82. Johnson, A. P. (2003) Antibiotic resistance in the intensive care unit setting. *Expert Rev. Anti. Infect. Ther.* **1,** 253–260.
83. Lowry, F. D. (2003) Antimicrobial resistance: the example of *Staphylococcus aureus*. *J. Clin. Invest.* **111,** 1265–1273.
84. Kampf, G., Lecke, C., Cimbal, A. K., Weist, K., and Ruden, H. (1999) Evaluation of the BBL Crystal MRSA ID System for detection of oxacillin resistance in *Staphylococcus aureus*. *J. Clin. Pathol.* **52,** 225–227.
85. Arbique, J., Forward, K., Haldane, D., and Davidson, R. (2001) Comparison of the Velogene Rapid MRSA Identification Assay, Denka MRSA-Screen Assay, and BBL Crystal MRSA ID System for rapid identification of methicillin-resistant *Staphylococcus aureus*. *Diagn. Microbiol. Infect. Dis.* **40,** 5–10.
86. Louie, L., Matsumura, S. O., Choi, E., Louie, M., and Simor, A. E. (2000) Evaluation of three rapid methods for detection of methicillin resistance in *Staphylococcus aureus*. *J. Clin. Microbiol.* **38,** 2170–2173.
87. Levi, K. and Towner, K. J. (2003) Detection of methicillin-resistant *Staphylococcus aureus* (MRSA) in blood with the EVIGENE MRSA detection kit. *J. Clin. Microbiol.* **41,** 3890–3892.
88. Maira-Litran, T., Kropec, A., Goldmann, D., and Pier, G. B. (2004) Biologic properties and vaccine potential of the staphylococcal poly-N-acetyl glucosamine surface polysaccharide. *Vaccine* **22,** 872–879.

17

Treatment of Infections Caused by Resistant *Staphylococcus aureus*

Gregory M. Anstead, Gabriel Quinones-Nazario, and James S. Lewis II

Summary

We review data on the treatment of infections caused by drug-resistant *Staphylococcus aureus*, particularly methicillin-resistant *S. aureus* (MRSA). In this review, we cover findings reported in the English language medical literature up to February 2006. Despite the emergence of resistant and multidrug resistant *S. aureus*, five effective drugs for which little resistance has been observed are in clinical use: vancomycin, quinupristin-dalfopristin, linezolid, tigecycline, and daptomycin. However, vancomycin is less effective for infections with MRSA isolates that have a high minimum inhibitory concentration in the susceptible range. Linezolid looks promising in the treatment of MRSA pneumonia and skin and soft-tissue infections (SSTIs). Daptomycin displays rapid bactericidal activity in vitro, and it has been shown to be noninferior to comparator agents in the treatment of SSTIs and bacteremia. Tigecycline was also noninferior to comparator drugs in the treatment of SSTIs. Clindamycin, trimethoprim-sulfamethoxazole, doxycycline, and minocycline are oral antistaphylococcal agents that may have utility in the treatment of SSTIs and osteomyelitis, but the clinical data for their efficacy is limited. There are four drugs with broad-spectrum activity against Gram-positive organisms at an advanced stage of clinical testing: ceptobiprole and three new glycopeptides with potent bactericidal activity, oritavancin, dalbavancin, and telavancin. Thus, there are currently many effective drugs to treat resistant *S. aureus* infections and many promising agents in the pipeline. Nevertheless, *S. aureus* remains a formidable adversary against which there are frequent treatment failures. The next goals are to determine the most appropriate indications and cost-effectiveness of each of these drugs in the treatment strategy against *S. aureus*.

Key Words: Methicillin-resistant *S. aureus*; vancomycin; linezolid; daptomycin; clindamycin; fusidic acid; fosfomycin.

1. Introduction

Staphylococcus aureus is a common pathogen, causing a variety of infections that range from minor skin lesions to serious and life-threatening conditions,

From: *Methods in Molecular Biology: MRSA Protocols*
Edited by: Y. Ji © Humana Press Inc., Totowa, NJ

such as osteomyelitis, bacteremia, endocarditis, and pneumonia. With its genetic plasticity, *S. aureus* has repeatedly thwarted efforts to control it with antibiotics. This propensity to evade antibiotic therapy, and to elaborate various virulence factors and adhesins, which mediate its adherence to both biological and prosthetic surfaces, has deservedly earned *S. aureus* the moniker "superbug," with a "super genome" *(1)*. In the 1990s, nosocomial infections caused by methicillin-resistant *S. aureus* (MRSA) became a problem of global magnitude. During the past half decade, MRSA infections have also become commonplace in the community, exceeding the number of methicillin-susceptible *S. aureus* (MSSA) infections in many locales *(2)*. Thus, MRSA now exists as two general entities: hospital-associated MRSA (HA-MRSA) and community-associated MRSA (CA-MRSA). The Centers for Disease Control and Prevention criteria for a CA-MRSA infection include (1) diagnosis in an outpatient setting or within 48 h of hospital admission; and (2) the absence of any of the following in the year prior to infection: hospitalization; admission to a nursing home, skilled nursing facility, or hospice; dialysis; surgery; indwelling catheters or medical devices *(3)*. Methicillin resistance is encoded by the *mecA* gene, which is part of the staphylococcal cassette chromosome (SCC). Five types of SCC*mec* have been identified; types I, II, and III, frequently found in HA-MRSA isolates, impart resistance to other antibiotics in addition to β-lactams. By contrast, CA-MRSA strains contain the type IV or V SCC*mec*, which does not confer multidrug resistance (MDR). Therefore, additional options may be available for the treatment of CA-MRSA infections, compared with those cause by the multidrug resistant HA-MRSA. Nevertheless, the CA-MRSA isolates often harbor specific virulence genes lacking in the HA-MRSA isolates *(3)*. Standard treatment regimens for various MRSA infections are provided in **Table 1**. Most of the recommendations come from that little bedside bible of infectology, the *Sanford Guide to Antimicrobial Therapy*, found tucked in the laboratory coat pocket of resident and attending physician alike *(4)*. However, the quality of data on which the recommendations are based, and the efficacy of the recommended treatments, vary widely. Furthermore, with the burgeoning incidence of MRSA cases and the intense concern that this organism has elicited, recommendations regarding optimal treatments are in continual flux.

In addition to MRSA, a number of other antibiotic-resistant phenotypes of *S. aureus* have recently been recognized. In 1997, the first vancomycin-intermediate *S. aureus* (VISA) strains were reported (at that time defined as a minimum inhibitory concentration [MIC] of 8–16 μg/mL). These isolates had thickened cell walls compared with non-VISA isolates *(9)*. There have also been three reports of vancomycin-resistant *S. aureus* (VRSA) (at that time defined as an MIC of ≥ 32μg/mL) *(10)*. These isolates contained the *vanA* gene, which was transferred on a plasmid from vancomycin-resistant *Enterococcus*

Table 1
Standard Treatment for Various Infections Caused by *S. aureus* (4)[a]

Infection	Antibiotic	Comments
iv Line infection (4,5)	Vanco 1 g iv q 12 h × 2 wk	Remove catheter; Use echocardiogram to rule out endocarditis
Osteomyelitis (4,6)	Vanco 1 g iv q 12 h; Clinda 600–900 mg iv q 8 h; TMP-SMX 8–10 mg/(kg/d) iv ÷ q 8 h; Linezolid 600 mg iv or po bid; 6 wk of therapy recommended	*Vertebral*: Perform MRI to evaluate for epidural abscess. *Sternal*: This may require debridement. *Contiguous with vascular insufficiency*: This may require revascularization. *Chronic*: Perform surgical debridement; remove hardware (if present)
Mastitis	Vanco 1 g iv q 12 h; Outpatient: TMP-SMX DS 2 tablets po bid	If abscess is present, discontinue nursing; do incision and drainage or needle aspiration
Endocarditis: native valve (7)	Vanco 1 g iv q 12 h × 4–6 wk; Dapto 6 mg/(kg/d) iv qd × 4–6 wk; Q/D only if isolate S to clindamycin; Others: TMP-SMX, linezolid	*Clinda:* There is a high rate of relapse in endocarditis-not recommended
Endocarditis: prosthetic valve (7)	(Vanco 1 g iv q 12 h + RIF 300 mg po q 8 h × 6 wk) + Gent 1 mg/kg iv q 8 h × 2 wk	Engage in early surgical consultation to consider valve removal
Septic arthritis (4,6)	See osteomyelitis recommendations; 2–4 wk of iv antibiotics, followed by oral antibiotics to complete 4–6 wk	Perform open or arthroscopic lavage or daily simple needle aspiration
Septic bursitis (4)	Vanco 1 g iv q 12 h × 2–3 wk	Initially, aspirate daily
Infected prosthetic joint (4,6)	Vanco 1 g iv q 12 h + RIF × 6 wk	This may require removal of hardware
Bacteremia (8)	Vanco 1 g iv q 12 h × 2 wk; Dapto 6 mg/(kg/d) iv qd × 2 wk	Echocardiogram is recommended; Observe for foci of infection and metastatic complications

[a]Vanco = vancomycin; Dapto = daptomycin; Clinda = clindamycin; Q/D = quinupristin/dalfopristin; RIF = rifampin; Gent = gentamicin.

faecalis (VRE) *(9)*. This is not Armageddon just yet, because the total number of cases of VRSA is still small and these isolates were susceptible to other antibiotics *(11)*. Nevertheless, the emergence of VRSA demonstrates both the incredible adaptability of *S. aureus* and a terrifying biological interaction between two resistant organisms, MRSA and VRE.

One mechanism of macrolide resistance in MRSA is ribosomal target modification, which affects susceptibility to macrolides, lincosamides, and the type B streptogramins (MLS$_B$ phenotype). MRSA strains with constitutive resistance (MLS$_B$c) show in vitro resistance to these three groups of antibiotics, including resistance to clindamycin. In isolates with the inducible MLS$_B$ phenotype (MLS$_B$i), resistance to erythromycin and susceptibility to clindamycin are observed in routine susceptibility testing *(12)*. However, during treatment with clindamycin, these MLS$_B$i strains can mutate to constitutive clindamycin resistance, leading to clinical failure *(13–15)*. Inducible resistance to clindamycin can be determined by the erythromycin induction test (also known as the D-test), in which an erythromycin disk is placed near a clindamycin disk on a Kirby-Bauer plate. If inducible clindamycin resistance is present, the zone of inhibition around clindamycin will be flattened in the direction of the erythromycin disk and will resemble the letter *D (12)*. Because of inducible resistance, clinicians should exercise caution when using clindamycin to treat *S. aureus* infections, especially if the local laboratory does not screen for the phenotype. Epidemiological data on the incidence of the MLS$_B$ phenotype are virtually nonexistent; in Western Australia, between 1995 and 2003, 56% of CA-MRSA isolates had the MLS$_B$ phenotype *(13)*. Most MRSA isolates retain susceptibility to trimethoprim-sulfamethoxazole (TMP-SMX), but recently a small colony variant with decreased susceptibility has been described *(16)*.

Although the new antibiotic pipeline in general is running at a slow trickle, three agents for the treatment of infections owing to Gram-positive organisms have entered clinical service in the new millennium: linezolid, daptomycin, and tigecycline. Furthermore, other drugs in the advanced stages of development may also soon be added to the arsenal of weapons to fight Gram-positive bugs, including dalbavancin, oritavancin, telavancin, and ceftobiprole.

2. Vancomycin: The Old Standard Bearer—Not Exactly the Magic Bullet

Vancomycin has been in widespread clinical use for the treatment of MRSA infections since the 1980s *(17)*. Unfortunately, there is ample clinical experience to indicate that vancomycin is a less than optimal drug for the treatment of *S. aureus* infections. In clinical trials in which vancomycin was compared to a β-lactam in the treatment of MSSA endocarditis, vancomycin was consistently outperformed by nafcillin, with failure rates of 37–50% for the former drug and 1.4–26% for the latter agent *(18,19)*. In the treatment of MRSA pneumonia, vancomycin treatment is successful in only 35–57% of patients *(18)*. Patients in intensive care units (ICUs) sometimes even develop MRSA pneumonia while being treated with vancomycin for other indications *(20)*. Vancomycin has low bone penetration and also performs poorly in the treatment of *S. aureus* osteomyelitis *(21)*. Chambers *(22)* assembled a series of cases of serious *S. aureus* infections in which vancomycin was used and found an overall cure rate of only 76%.

Kollef and Micek *(18)* have recently enumerated the various deficiencies of vancomycin. For example, although it is -cidal for *S. aureus*, vancomycin displays ponderously slow killing kinetics compared to β-lactams or daptomycin. Another emerging problem with vancomycin is "MIC creep," the evolution of increasing MICs for vancomycin by various MRSA strains. (However, Jones *[23]* has provided data that such MIC drift is not occurring.) Finally, as a cell wall active agent, vancomycin does not affect MRSA toxin production (as opposed to the antibiotics that inhibit protein synthesis), which may limit clinical improvement until the slow -cidality of the antibiotic is brought to bear *(18)*.

To address these deficiencies, techniques to increase vancomycin exposure, such as continuous infusion and higher dosing of vancomycin (to attain trough levels ≥15 μg/mL), have been applied *(24)*. Recent guidelines issued by the American Thoracic Society and Infectious Diseases Society of America advocate vancomycin trough concentrations of 15–20 μg/mL during the treatment of MRSA nosocomial pneumonia *(25)*, although there are no specific clinical data to support this recommendation. In Australia, recent guidelines recommend a trough concentration of 10–20 μg/mL *(26)*. However, this strategy of increasing the trough concentration of vancomycin to enhance its therapeutic efficacy may still be unsuccessful if the MRSA isolate has a higher MIC within the previously accepted susceptibility range. In a study by Hidayat et al. *(27)*, patients with MRSA infection were treated with vancomycin to attain a trough ≥15 μg/mL; in the group of patients with MRSA isolates with an MIC ≥2 μg/mL, only 62% were clinical responders, vs 85% for the group of patients with isolates with an MIC <2 μg/mL *(27)*. Those *S. aureus* isolates with an MIC ≥4 μg/mL have been termed *S. aureus* with reduced vancomycin susceptibility (SA-RVS), and a previous study found that patients with SA-RVS infections have higher in-hospital mortality than other MRSA patients *(28)*.

3. Newer Agents: Linezolid (Zyvox)

Linezolid is an oxazolidinone antibiotic that blocks assembly of the initiation complex required for protein synthesis. This is a unique mechanism of action and, thus, there is no cross-resistance between other antibiotics and linezolid. Linezolid has broad activity against Gram-positive bugs, but no Gram-negative activity *(29)*. A strong suit of linezolid is its high oral bioavailability (approx 100%), allowing seamless iv to oral switch therapy and obviating the need for iv access and home health services, thereby offsetting the high costs of drug acquisition.

The most exciting, albeit controversial, clinical data to emerge about linezolid concern the treatment of nosocomial and ventilator-associated pneumonia (VAP) caused by MRSA *(30,31)*. Kollef et al. *(31)* retrospectively analyzed two randomized double-blind studies of VAP that enrolled 264 patients with Gram-positive infections and 91 with MRSA infection *(31)*. The treatment arms were linezolid or vancomycin, plus aztreonam. The use of linezolid was an independent predictor of clinical cure and survival, with an odds ratio of 2.4 and 2.6 for Gram-positive VAP, and 20 and 4.6 for MRSA VAP.

In an accompanying editorial, Ioanas and Lode *(32)* proclaimed, "There is sufficient evidence that vancomycin is no longer a recommendable therapeutic option for pulmonary infections, especially when MRSA is involved". The rationale for the superior efficacy of linezolid in pneumonia is its greater intrapulmonary penetration vis-à-vis vancomycin *(33)*. The existing data for the use of linezolid as first-line treatment in Gram-positive VAP are compelling. Nevertheless, important questions remain: Why does treatment with linezolid impart a significant survival advantage in MRSA but not for MSSA VAP? In addition, if vancomycin were given at a higher dose would the survival advantage of linezolid disappear? A larger trial of linezolid in VAP is currently under way.

More recently, the use of linezolid or clindamycin has been advocated for the treatment of necrotizing pneumonia caused by MRSA strains secreting the Panton-Valentine leukocidin (PVL) virulence factor. In a case series of four patients, three patients who were clinically failing vancomycin were successfully treated with linezolid or clindamycin *(34)*. The rationale is that antibiotics that inhibit protein synthesis, such as linezolid or clindamycin, block toxin production, whereas vancomycin does not. A previous case series of four patients with necrotizing pneumonia caused by PVL-producing MRSA had less favorable results *(35)*. One patient, treated with vancomycin, gatifloxacin, and meropenem, died. Two patients survived but suffered lower-extremity necrosis requiring amputation (one of these patients received vancomycin and clindamycin; the other, vancomycin, linezolid, and rifampin). A fourth patient survived after a 4-wk stay in the ICU; this patient initially received vancomycin, gatifloxacin, and meropenem, which was later changed to gatifloxacin and rifampin. Because of the rarity of pneumonia induced by PVL-producing strains, there will probably not be a clinical trial comparing antibiotic therapy with and without a protein synthesis inhibitor; nevertheless, this does make mechanistic sense and this strategy of using a protein synthesis inhibitor has been previously applied to treat another death-dealing infection, necrotizing fasciitis caused by *Streptococcus pyogenes*. Recently, new toxin-mediated clinical entities resulting from MRSA, such as necrotizing fasciitis and purpura fulminans, have been described *(36)* and, thus, therapy directed specifically at toxin production merits further investigation. Subinhibitory concentrations of linezolid are known to impair the production of virulence factors by *S. aureus (37,38)*. However, one caveat is that in the case of MRSA isolates with clindamycin resistance, clindamycin is unable to abort toxin production *(14)*.

MRSA osteomyelitis is an infection for which an effective oral treatment would be extremely useful. Linezolid was evaluated in a compassionate use trial of 55 patients with Gram-positive osteomyelitis; greater than 60% of the patients with MRSA osteomyelitis were cured (comparable with historical controls treated with vancomycin) *(39)*. Many of the patients had already failed protracted courses of vancomycin and quinupristin-dalfopristin (QD) or had retained hardware.

In another study, the bone and joint tissue penetration of linezolid was determined in 13 patients scheduled for debridement of MRSA-infected bone and soft tissue *(40)*. Linezolid rapidly reached mean levels >10 µg/mL in infected joint tissue (plasma concentration of >11 µg/mL). However, the mean linezolid concentration in the bone was only 3.9 ± 2.0 mg/L; thus, intrabone concentrations of linezolid below the MIC_{90} occurred for some isolates. The variable penetration of linezolid was probably owing to poor perfusion that may exist in infected bone *(40)*. However, everything is relative; the issue is how well linezolid will perform in well-designed comparator trials in which appropriate surgical intervention is also performed. Unfortunately, what is sorely lacking is an osteomyelitis trial in which linezolid is pitted against its archrival, vancomycin, or one of the youthful upstarts, daptomycin or tigecycline. Furthermore, because osteomyelitis requires prolonged therapy, the possible toxicity of linezolid with long-term use (myelosuppression, optic and peripheral neuropathy) requires further evaluation and careful clinical surveillance.

Another trial challenges the conventional wisdom that diabetic foot infections require broad-spectrum coverage for successful treatment owing to their polymicrobial nature. Linezolid was compared to ampicillin-sulbactam followed by amoxicillin-clavulanate in the treatment of diabetic foot infection, including infected ulcer, cellulitis, deep soft-tissue infection, paronychia, abscess, and osteomyelitis *(41)*. Vancomycin was added to the aminopenicillin-β-lactamase inhibitor arm when MRSA was suspected or confirmed. No difference was found in the overall clinical cure rate in the two groups, but linezolid outperformed the comparator group in the subsets of patients with ulcers and without osteomyelitis. In a series from the pediatric literature, linezolid was comparable to vancomycin in the treatment of complicated skin and soft-tissue infections (SSTIs), pneumonia, and bacteremia *(42)*.

Shorr et al. *(43)* analyzed five randomized studies comparing linezolid to vancomycin and extracted data on 53 patients with secondary MRSA bacteremia;

they found that linezolid was noninferior to vancomycin with respect to clinical, microbiological, and survival outcomes. Problems with the study were the small number of patients, the absence of strict monitoring of vancomycin levels, and the predominance of patients in the studies with SSTIs, which decreased the extent of illness severity. In addition, none of the patients had endocarditis *(43)*. Thus, a prospective comparative trial is still needed to determine the relative efficacy of linezolid in bacteremia and endocarditis.

Another study compared oral linezolid to vancomycin in the treatment of lower-extremity SSTIs caused by MRSA *(44)*. The use of linezolid was associated with greater rates of clinical cure and improvement (97% for linezolid, 43% for vancomycin) and a 3-d shorter median length of hospital stay. Conversely, the use of vancomycin was associated with more treatment failures (57%, vs 3% for linezolid) and, consequently, a higher rate of amputations (7%, vs 0% for linezolid).

Linezolid was also compared to vancomycin in the treatment of surgical-site infections in an open-label randomized trial; rates of clinical success were similar for the two agents. However, for those patients with MRSA infection, linezolid outperformed vancomycin: 87% of patients microbiologically cured vs 48% for those treated with vancomycin *(45)*. In a similar study of SSTIs, 92.2% of patients were cured in the linezolid group, vs 88.5% for vancomycin. But once again, in the subset of patients with MRSA infection, outcomes were superior for linezolid (88.6% cured vs 66.9% for vancomycin) *(46)*.

Several studies document greater efficacy for linezolid vs vancomycin for infections caused by MRSA, but not MSSA. Why is this? There may be two reasons for this phenomenon *(47)*. First, MRSA strains tend to have higher MICs to vancomycin than MSSA isolates. Even though their MICs are still in the susceptible range, infections resulting from these higher-MIC MRSA isolates may be more refractory to vancomycin treatment, owing to vancomycin's poor tissue penetration. One study found that clinical failure with vancomycin was directly correlated with the MIC of the isolate, even in the susceptible range *(48)*. Thus, break points for susceptibility to vancomycin were officially decreased by one dilutional factor on January 1, 2006; the new break points are as follows: susceptible is ≤2 µg/mL, intermediate is 4–8 µg/mL, and resistant is ≥16 µg/mL *(49)*.

Second, there is a recently described ruse of MRSA called heteroresistance; these strains appear susceptible to vancomycin (MIC ≤4 µg/mL) by routine testing, but they contain subpopulations of organisms at a frequency of 10^{-6} for which the MIC of vancomycin is in the intermediate range *(8)*. Infections with such hetero-VISA (hVISA) organisms often fail treatment with vancomycin. The proportion of hVISA strains determined from sites all around the globe (Spain, France, Germany, Italy, Belgium, Brazil, Japan, and England) has varied from 0 to 65% *(50)*. This great variation is owing to the lack of standardized methods for detecting heteroresistance. In a carefully conducted study from a Turkish hospital, 18% of 256 isolates obtained between January 1998 and January 2002 were hVISA. Furthermore, the investigators found an increase in the prevalence of hVISA from 1.6% in 1998 to 36% in 2001. These are very worrisome data indeed. At this point in time, the epidemiology of heteroresistance in North America is a total mystery and, furthermore, there is currently no practical means to routinely test for heteroresistance in the hospital laboratory. Surveillance data collected from diverse sites from 1997 to 2001 showed that only 1.67% of *S. aureus* strains displayed heteroresistance *(51)*. However, considering the intense use of vancomycin in American hospitals, it is essential to determine the current prevalence of heteroresistance to provide appropriate therapy in the cases of MRSA infection that involve these strains.

One issue that has haunted the use of linezolid is myelosuppression, especially thrombocytopenia. However, in studies by Nasraway et al. *(52)* and Rao et al. *(53)*, linezolid and vancomycin were found to have comparable rates of thrombocytopenia. These studies establish a new comfort level for using linezolid without excessive fear of myelosuppression.

Adverse effects of linezolid recognized in postmarketing surveillance include serotonin syndrome and optic and peripheral neuropathy. There have been nine published reports of serotonin syndrome, which manifests as hyperpyrexia, changes in mental status, and neuromuscular symptoms (tremors, incoordination, myoclonus, and hyperreflexia), when linezolid has been used with various antidepressants. In the reported cases of serotonin syndrome, usually high doses of a serotonergic drug or multiple concomitant serotonergic drugs were involved *(54)*. For the neuropathies, in some cases, it was reversible on discontinuation of linezolid, but in other cases, it persisted *(55–59)*. The neuropathies have primarily occurred in patients treated for longer than the Food and Drug Administration (FDA)–approved period of 28 d, although some patients received shorter courses. The accumulated case reports cannot be used to calculate estimates of risk because the number of reports received may not reflect the actual number of cases. In a compassionate use program of 796 patients treated with linezolid, peripheral neuropathy was reported in three patients (0.4%) who received the drug for a mean of 95 d *(56)*. Because of the potential irreversible nature of the neuropathies, patients on long-term linezolid therapy should be advised regarding these possible adverse effects. A study is currently enrolling patients for long-term linezolid treatment to determine whether early warning signs of optic neuropathy are apparent with close clinical surveillance.

The cost of drug acquisition has probably been the major factor that has stopped clinicians from wholeheartedly jumping on the linezolid bandwagon; linezolid costs eight times that of vancomycin. Shorr et al. *(60)* determined that the cost per quality-adjusted life year for the use of linezolid in VAP is approx

$30,000, comparable with other critical-care procedures. Likewise, Machado et al. *(61)* calculated that linezolid is more cost-effective than vancomycin in the treatment of VAP. Another determinant of health-care cost is length of hospital stay (LOS). In a study comparing linezolid to vancomycin in the treatment of SSTIs caused by MRSA, patients in the linezolid group had an LOS 6.6 d less than those in the vancomycin group, owing to the switch to oral linezolid therapy *(62)*.

A second factor that has tempered the use of linezolid is the concern that overuse will promote resistance and that this ace-in-the-hole agent will be quickly squandered away. Certainly, the resistance of *Enterococcus faecium* to linezolid is already a concern; in a recent surveillance study, 3.6% of *E. faecium* isolates were not susceptible to linezolid *(63)*. However, resistance of *S. aureus* to linezolid is exceedingly rare so far. In recent global surveillance data, no linezolid resistance was detected in >2100 staphylococcal isolates *(64,65)*. Similarly, data from the LEADER 2004 Surveillance Program, which evaluated 2872 *S. aureus* isolates collected in American hospitals, also revealed no resistance to linezolid *(63)*. Nevertheless, there are a few reported cases of MRSA resistance to linezolid when the drug was used for a prolonged period *(66)*. Despite the limited resistance that has been observed so far, as a general principle of treating bacterial infections, it would be prudent to test all Gram-positive isolates for susceptibility to linezolid. It is potentially disastrous for individual patients and the hospital infections control situation to assume ongoing susceptibility to any antibiotic for this crafty group of bacteria.

4. Newer Agents: Daptomycin (Cubicin)

Daptomycin was approved by the FDA in September 2003 and is the first drug from the cyclic lipopeptide class of antimicrobials. Daptomycin promotes the efflux of potassium out of the bacterial cell, leading to cell death. With this unique mechanism of action, there is no cross-resistance between daptomycin and other antimicrobials. Daptomycin also displays very rapid in vitro time-kill kinetics against Gram-positive organisms *(67)*. Other attractive features of daptomycin include once daily dosing, minimal adverse effects, and few drug interactions.

Two clinical trials have compared daptomycin to vancomycin or a β-lactam for the treatment of SSTIs *(68,69)*, its only FDA-approved indication. In both trials, daptomycin was clinically equivalent to the comparators. However, a prospective trial involving 740 patients with pneumonia was terminated early when daptomycin was found to be less effective than ceftriaxone *(67)*. Daptomycin is not effective in pneumonia because the drug is bound by pulmonary surfactant.

Daptomycin is generally well tolerated; however, myopathy has been reported *(70)*, so creatine phosphokinase levels should be monitored during therapy. In addition, statins should be withheld while a patient is being treated with daptomycin, because of the potential for additive myopathic effects *(67)*.

The approved dosage of daptomycin is 4 mg/(kg/d) for soft-tissue infections. Higher dosages are being studied for endocarditis and bacteremia, up to 6–8 mg/(kg/d). At our institution, we typically use 6 mg/(kg/d) for serious infections. Drawbacks of daptomycin include its high cost, lack of an oral formulation, poor penetration into pulmonary and osseous tissues, and myopathy as a potential adverse effect. Another concern with the use of daptomycin is the acquisition of resistance while the patient is on therapy. There are several reported cases of MRSA resistance to daptomycin arising during therapy (*[71]*; see **Subheading 13.3.**).

Although daptomycin is FDA-approved only for the treatment of SSTIs, with its rapid bactericidal actions, clinicians remain hopeful that this agent may prove to be superior to vancomycin in treating serious infections, such as endocarditis and bacteremia, in which a -cidal agent may be preferred. However, our initial hopes were dashed when a paper presented at the 2005 Interscience Conference on Antimicrobial Agents and Chemotherapy revealed that daptomycin was merely noninferior to vancomycin or a semisynthetic penicillin in the treatment of *S. aureus* bacteremia, including endocarditis *(72)*. The median time to clearance of bacteremia was 5 d in the daptomycin group vs 4 d in the comparator arm. Persistence or relapse occurred in 15.8% of the daptomycin group vs 9.6% in the comparator group. Six of 19 daptomycin-treated patients failed with increased MICs to daptomycin (0.25–0.5 µg/mL to 2–4 µg/mL). In the MRSA subgroup, at the test of cure, 44.4% of patients had successful outcome in the daptomycin group vs 31.8% in the vancomycin group, but this difference was not statistically significant. Thus, until further studies are available, daptomycin should be reserved for patients with resistant Gram-positive infections who cannot tolerate or who fail vancomycin. Despite this initial discouraging result, it is hoped that the developer of daptomycin, Cubicin, will embark on additional larger trials to study the use of daptomycin in MRSA bacteremia.

Several studies indicate synergy between daptomycin and other classes of antibiotics. In a rat model of MRSA aortic valve endocarditis, daptomycin was compared to vancomycin, vancomycin plus rifampin, rifampin alone, and daptomycin plus rifampin. The relative efficacy of the five regimens in decreasing vegetation bacterial densities was daptomycin + rifampin > vancomycin + rifampin > rifampin > daptomycin > vancomycin *(73)*. In an in vitro model of simulated endocardial vegetations, there were synergistic effects on the bactericidal activity of both daptomycin and vancomycin by the addition of gentamicin *(74)*. In another in vitro study, synergy was demonstrated between daptomycin and oxacillin *(75)*. It is hoped that this synergy between daptomycin and other antibiotics can be exploited in the treatment of refractory cases of MRSA bacteremia.

5. Quinupristin-Dalfopristin (Synercid)

With the dearth of treatment options for VRE infections in the late 1990s, the approval of QD in 1999 was greeted with much jubilation. QD is a bactericidal drug composed of two streptogramins, each of which binds to a different site on the ribosomal subunit and which act synergistically to block protein synthesis. Staphylococcal isolates have remained highly susceptible to the drug *(63,64)*. However, QD has never shown superiority over vancomycin or a β-lactam in any clinical trial *(76)*. In isolates with the $MLS_B c$ phenotype, quinupristin is inactive, and QD is then bacteriostatic instead of -cidal *(3)*. With the advent of linezolid and daptomycin, this agent has been relegated to benchwarmer status, because of its high cost and adverse effect profile (myalgias, arthalgias, and thrombophlebitis).

Recently, a new streptogramin antibiotic (XRP 2868) has been developed with improved potency against MRSA compared to QD and increased aqueous solubility, which may permit oral administration *(77)*. If this drug sees the clinical light of day, it will represent a real improvement over QD and will be a genuine advance in the crusade against MRSA.

6. Newer Agents: Tigecycline (Tigacyl)

Currently, the overall use of antibiotics of the tetracycline family has decreased because of resistance, which is mediated by efflux pumps and ribosomal protection. The glycylcyclines are analogs of tetracyclines that display activity against these resistant organisms. The most potentially useful glycylcycline is tigecycline. In recent in vitro studies, 98.9% of 348 global MRSA isolates were susceptible to tigecycline, with an average MIC_{90} of 0.25 µg/mL *(65)*. In addition to its potent activity against resistant Gram-positive organisms, it is active against certain resistant Gram-negative bacteria, such as members of the Enterobacteriaceae that sport extended spectrum β-lactamases.

Tigecycline was compared to vancomycin + aztreonam in a randomized, double-blind, controlled trial of 546 patients with SSTIs *(78)*. The clinical responses in the tigecycline and the vancomycin + aztreonam groups were similar. Among patients infected with MRSA and MSSA, the clinical and the microbiological success rates were 83 and 87%, respectively, for tigecycline-treated patients and 50 and 95%, respectively, for the vancomycin + aztreonam–treated patients. However, the number of MRSA infections was very small (six in each group), so it is hard to draw conclusions at this time about the efficacy of tigecycline in SSTIs caused by MRSA. The numbers of patients reporting adverse events were similar in the two groups, with increased nausea and vomiting rates in the tigecycline group and an increased incidence of rash and increases in liver transaminase levels in the vancomycin + aztreonam group.

Ultimately, tigecycline may prove useful as single-agent therapy in mixed infections, such as SSTIs, diabetic foot infections, and nosocomial peritonitis, in which resistant organisms are present. A trial of tigecycline in nosocomial pneumonia is under way *(32)*. Overall enthusiasm for tigecycline is tempered by the lack of an oral formulation, the necessity for twice daily dosing, the relatively high cost, and the high rate of nausea and vomiting.

7. Agents in Development: Oritavancin

Oritavancin (LY333328) is a semisynthetic glycopeptide that has completed phase III trials for complicated SSTIs and will be evaluated by the FDA in the near future. Susceptibility to oritavancin is not affected by the *VanA*-, *VanB*-, and *VanC*-encoded alterations in the bacterial cell wall that impart vancomycin resistance *(79)*. Oritavancin is also more rapidly bactericidal than vancomycin *(80)*. The increased activity of oritavancin against vancomycin-resistant organisms is owing to its ability to dimerize, allowing more favorable interactions with the bacterial peptidoglycan chain. In addition, oritavancin inhibits cell wall formation by blocking the transglycosylation step in peptidoglycan biosynthesis *(79)*. Oritavancin is active in vitro against glycopeptide-sensitive and -resistant Gram-positive bacteria, including VISA and VRSA *(10)*.

Two phase III studies have been completed for the use of oritavancin in the treatment of SSTIs *(81,82)*. Both studies compared 3–7 d of oritavancin treatment to 3–7 d of vancomycin treatment followed by oral cephalexin to complete 10–14 d of therapy. In both studies, there was no difference in efficacy between the groups. However, oritavancin's placement will unlikely be for SSTIs because a number of other antimicrobials can fulfill this role. Oritavancin may provide an additional weapon against resistant Gram-positive infections caused by VISA, VRSA, and VRE. Additionally, with its rapid bactericidal activity, it would be interesting to see how this drug compares to vancomycin, linezolid, and daptomycin for the treatment of MRSA/VRE endocarditis, bacteremia, and osteomyelitis. A phase II trial for the treatment of bacteremia is currently under way.

Oritavancin's half-life is about 6 d, which may be both beneficial and potentially problematic. The long half-life allows for once daily dosing, but what happens if there is an adverse reaction? Fortunately, few serious adverse reactions have been reported for oritavancin thus far. Another important question that remains to be answered is: Once therapy is completed, will low, subtherapeutic levels of oritavancin linger on, allowing genetically resourceful organisms such as *Enterococcus* and *Staphylococcus* time to acquire resistance mechanisms?

8. Agents in Development
8.1. Dalbavancin

Dalbavancin (BI397) is a second glycopeptide that will be evaluated by the FDA in the near future for the treatment of resistant Gram-positive infections. Its spectrum of activity is similar to that of oritavancin, but with less activity against VRE *(82,83)*. The half-life of dalbavancin (9–12 d) is even longer than that of oritavancin, which permits once weekly administration. Bactericidal activity against *S. aureus* was maintained 7 d after an initial 500-mg iv dose in healthy volunteers *(84)*.

A phase II randomized, controlled, open-label trial comparing a single infusion of dalbavancin (1100 mg), with 1000 mg on d 1 and 500 mg on d 8, and standard care (various agents) for the treatment of SSTIs was recently published *(85)*. The trial was small (62 patients), so it was not possible to draw statistical conclusions. However, there was a trend toward improved response with the two-dose arm of the trial. Adverse reactions were similar in all arms, and there were no discontinuations of study drug owing to adverse events.

In August 2004, Vicuron Pharmaceuticals (the developer of dalbavancin) announced the results of a phase III trial of more than 1500 patients with SSTIs in which dalbavancin was found to be noninferior compared to linezolid, cefazolin followed by oral cephalexin, or vancomycin *(86)*. Dalbavancin has also been compared to linezolid in a randomized, double-blind trial for the treatment of SSTIs. In that trial, 1000- and 500-mg infusions of dalbavancin were administered on d 1 and 8, respectively, vs 600 mg of linezolid twice daily for 14 d. The drugs had comparable efficacy *(87)*.

The long half-life of dalbavancin will permit once weekly dosing and may allow earlier hospital discharge without the need for long-term venous access. Once weekly administration of a parenteral antibiotic represents a new treatment paradigm, and it will be interesting to see the ways in which clinicians may exploit this pharmacokinetic property. It is anticipated that dalbavancin will become available for use in 2007.

8.2. Ceftobiprole

Ceftobiprole is a parenteral β-lactam antibiotic that combines the spectrum of a fourth-generation cephalosporin (i.e., cefepime) with activity against MRSA. Prolonged serial transfer of staphylococci in the presence of subinhibitory concentrations of this drug did not lead to resistance. This drug looks extremely promising, especially as empiric therapy against both community- and hospital-acquired pathogens *(88)*. The development of other β-lactams with activity against MRSA is under way *(89)*.

8.3. Telavancin

Telavancin (TD-6424) is a new lipoglycopeptide that affects both cell wall synthesis and cell membrane integrity to exert potent and rapid bactericidal activity against Gram-positive organisms. Recently, Reyes et al. *(90)* compared this drug to linezolid and vancomycin in a neutropenic mouse model of MRSA pneumonia. Telavancin handily outperformed both rivals, with the proportion

Table 2
Cost of Various Items and Services Associated with Treatment of MRSA Infections[a]

Item	Cost
PICC line insertion	$345
CXR (PA)- to verify PICC line placement	$140
Complete blood count[b]	$37
Vancomycin level	$65
Creatinine Kinase level[c]	$37
Chemistry panel 7[d]	$85
Home health nursing visit[e]	$112

[a]Based on charges at University Hospital, San Antonio, TX, November 2005.
[b]Monitor during long-term linezolid or vancomycin therapy.
[c]Monitor during daptomycin therapy.
[d]Monitor creatinine (renal function) during vancomycin therapy.
[e]Based on reported charges from one large San Antonio home health-care service provider for a 2-h home visit for setting up iv antibiotics. The number of home health visits will be determined by the antibiotic dosing, the number of iv line dressing changes, and the ability of the patient to self-administer the antibiotic.

of survivors in the control, telavancin, vancomycin, and linezolid groups being 14, 86, 73, and 50%, respectively. The investigators proposed that the superior efficacy of telavancin was owing to its rapid bactericidal activity. The drug will be available only in parenteral form, probably for once daily administration. Obviously, it looks promising. As of January 2007, phase III studies on the treatment of pneumonia and SSTIs were under way, and the FDA had granted the drug fast-track status.

9. Oral vs IV: The Pharmacoeconomics of Defeating *S. aureus*

As one advertisement says, the best thing a patient can hear is "welcome home." Hospital administrators also like patients to hear those words, because getting patients out of the hospital not only is good for patients' mental health, but also saves money—big money. Several studies have examined the cost savings associated with either switching from iv vancomycin to oral linezolid or using oral linezolid initially instead of vancomycin for the management of MRSA infections, which allows early hospital discharge. This strategy can decrease the LOS by 3–8 d, with the cost of each day of hospitalization estimated to be $1,689 *(91,92)*.

In addition to drug acquisition costs, another economic factor is the cost associated with the use of iv catheters *(93)* (*see* **Table 2**). At our clinical site (University Hospital, San Antonio, TX), the charge for placement of a peripherally inserted central venous catheter (PICC line) is $345 and the chest X-ray to verify its placement is a further $140. And then there is the cost when problems from the catheter arise (infection and thrombophlebitis); rates of catheter-related adverse events run from 1.1 to 12.1 per 1000 catheter days, at a cost of $4,300–$54,000 per event *(91)*! Even in the hospital, nursing time is greater for parenteral agents compared to oral therapy *(91)*. For those patients who continue iv therapy at home, there is then the cost of home health nursing care to administer the agent ($112 per 2-h visit). If vancomycin is used for a prolonged period, there may be additional costs to monitor vancomycin levels ($65 per test) and creatinine ($85 for a serum chemistry 7 panel). Sharpe et al. *(44)* calculated that the daily cost of outpatient therapy was $97 less for oral linezolid compared to iv vancomycin, even without considering the cost of catheter placement and pharmacokinetic monitoring.

Considering the cost and morbidity associated with the use of iv lines, the use of oral agents to treat MRSA infections would be preferable. In the United States, there are many potential oral agents for the treatment of MRSA infections: linezolid, clindamycin, minocycline, doxycycline, rifampin, the fluoroquinolones (FQs), and TMP-SMX. The daily cost of these oral antibiotics varies considerably, from $0.12/d for doxycycline to $105.88/d for oral linezolid (**Table 3**), so it behooves researchers to determine how these drugs can be used against MRSA in the most efficacious and cost-effective manner.

Linezolid is the oral antistaphylococcal agent that has the greatest weight of evidence for efficacy in a variety of MRSA infections, including SSTIs, pneumonia, diabetic foot infections, and surgical-site infections (*vide supra*), because it is backed by the clinical trial muscle of pharmaceutical titan Pfizer. Unfortunately, the other oral medications for use in MRSA infections lack the backing of the pharmaceutical industry for comparative trials and, therefore, the data for their efficacy are limited.

Although most nosocomial MRSA isolates are resistant to clindamycin, most community-acquired isolates still retain susceptibility. Clindamycin has high oral bioavailability and achieves levels in abscess fluid and bone at about 50% of the serum concentration. In neutrophils, it is concentrated 10X compared to the extracellular milieu *(22)*.

Nevertheless, the use of clindamycin has several potential drawbacks. First is the phenomenon of inducible clindamycin resistance (*vide supra*). Clindamycin should not be used to treat serious staphylococcal infections if the isolate is resistant to erythromycin unless the D-test is performed *(14)*. A second drawback is its short half-life, necessitating three or four times daily dosing, which may decrease patient compliance. There are no clinical data comparing clindamycin to other agents in the treatment of staphylococcal infections. In the opinion of Chambers *(22)*, clindamycin probably has efficacy similar to that of vancomycin. Furthermore, the incidence of *Clostridium difficile* colitis may be higher with clindamycin than with other antibiotics.

Table 3
Daily Cost of Various Antibiotics to Treat *S. aureus* (93)[a]

Drug	Dosing	Daily cost
Vancomycin	1 g iv q 12 –h	$11.24
Linezolid	600 mg iv q 12 h	$133.76
Linezolid	600 mg po bid	$105.88
Quinupristin-Dalfopristin	500 mg iv q 8 h	$317.88
Daptomycin	500 mg iv q 24 h	$149.69 (based on 70-kg patient)
Daptomycin	6 mg/kg iv qd	$143.70 (based on 70-kg patient)
Tigecycline	50 mg iv q 12 –h	$84.00
Minocycline	100 mg po q 12 h	$0.28
Doxycycline	100 mg po q 12 h	$0.12
Clindamycin	300 mg po q 6 h	$14.87
TMP-SMX	800/160 (DS) 2 tablets po q 12 h	$0.20

[a]Prices may vary according to specific pricing negotiations between a pharmacy and a supplier.

The long-acting tetracyclines, doxycycline and minocycline, offer excellent oral bioavailability, low cost, few adverse effects, and the availability of oral and iv preparations. Additionally, the majority of CA-MRSA isolates have retained susceptibility to these agents. Even if tetracycline resistance is present, MRSA isolates typically retain susceptibility to minocycline *(94,95)*. However, clinical data for the use of the long-acting tetracyclines in MRSA infections are sparse. In a retrospective study of 17 patients with staphylococcal infections (four were MRSA), 15 patients responded to minocycline therapy *(95)*. In one recent case series of the use of long-acting tetracyclines in the treatment of 24 patients with MRSA infections, there was an overall 83% success rate (93% for SSTIs) *(96)*. In a rabbit model of MRSA endocarditis, minocycline was equivalent to vancomycin in the reduction of bacterial densities in the vegetations *(97)*. Prospective comparative studies using minocycline and doxycycline in the treatment of MRSA infections are needed. Tetracyclines should not be used in children less than 8 yr old, because of the risk of dental staining.

Rifampin is a bactericidal antibiotic that inhibits chain initiation of bacterial DNA-dependent RNA polymerase. It is available in oral and iv preparations. Resistance to rifampin is rare (<3%) among American *S. aureus* isolates *(63)*. Unfortunately, when rifampin is used against *S. aureus* as monotherapy, resistance occurs rapidly via a one-step target-site mutation in the RNA polymerase *(26)*. Rifampin is often added to another antibiotic as an act of faith that it will improve outcomes in staphylococcal infections, despite the paucity of convincing clinical data for such an approach. In fact, enhanced activity, indifference, and even antagonism have all been reported for rifampin combinations *(22)*. In a study comparing vancomycin to vancomycin + rifampin in the treatment of

MRSA endocarditis, clinical outcome was identical for the two regimens, but patients in the rifampin arm had a longer duration of bacteremia *(98)*. In a rabbit model of MRSA endocarditis, rifampin did not contribute to the antibacterial effect of linezolid *(99)*. There is also concern that rifampin decreases serum levels of linezolid *(100)*.

Krut et al. *(101)* examined the in vitro activity of rifampin and other antibiotics in inhibiting *S. aureus* sequestered in nonphagocytic cells. Clindamycin, azithromycin, and linezolid exerted significant bacteriostatic activity against intracellular *S. aureus*, but only rifampin eradicated the intracellular bacteria. The clinical relevance of this observation awaits further investigation. The practicality of using rifampin is also hampered by its plethora of drug interactions *(26)*.

Shopsin et al. *(102)* have proposed that the newer FQs, such as gatifloxacin and moxifloxacin, may also be useful in the treatment of MRSA infections, because unlike older FQs, such as ciprofloxacin, these drugs maintain serum concentrations well above those needed to block the growth of the least susceptible mutant. This hypothesis has yet to be tested in the clinical arena. Recent surveillance data from American hospitals indicate that 55.6% of *S. aureus* isolates were susceptible *(63)* to levofloxacin. Unfortunately, owing to the phenomenon of MDR in MRSA, far fewer of these isolates are susceptible to FQs; at our institution in 2004, 82% of all *S. aureus* isolates were susceptible to gatifloxacin, whereas only 30% of the MRSA isolates were susceptible. There is also concern about the emergence of resistance when these agents are used as monotherapy against *S. aureus*, although in vitro gatifloxacin and moxifloxacin both appear to be less prone to select for resistance as rapidly as the older FQs. However, in one observational study, only 56% of patients with MRSA skin infections responded to moxifloxacin *(103)*. The prospect of using the newer FQs in the treatment of certain MRSA infections is enticing, even if only in select patients, such as those with osteomyelitis, because of their convenient once daily oral dosing and excellent bone and soft-tissue penetration. Thus, the acquisition of clinical data on the efficacy of the new FQs, perhaps in combination with a second agent to delay the emergence of resistance, would be very useful.

TMP-SMX offers some useful features in the battle against MRSA: the availability of iv and oral preparations, good oral bioavailability, -cidal activity, low cost, and retention of activity in vitro against *S. aureus* despite almost three decades of use (recent American surveillance data indicate <3% resistance) *(63)*. Liabilities of the drug are a high rate of hypersensitivity reactions, myelosuppression, and electrolyte disturbances. Unfortunately, a recent review reveals just how skimpy the data are on the use of TMP-SMX in MRSA infections *(104)*: only one clinical trial, one observational study, and a few anecdotal reports. TMP-SMX was outperformed by vancomycin in the only randomized trial of the treatment of staphylococcal infections, achieving a cure rate of 86%, vs 98% for vancomycin *(105)*. Blood cultures remained positive for 2 to 3 d longer in the patients treated with TMP-SMX. Furthermore, in a rabbit model of *S. aureus* endocarditis, vancomycin was also more effective than TMP-SMX *(106)*. These studies suggest that TMP-SMX is not as efficacious as vancomycin in the treatment of *S. aureus* infections. Nevertheless, Shams et al. *(107)* reported a case of a patient with persistent bacteremia caused by an MRSA-infected left-ventricular assist device in which TMP-SMX treatment was successful after vancomycin, linezolid, and QD all failed *(107)*. This may have been the result of poor penetration of the other antibiotics into the MRSA biofilm on the device compared to TMP-SMX. When oral TMP-SMX is used as an anti-MRSA agent, it should be dosed as two double-strength tablets twice a day *(108)*. One concern about the empiric use of TMP-SMX in SSTIs is the frequent resistance observed in *S. pyogenes* and the drug's limited activity against *Streptococcus agalactiae (108)*. The TMP-SMX + rifampin combination is often bandied about as a treatment for *S. aureus* infections *(109)*, but there is little published information on its efficacy. In one small series of six patients with cutaneous MRSA infections, treatment with TMP-SMX + rifampin was uniformly successful *(103)*.

One encouraging study used high-dose oral TMP-SMX (20 mg/[kg/d] of the TMP component) for 6–9 mo in the treatment of orthopedic implants infected with MRSA *(110)*. Although here was a high rate (21%) of discontinuation of the drug owing to adverse effects (rash, vomiting, diarrhea), the success rate was 86.7% in those who completed therapy, which is quite high considering the difficult nature of this type of infection. Furthermore, in 65.4% of the successfully treated patients, there was no removal of orthopedic material.

10. Other Antistaphylococcal Drugs Available Outside of the United States

The United States may have nearly every conceivable consumer good, but when it comes to antistaphylococcal antibiotics it is at a distinct disadvantage compared with some other countries. Fusidic acid has been available in Europe and Australia since the 1960s and is available in oral, iv, and topical forms *(111)*. In Australia, 95% of even MDR isolates of MRSA retain susceptibility to fusidic acid *(112)*. To avoid emergence of resistance, fusidic acid is usually combined with a second antibiotic, such as rifampin or a β-lactam. Clinical trial data support the use of the oral form in combination with rifampin to treat SSTIs and bone/joint infections after initial treatment with vancomycin. It is also used to treat bacteremia and endocarditis, but there is less evidence to support these applications *(111)*.

Another oral antistaphylococcal antibiotic available in Europe is the streptogramin pristinamycin. This antibiotic acts synergistically with tetracyclines and has been used primarily to treat skin and soft-tissue, bone, and joint infections. The cost of pristinamycin is about 10% that of linezolid *(26)*.

Fosfomycin is a bactericidal broad-spectrum antibiotic available in parenteral form to treat systemic infections that has been used for decades in Europe, Japan, Brazil, and South Africa. It has a unique structure and mechanism of action, so cross-resistance with other classes of antibiotics is not a problem. Fosfomycin itself has poor oral bioavailability, although its tromethamine salt has 34–41% oral bioavailability and has an FDA indication for the treatment of urinary tract infections. Fosfomycin shows synergy against MRSA with linezolid, QD, ciprofloxacin, and rifampin *(113,114)*. The question is: Why is this drug not available in the United States in parenteral form? It would be potentially useful not only for MRSA infections, but also against *Pseudomonas aeruginosa*.

11. Bactericidal vs Bacteriostatic: Does It Make a Difference?

Among the drugs described herein, vancomycin, QD, oritavancin, dalbavancin, televancin, daptomycin, TMP-SMX, fosfomycin, and ceftibiprole are bactericidal, whereas linezolid, doxycycline, clindamycin, and tigecycline are considered bacteriostatic. Are the bactericidal agents intrinsically more efficacious? Pankey and Sabath *(115)* readdressed this issue and concluded that the superiority of bactericidal action over bacteriostatic is of little clinical relevance and that pharmacokinetic/pharmacodynamic parameters of the agents may be more important factors in clinical success. Linezolid, a static agent, has never been outperformed in a single trial by -cidal vancomycin or β-lactams. Further trials are needed to determine whether rapidly bactericidal agents such as daptomycin or telavancin will be clinically more efficacious in serious conditions such as bacteremia and endocarditis.

12. Are Antibiotics Always Necessary?

If limiting overall antibiotic use is a noble goal in itself, are there minor MRSA infections that do not require antibiotic treatment? Lee et al. *(116)* reported that incision and drainage without antibiotic treatment was effective in the treatment of MRSA skin abscesses <5 cm in diameter in immunocompetent children. Thus, for smaller abscesses, clinicians may consider withholding antibiotics after draining the swamp.

13. Three Illustrative Cases of Difficult MRSA Infection
13.1. Case #1

A 21-yr-old Hispanic male presented with pain and swelling over the left side of his face that started 6 d prior to medical evaluation. The patient's symptoms started after self-manipulation of a pustule on his chin. On examination, the patient was febrile with swelling of the left side of his face involving his chin, lips, cheek, preauricular area, and lower eyelid. The patient also had an abscess on his chin. He reported pain on palpation of teeth #23–25. Initial laboratory results were only remarkable for a mild leukocytosis (white blood cell [WBC] = 17,000 cells/mm^3). An infection with mixed oral flora was suspected, and the patient was started on 400 mg of iv gatifloxacin daily and 900 mg of iv clindamycin every 8 h. The chin abscess was surgically drained, and teeth #23–25 were extracted, after radiographically documented periapical involvement. However, the patient remained febrile and the facial swelling worsened. MRSA was isolated from multiple blood cultures and cultures of the abscess. The patient's antibiotic was switched to 1 g of iv vancomycin every 8 h. A computerized tomographic (CT) scan of the head and neck revealed extensive soft-tissue swelling of the face and submandibular area and bilateral apical pulmonary nodules. A CT scan of the chest showed multiple bilateral nodular opacities with cavitation, suggestive of septic emboli. A CT scan of the neck revealed a thrombus of the left internal jugular vein as the probable source of the septic emboli. A transthoracic echocardiogram (TEE) showed a partially occluding intraluminal thrombus that extended down to the junction with the right atrium; no vegetations were noted. The patient was started on enoxaparin for pulmonary emboli. MRSA continued to be repeatedly isolated from blood cultures during the following days; all isolates were susceptible to TMP-SMX and doxycycline and resistant to erythromycin and clindamycin. In view of the persistent bacteremia and difficulty reaching a therapeutic trough level with vancomycin, the dose was increased to 2 g every 8 h, but levels remained <10 µg/mL. Trough levels within the 15- to 20-µg/mL range were consistently attained when the dose of vancomycin was increased to 2 g every 6 h. Clindamycin was resumed because of concern about infection by anaerobes in the setting of apparent Lemierre's syndrome. Despite these measures, the patient continued to have fever, prominent facial swelling, and persistent bacteremia. Intravenous linezolid at 600 mg every 12 h was then added to 2 g of vancomycin every 8 h and 600 mg of clindamycin every 8 h. Clinical improvement (defervescence and resolving leukocytosis) was observed within 48 h after the addition of linezolid to the treatment regimen. After 15 d of sustained bacteremia, the patient finally showed negative blood cultures on linezolid, vancomycin, and clindamycin. Clindamycin was discontinued after 13 d and linezolid was switched to its oral formulation by d 15 of therapy. Marked improvement became evident during the next few days, and the patient was discharged after 25 d of hospitalization. The patient received an additional 4-wk course of 2 g of iv vancomycin every 12 h owing to funding issues for prolonged therapy with linezolid.

This case illustrates two problems with the use of vancomycin: difficulty obtaining therapeutic levels and clinical failure despite supratherapeutic levels.

In this case, it is uncertain whether the addition of linezolid to the regimen or the cumulative intensity of antibiotic treatment was the important factor in achieving resolution of infection.

13.2. Case #2

A 26-yr-old Hispanic male presented with a productive cough, a fever, chills, and a chest X-ray with a right lower lobe infiltrate. He was treated for community-acquired pneumonia with oral gatifloxacin. After 4 d, he returned owing to dyspnea, pleuritic chest pain, and swelling and pain of the left side of his face that appeared after self-manipulation of a pustule. On examination, the patient was febrile, mildly tachycardic, and had moderate oxygen desaturation on minimal ambulation. He had swelling of the left face, periorbital area, and a draining abscess in the left nare. Laboratory tests showed marked leukocytosis (WBCs = 24,000 cells/mm^3) and elevated C-reactive protein of 42 µg/dL. A chest X-ray showed a right pleural effusion and a right lower lobe infiltrate. Thoracentesis afforded a bloody exudate. The abscess in the left nare was drained and iv cefotaxime, azithromycin, and clindamycin were started. MRSA was isolated from cultures of the blood, exudates, and pleural effusion, and vancomycin was added. Because of loculation of the pleural effusion, thoracotomy with decortication and chest tube drainage were necessary. The patient continued to be bacteremic on hospital d 5. After 1 wk of hospitalization, the patient remained febrile and continued to have significant leukocytosis. Vancomycin trough levels were <15 µg/mL. At that time, the patient was switched to linezolid. Defervescence and resolution of leukocytosis occurred during the following days, and the patient was discharged 13 d after starting treatment with linezolid. The patient then uneventfully completed a 4-wk course of oral linezolid.

This case illustrates clinical failure using vancomycin and successful salvage therapy with linezolid. The superior penetration of linezolid into pulmonary tissue compared to vancomycin may have been a factor in the clinical success observed with linezolid.

13.3. Case #3

A 51-yr-old African American male presented with a 5-d history of fever, chills, night sweats, and severe right lumbar pain with radiation to his right lower extremity associated with mild lower-limb weakness. The patient also reported polydipsia and polyuria. He had a history of MRSA endocarditis 1 yr before that was complicated by multiple septic emboli and was treated with a 6-wk course of vancomycin. However, the patient developed an interstitial nephritis while taking vancomycin at that time. The patient denied any recent use of iv drugs. Examination was remarkable for fever, a systolic ejection murmur loudest at the apex, and exquisite pain on palpation along the right lumbosacral spine area. Laboratory abnormalities on admission included leukocytosis (WBC = 27,000 cells/mm^3), hyperglycemia, and mild metabolic acidosis. Since the patient had a previous history of interstitial nephritis while taking vancomycin, he was initially administered daptomycin at 4 mg/(kg/d). All blood cultures grew MRSA susceptible to daptomycin with an MIC ≤0.5 µg/mL. A TEE was negative for vegetations. The patient remained febrile and had persistent bacteremia despite antibiotic therapy. The dosage of daptomycin was then increased to 6 mg/(kg/d). A transesophageal echocardiogram did not show valvular lesions at that time. Despite 10 d of daptomycin therapy, the patient remained bacteremic with MRSA; piperacillin/tazobactam was then added based on the reported synergy of β-lactams and daptomycin. An abdominal CT scan revealed a right psoas abscess, but attempts at percutaneous drainage were unsuccessful. On hospital d 15, the patient developed acute paraparesis and required urgent laminectomy with debridement, as a result of L5 osteomyelitis with a soft-tissue process involving the conus medularis and cauda equina. With the persistence of MRSA bacteremia, the dosage of daptomycin was increased to 8 mg/(kg/d) and 600 mg of iv gentamicin every day was added. Subsequent testing revealed that the MICs of the MRSA strain for daptomycin had progressively increased from initial values of 0.5–1 to 2 µg/mL. The patient showed clearance of bacteremia 7 d after the addition of gentamicin. A third echocardiogram showed a vegetation on the aortic valve, and piperacillin-tazobactam was discontinued and 300 mg of rifampin orally every 8 h was started. One week later, while on combination therapy of daptomycin, rifampin, and gentamicin, the patient developed a marked increase in creatine phosphokinase levels and a moderate increase in serum creatinine level. These three antibiotics were discontinued, and 600 mg of linezolid orally every 12 h was then started. Percutaneous drainage of the psoas abscess was eventually achieved but the patient suffered paraparesis. An additional 4-wk course of oral linezolid was completed.

This case illustrates the acquisition of daptomycin resistance by MRSA during daptomycin therapy and the adverse effect of myositis when the dosage of daptomycin was increased to high levels.

14. Conclusion

Four agents, QD, linezolid, daptomycin, and tigecycline, are now available as alternatives to vanomycin for the treatment of serious MRSA infections. These agents will also be effective against the next assault currently being launched by *S. aureus* against the human race—isolates with reduced susceptibility to vancomycin (VISA, VRSA). Linezolid, with its high oral bioavailability and tissue penetration, resilience to the development of resistance, and favorable results in a relatively large number of clinical trials and pharmacoeconomic analyses, is the most versatile of the currently available agents against MRSA.

However, some safety issues when linezolid is used for longer-term treatment still need to be resolved. Furthermore, there are four other agents (oritavancin, dalbavancin, telavancin, ceftobiprole) that will probably receive FDA approval in the next few years. Despite these promising developments, high-quality clinical data for some of the vexing problems caused by MRSA are woefully lacking. Clinical trial data that are needed to combat MRSA more effectively include efficacy data for the use of inexpensive oral drugs (doxycycline, minocycline, TMP/SMX, and FQ plus rifampin) for the treatment of SSTIs; comparator trials for MRSA osteomyelitis, especially comparing oral linezolid to the cheap oral drugs; and additional prospective comparator trials for the newer drugs in MRSA bacteremia and endocarditis (investigators should abandon the prejudice against trying linezolid in this role just because of its in vitro bacteriostatic activity). Additionally, although there are a multitude of in vitro studies on antimicrobial synergy against MRSA, there is a paucity of clinical data regarding whether this property can be exploited to improve outcomes, especially in difficult-to-treat MRSA infections, such as chronic osteomyelitis, pneumonia, and refractory bacteremia/endocarditis.

References

1. Lindsay, J. A. and Holden, M. T. G. (2004) *Staphylococcus aureus*: superbug, super genome? *Trends Microbiol.* **12**, 378–385.
2. Naimi, T. S., LeDell, K. H., Como-Sabetti, K., et al. (2003) Comparison of community- and health care-associated methicillin-resistant *Staphylococcus aureus* infection. *JAMA* **290**, 2976–2984.
3. Rybak, M. J. and LaPlante, K. L. (2005) Community-associated methicillin-resistant *Staphylococcus aureus*: a review. *Pharmacotherapy* **25**, 74–85.
4. Gilbert, D. N., Moellering, R. C., Eliopoulos, G. M., and Sande, M. A. (eds.). (2005) *The Sanford Guide to Antimicrobial Therapy 2005*, 35th ed., Antimicrobial Therapy, Hyde Park, VT.
5. Gosbell, I. B. (2005) Diagnosis and management of catheter-related bloodstream infections due to *Staphylococcus aureus*. *Intern. Med. J.* **35**, S45–S62.
6. Davis, J. S. (2005) Management of bone and joint infections due to *Staphylococcus aureus*. *Intern. Med. J.* **35**, S79–S96.
7. Murray, R. J. (2005) *Staphylococcus aureus* infective endocarditis: diagnosis and management guidelines. *Intern. Med. J.* **35**, S25–S44.
8. Mitchell, D. H. and Howden, B. P. (2005) Diagnosis and management of *Staphylococcus aureus* bacteraemia. *Intern. Med. J.* **35**, S17–S24.
9. Cosgrove, S. E., Carroll, K. C., and Perl, T. M. (2004) *Staphylococcus aureus* with reduced susceptibility to vancomycin. *Clin. Infect. Dis.* **39**, 539–545.
10. Anonymous. (2004) Vancomycin-resistance *Staphylococcus aureus*—New York, 2004. *MMWR Morbid. Mortal. Wkly. Rep.* **53**, 322, 323.
11. Bozdogan, B., Esel, D., Whitener, C., Browne, F. A., and Appelbaum, P. C. (2004) Antibacterial susceptibility of a vancomycin-resistant *Staphylococcus aureus*

strain isolated at the Hershey Medical Center. *J. Antimicrob. Chemother.* **52,** 864–868.
12. Fiebelkorn, K. R., Crawford, S. A., McElmeel, M. L., and Jorgensen, J. H. (2003) Practical disk diffusion method for detection of inducible clindamycin resistance in *Staphylococcus aureus* and coagulase-negative staphylococci. *J. Clin. Microbiol.* **41,** 4740–4744.
13. O'Brien, F. G., Zainin, Z., Coombs, G. W., Pearson, J. C., Christiansen, K., and Grubb, W. B. (2005) Macrolide, lincosamide and streptogramin B resistance in a dominant clone of Australian community methicillin-resistant *Staphylococcus aureus*. *J. Antimicrob. Chemother.* **56,** 985, 986.
14. Lewis, J. S. II and Jorgensen, J. H. (2005) Inducible clindamycin resistance in staphylococci: should clinicians and microbiologists be concerned? *Clin. Infect. Dis.* **40,** 280–285.
15. Siberry, G. K., Tekle, T., Carroll, K., and Dick, J. (2003) Failure of clindamycin treatment of methicillin-resistant *Staphylococcus aureus* expressing inducible clindamycin resistance in vitro. *Clin. Infect. Dis.* **37,** 1257–1260.
16. von Eiff, C., Lubritz, G., Heese, C., Peters, G., and Becker, K. (2004) Effect of trimethoprim-sulfamethoxazole in AIDS patients on the formation of the small colony variant phenotype of *Staphylococcus aureus*. *Diagn. Microbiol. Infect. Dis.* **48,** 191–194.
17. Levine, D. P. (2006) Vancomycin: a history. *Clin. Infect. Dis.* **42,** S5–S12.
18. Kollef, M. H. and Micek, S. T. (2005) *Staphylococcus aureus* pneumonia: a "superbug" infection in community and hospital settings. *Chest* **128,** 1093–1096.
19. Stevens, D. L. (2006) The role of vancomycin in the treatment paradigm. *Clin. Infect. Dis.* **42,** S51–S57.
20. Rello, J., Torres, A., Ricart, M., Valles, J., Gonzalez, J., Artigas, A., and Rodriguez-Roisin, R. (1994) Ventilator-associated pneumonia by *Staphylococcus aureus*: comparison of methicillin-sensitive and methicillin-resistant episodes. *Am. J. Respir. Crit. Care Med.* **150,** 1545–1549.
21. Tice, A. D., Hoagland, P., and Shoultz, D. A. (2003) Outcomes of osteomyelitis among patients treated with outpatient parenteral antimicrobial therapy. *Am. J. Med.* **114,** 723–728.
22. Chambers, H. F. (1997) Parenteral antibiotics for the treatment of bacteremia and other serious staphylococcal infections, in *The Staphylococci in Human Disease* (Crossley, K. B., and Archer, G. L., eds.), Churchill-Livingstone, New York, pp. 583–601.
23. Jones, R. N. (2006) Microbiologic features of vancomycin in the 21st century: minimum inhibitory concentration creep, bactericidal/static activity, and applied breakpoints to predict clinical outcomes or detect resistant strains. *Clin. Infect. Dis.* **42,** S13–S24.
24. Rello, J. N., Sole-Violan, J., Sa-Borges, M., et al. (2005) Pneumonia caused by oxacillin-resistant *Staphylococcus aureus* treated with glycopeptides. *Crit. Care Med.* **33,** 1983–1987.
25. American Thoracic Society, Infectious Diseases Society of America. (2005) Guidelines for the management of adults with hospital-acquired, ventilator-

associated, and healthcare-associated pneumonia. *Am. J. Respir. Crit. Care Med.* **171**, 388–416.
26. Rayner, C. and Munckhof, W. J. (2005) Antibiotic currently used in the treatment of infections caused by *Staphylococcus aureus*. *Intern. Med. J.* **35**, S3–S16.
27. Hidayat, L. K., Hsu, D. I., Quist, R., Shriner, K. A., and Wong-Beringer, A. (2006) High-dose vancomycin therapy for methicillin-resistant *Staphylococcus aureus* infections: efficacy and toxicity. *Arch. Intern. Med.* **166**, 2138–2144.
28. Fridkin, S. K., Hageman, J., McDougal, L. K., Mohammed, J., Jarvis, W. R., Perl, T. M., and Tenover, F. C. (2003) Vancomycin-Intermediate *Staphylococcus aureus* Epidemiology Study Group. Epidemiological and microbiological characterization of infections caused by *Staphylococcus aureus* with reduced susceptibility to vancomycin, United States, 1997–2001. *Clin. Infect. Dis.* **36**, 429–439.
29. Livermore, D. M. (2003) Linezolid in vitro: mechanism and antibacterial spectrum. *J. Antimicrob. Chemother.* **51(Suppl. S2)**, ii9–ii16.
30. Wunderink, R. G., Bello, J., Cammarata, S. K., Croos-Dadrera, R. V., and Kollef, M. H. (2003) Linezolid vs vancomycin: analysis of two double-blind studies of patients with methicillin-resistant *Staphylococcus aureus* nosocomial pneumonia. *Chest* **124**, 1789–1797.
31. Kollef, M. H., Rello, J., Cammarata, S. K., Croos-Dabrera, R. V., and Wunderink, R. G. (2004) Clinical cure and survival in Gram-positive ventilator-associated pneumonia: retrospective analysis of two double-blind studies comparing linezolid with vancomycin. *Intensive Care Med.* **30**, 388–394.
32. Ioanas, M. and Lode, H. (2004) Linezolid in VAP by MRSA: better choice? *Intensive Care Med.* **30**, 343–346.
33. Honeybourne, D., Tobin, C., Jevons, G., Andrews, J., and Wise, R. (2003) Intrapulmonary penetration of linezolid. *J. Antimicrob. Chemother.* **51**, 1431–1434.
34. Micek, S. T., Dunne, M., and Kollef, M. H. (2005) Pleuropulmonary complications of Panton-Valentine leukocidin-positive community-acquired methicillin-resistant *Staphylococcus aureus*: importance of treatment with antimicrobials inhibiting exotoxin production. *Chest* **128**, 2732–2738.
35. Francis, J. S., Doherty, M. C., Lopatin, U., et al. (2005) Severe community-onset pneumonia in healthy adults caused by methicillin-resistant *Staphylococcus aureus* carrying the Panton-Valentine Leukocidin genes. *Clin. Infect. Dis.* **40**, 100–107.
36. Shukla, S. K. (2005) Community-associated methicillin-resistant *Staphylococcus aureus* and its emerging virulence. *Clin. Med. Res.* **3**, 57–60.
37. Gemmel, C. G. and Ford, C. W. (2002) Virulence factor expression by Gram-positive cocci exposed to subinhibitory concentrations of linezolid. *J. Antimicrob. Chemother.* **50**, 667–672.
38. Bernardo, K., Pakulat, N., Fleer, S., et al. (2004) Subinhibitory concentrations of linezolid reduce *Staphylococcus aureus* virulence factor expression. *Antimicrob. Agents Chemother.* **48**, 544–546.
39. Rayner, C. R., Baddour, L. M., Birmingham, M. C., Norden, C., Meagher, A. K., and Schetag, J. J. (2004) Linezolid in the treatment of osteomyelitis: results of a compassionate use experience. *Infection* **32**, 8–14.

40. Kutscha-Lissberg, F., Hebler, U., Muhr, G., and Koller, M. (2003) Linezolid penetration into bone and joint tissues infected with methicillin-resistant staphylococci. *Antimicrob. Agents Chemother.* **47,** 3964–3966.
41. Lipsky, B. A., Itani, K., and Norden, C. (2004) Linezolid Diabetic Foot Infection Study Group. Treating foot infections in diabetic patients: a randomized, multicenter, open-label trial of linezolid versus ampicillin-sulbactam/amoxicillin-clavulanate. *Clin. Infect. Dis.* **38,** 17–24.
42. Kaplan, S. L., Afghani, B., Lopez, P., et al. (2003) Linezolid for the treatment of methicillin-resistant *Staphylococcus aureus* in children. *Pediatr. Infect. Dis. J.* **22,** S178–S185.
43. Shorr, A. F., Kunkek, M. J., and Kollef, M. (2005) Linezolid versus vancomycin for *Staphylococcus aureus* bacteraemia: pooled analysis of randomized studies. *J. Antimicrob. Chemother.* **56,** 923–929.
44. Sharpe, J. N., Shively, M. D., and Polk, H. C. Jr. (2005) Clinical and economic outcomes of oral linezolid versus intravenous vancomycin in the treatment of MRSA-complicated, lower-extremity skin and soft-tissue infections caused by methicillin-resistant *Staphylococcus aureus. Am. J. Surg.* **189,** 425–428.
45. Weigelt, J., Kaafarani, H. M. A., Itani, K. M. F., and Swanson, R. N. (2004) Linezolid eradicates MRSA better than vancomycin from surgical-site infections. *Am. J. Surg.* **188,** 760–766.
46. Weigelt, J., Itani, K., Stevens, D., Lau, W., Dryden, M., Knirsch, C., and Linezolid CSSTI Study Group. (2005) Linezolid versus vancomycin in treatment of complicated skin and soft tissue infections. *Antimicrob. Agents Chemother.* **49,** 2260–2266.
47. Howden, B. P., Charles, P. G. P., Johnson, P. D. R., Ward, P. B., and Grayson, M. L. (2005) Improved outcomes with linezolid for methicillin-resistant *Staphylococcus aureus* infections: better drug or reduced vancomycin susceptibility? *Antimicrob. Agents Chemother.* **49,** 4816, 4817.
48. Moise-Broder, P. A., Sakoulas, G., Eliopoulos, G. M., Schentag, J. J., Forrest, A., and Moellering, R. C. Jr. (2004) Accessory gene regulatory group II polymorphism in methicillin-resistant *Staphylococcus aureus* is predictive of failure of vancomycin therapy. *Clin. Infect. Dis.* **38,** 1700–1705.
49. Clinical and Laboratory Standards Institute/NCCLS. (2006) *Performance Standards for Antimicrobial Susceptibility Testing,* Supplement M100-S16, Clinical and Laboratory Standards Institute, Wayne, PA.
50. Sancak, B., Ercis, S., Menemenlioglu, D., Çolakoglu S., and Hasçelik, G. (2005) Methicillin-resistant *Staphylococcus aureus* heterogeneously resistant to vancomycin in a Turkish university hospital. *J. Antimicrob. Chemother.* **56,** 519–523.
51. Liu, C. and Chambers, H. F. (2003) *Staphylococcus aureus* with heterogeneous resistance to vancomycin: clinical significance and critical assessment of diagnostic methods. *Antimicrob. Agents Chemother.* **47,** 3040–3045.
52. Nasraway, S. A., Shorr, A. F., Kuter, D. J., O'Grady, N., Le, V. H., and Cammarata, S. K. (2003) Linezolid does not increase the risk of thrombocytopenia in patients with nosocomial pneumonia: comparative analysis of linezolid and vancomycin use. *Clin. Infect. Dis.* **37,** 1609–1616.

53. Rao, N., Ziran, B. H., Wagener, M. M., Santa, E. R., and Yu, V. L. (2004) Similar hematologic effects of long-term linezolid and vancomycin therapy in a prospective observational study of patients with orthopedic devices. *Clin. Infect. Dis.* **38,** 1058–1064.
54. Bergeron, L., Boulé, M., and Perreault, S. (2005) Serotonin toxicity associated with concomitant use of linezolid. *Ann. Pharmacother.* **39,** 956–961.
55. Anonymous. (2004) Pfizer: Zyvox (linezolid). Safety-nervous-psychiatric. Reports of peripheral or optic neuropathy. Data on file.
56. Birmingham, M. C., Rayner, C. R., Meagher, A. K., Flavin, S. M., Batts, D. H., and Schentag, J. J. (2003) Linezolid for the treatment of multidrug-resistant, gram-positive infections: experience from a compassionate-use program. *Clin. Infect. Dis.* **36,** 159–168.
57. Legout, L., Senneville, E., Gomel, J. J., Yazdanpanah, Y., and Mouton, Y. (2004) Linezolid-induced neuropathy. *Clin. Infect. Dis.* **38,** 767, 768.
58. Frippiat, F., Bergiers, C., Michel, C., Dujardin, J. P., and Derue, G. (2004) Severe bilateral optic neuritis associated with prolonged linezolid therapy. *J. Antimicrob. Agents Chemother.* **53,** 114, 115.
59. Ferry, T., Ponceau, B., Simon, M., et al. (2005) Possibly linezolid-induced peripheral and central nervous system neurotoxicity: report of four cases. *Infection* **33,** 151–154.
60. Shorr, A. F., Susla, G. M., and Kollef, M. H. (2004) Linezolid for treatment of ventilator-associated pneumonia: a cost-effective alternative to vancomycin. *Crit. Care Med.* **32,** 137–143.
61. Machado, A. R., Arns Cda, C., Follador, W., and Guerra, A. (2005) Cost-effectiveness of linezolid versus vancomycin in mechanical ventilation-associated nosocomial pneumonia caused by methicillin-resistant *Staphylococcus aureus*. *Braz. J. Infect. Dis.* **9,** 191–200.
62. Li, J. Z., Willke, R. J., Rittenhouse, B. E., and Rybak, M. J. (2003) Effect of linezolid versus vancomycin on length of hospital stay in patients with complicated skin and soft tissue infections caused by known or suspected methicillin-resistant staphylococci: results from a randomized clinical trial. *Surg. Infect.* **4,** 45–58.
63. Draghi, D. C., Sheehan, D. J., Hogan, P., and Sahm, D. F. (2005) In vitro activity of linezolid against key gram-positive organisms isolated in the United States: results of the LEADER 2004 Surveillance Program. *Antimicrob. Agents Chemother.* **49,** 5024–5032.
64. Streit, J. M., Jones, R. N., Sader, H. S., and Fritsche, T. R. (2004) Assessment of pathogen occurrence and resistance profiles among infected patients in the intensive care unit report from the SENTRY Antimicrobial Surveillance Program (North America, 2001). *Int. J. Antimicrob. Agents* **24,** 11–18.
65. Hoban, D. J., Bouchillon, S. K., Johnson, B. M., Johnson, J. L., and Dowzicky, M. J. (2005) In vitro activity of tigecycline against 6792 Gram-negative and Gram-positive clinical isolates from global Tigecycline Evaluation and Surveillance Trial (TEST Program, 2004). *Diagn. Microbiol. Infect. Dis.* **52,** 215–227.
66. Lomaestro, B. M. (2003) Resistance to linezolid. Are we surprised? How hard should we look? *Ann. Pharmacother.* **37,** 909–911.

67. Fenton, C., Keating, G. M., and Curran, M. P. (2004) Daptomycin. *Drugs* **64**, 445–455.
68. Arbeit, R. D., Maki, D., Tally, F. P., Campanaro, E., Eisenstein, B. I., and Daptomycin 98-01 and 99-01 Investigators. (2004) The safety and efficacy of daptomycin for the treatment of complicated skin and skin-structure infections. *Clin. Infect. Dis.* **38**, 1673–1681.
69. Alder, J. D., Thorne, G., Luperchio, S., Anastasiou, D., Eisenstein, B., and Tally, F. (2003) Daptomycin compared to semi-synthetic penicillins or vancomycin for treatment of complicated skin and skin structure infections (cSSSI) caused by *S. aureus* [abstract 296], in *Program and Abstracts of the 41st Annual Meeting of the Infectious Diseases Society of America*, (Sears, C. L., Karp, C. L. (eds.), Alexandria, VA. p. 83.
70. Echevarria, K., Datta, P., Cadena, J., and Lewis, J. S. II (2005) Severe myopathy and possible hepatotoxicity related to daptomycin. *J. Antimicrob. Chemother.* **55**, 599, 600.
71. Hayden, M. K., Rezai, K., Hayes, R. A., Lolans, K., Quinn, J. P., and Weinstein, R. A. (2005) Development of daptomycin resistance in vivo in methicillin-resistant *Staphylococcus aureus*. *J. Clin. Microbiol.* **43**, 5285–5287.
72. Fowler, V. G. Jr., Boucher, H. W., Corey, G. R., et al., *S. aureus* Endocarditis and Bacteremia study Group. (2006) Daptomycin versus standard therapy for bacteremia and endocarditis caused by *Staphylococcus aureus*. *New Eng. J. med.* **355**, 653–665.
73. Sakoulas, G., Eliopoulos, G. M., Alder, J., and Eliopoulos, C. T. (2003) Efficacy of daptomycin in experimental endocarditis due to methicillin-resistant *Staphylococcus aureus*. *Antimicrob. Agents Chemother.* **47**, 1714–1718.
74. Tsuji, B. T. and Rybak, M. J. (2005) Short-course gentamicin in combination with daptomycin or vancomycin against *Staphylococcus aureus* in an in vitro pharmacodynamic model with simulated endocardial vegetations. *Antimicrob. Agents Chemother.* **49**, 2735–2745.
75. Rand, K. H. and Houck, H. J. (2004) Synergy of daptomycin with oxacillin and other β-lactams against methicillin-resistant *Staphylococcus aureus*. *Antimicrob. Agents Chemother.* **48**, 2871–2875.
76. Klastersky, J. (2003) Role of quinupristin/dalfopristin in the treatment of Gram-positive nosocomial infections in haematological or oncological patients. *Cancer Treat. Rep.* **29**, 431–440.
77. Mabe, S. and Champney, W. S. (2005) A comparison of a new oral streptogramin XRP 2868 with quinupristin-dalfopristin against antibiotic-resistant strains of *Haemophilus influenzae*, *Staphylococcus aureus*, and *Streptococcus pneumoniae*. *Curr. Microbiol.* **51**, 363–366.
78. Breedt, J., Teras, J., Gardovskis, J., et al., and Tigecycline 305 cSSSI study Group. (2005) Safety and efficacy of tigecycline in treatment of skin and skin structure infections: results of a double-blind phase 3 comparison study with vancomycin-aztreonam. *Antimicrob. Agents Chemother.* **49**, 4658–4666.
79. Allen, N. E. and Nicas, T. I. (2003) Mechanism of action of oritavancin and related glycopeptide antibiotics. *FEMS Microbiol. Rev.* **26**, 511–532.
80. Mercier, R.-C., Houlihan, H. H., and Rybak, M. J. (1997) Pharmacodynamic evaluation of a new glycopeptide, LY333328, and in vitro activity against

Staphylococcus aureus and *Enterococcus faecium*. *Antimicrob. Agents Chemother.* **41,** 1307–1312.
81. Giamarellou, H., O'Riordan, W., Harris, H., Owen, S., Porter, S. B., and Loutit, J. S. (2003) Phase 3 trial comparing 3–7 days of oritavancin vs. 10–14 days of vancomycin/cephalexin in the treatment of patients with complicated skin and skin structure infections (cSSSI) [abstract L-739a], in *Program and Abstracts of the 43rd Interscience Conference of Antimicrobial Agents and Chemotherapy,* American Society for Microbiology, Washington, DC.
82. Wasilewski, M., Disch, D., McGill, J., Harris, H., O'Riordan, W., and Zeckel, M. (2001) Equivalence of shorter course therapy with oritavancin vs. vancomycin/ cephalexin in complicated skin-skin structure infections (cSSSI) [abstract UL-18], in *Program and Abstracts of the 41st Interscience Conference on Antimicrobial Agents and Chemotherapy,* American Society for Microbiology, Washington, DC.
83. van Bambeke, F., van Laethem, Y., Courvalin, P., and Tulkens, P. (2004) Glycopeptide antibiotics: from conventional molecules to new derivatives. *Drugs* **64,** 913–936.
84. Leighton, A., Gottlieb, A. B., Dorr, M. B., et al. (2004) Tolerability, pharmacokinetics, and serum bactericidal activity of intravenous dalbavancin in healthy volunteers. *Antimicrob. Agents Chemother.* **48,** 940–945.
85. Seltzer, E., Dorr, M. B., Goldstein, B. P., Perry, M., Dowell, J. A., Henkel, T., and Dalbavancin Skin and Soft-Tissue Infection Study Group. (2003) Once-weekly dalbavancin versus standard-of-care antimicrobial regimens for treatment of skin and soft-tissue infections. *Clin. Infect. Dis.* **37,** 1298–1303.
86. Vicuron Pharmaceuticals. (2004) Press release. Vicuron Pharmaceuticals announces positive pivotal phase III results for dalbavancin in skin and soft tissue infections. Accessed August 12, 2004. http://biz.yahoo.com/prnews/040812/sfth039_1.html.
87. Jauregui, L. E., Babazadeh, S., Seltzer, E., et al. (2005) Randomized, double-blind comparison of once-weekly dalbavancin versus twice-daily linezolid therapy for the treatment of complicated skin and skin structure infections. *Clin. Infect. Dis.* **41,** 1407–1415.
88. Bogdanovich, T., Ednie, L. M., Shapiro, S., and Appelbaum, P. C. (2005) Antistaphylococcal activity of ceftobiprole, a new broad-spectrum cephalosporin. *Antimicrob. Agents Chemother.* **49,** 4210–4219.
89. Guignard, B., Entenza, J. M., and Moreillon, P. (2005) β-lactams against methicillin-resistant *Staphylococcus aureus*. *Curr. Opin. Pharmacol.* **5,** 479–489.
90. Reyes, N., Skinner, R., Kaniga, K., et al. (2005) Efficacy of telavancin (TD-6424), a rapidly bactericidal lipoglycopeptide with multiple mechanisms of action, in a murine model of pneumonia induced by methicillin-resistant *Staphylococcus aureus*. *Antimicrob. Agents Chemother.* **49,** 4344–4346.
91. McCollum, M., Rhew, D. C., and Parodi, S. (2003) Cost analysis of switching from IV vancomycin to po linezolid for the management of methicillin-resistant *Staphylococcus* species. *Clin. Ther.* **25,** 3173–3189.
92. Itani, K. M. F., Weigelt, J., Li, J. Z., and Duttagupta, S. (2005) Linezolid reduces length of stay and duration of intravenous treatment compared with

vancomycin for complicated skin and soft tissue infections due to suspected or proven methicillin-resistant *Staphylococcus aureus*. *Int. J. Antimicrob. Agents* **26**, 442–448.
93. Jorgensen, J. H. and Fiebelkorn, K. (2006) *Susceptibility of Common Organisms. January–December 2005*, University Health System, San Antonio, TX.
94. Reynolds, R., Potz, N., Colman, M., Williams, A., Livermore, D., MacGowan, A., and BSAC Extended Working Party on Bacteraemic Resistance Surveillance. (2004) Antimicrobial susceptibility of the pathogens of bacteraemia in UK and Ireland 2001–2002: the BSAC Bacteraemia Resistance Surveillance Programme. *J. Antimicrob. Chemother.* **53**, 1018–1032.
95. Yuk, J. H., Dignani, M. C., Harris, R. L., Bradshaw, M. W., and Williams, T. W. Jr. (1991) Minocycline as an alternative antistaphylococcal agent. *Rev. Infect. Dis.* **13**, 1023, 1024.
96. Ruhe, J. J., Monson, T., Bradshear, R. W., and Menon, A. (2005) Use of long-acting tetracyclines for methicillin-resistant *Staphylococcus aureus* infections: case series and review of the literature. *Clin. Infect. Dis.* **40**, 1429–1434.
97. Nicolau, D. P., Freeman, C. D., Nightingale, C. H., Coe, C. J., and Quintiliani, R. (1994) Minocycline versus vancomycin for treatment of experimental endocarditis caused by oxacillin-resistant *Staphylococcus aureus*. *Antimicrob. Agents Chemother.* **38**, 1515–1518.
98. Levine, D. P., Fromm, B. S., and Reddy, B. R. (1991) Slow response to vancomycin or vancomycin plus rifampin in methicillin-resistant *Staphyloccoccus aureus* endocarditis. *Ann. Intern. Med.* **115**, 674–680.
99. Dailey, C. F., Pagano, P. J., Buchanan, L. V., Paquette, J. A., Haas, J. V., and Gibson, J. K. (2003) Efficacy of linezolid plus rifampin in an experimental model of methicillin-susceptible *Staphylococcus aureus* endocarditis. *Antimicrob. Agents Chemother.* **47**, 2655–2658.
100. Egle, H., Trittler, R., Kummerer, K., and Lemmen, S. W. (2005) Linezolid and rifampin: drug interaction contrary to expectations? *Clin. Pharmacol. Ther.* **77**, 451–453.
101. Krut, O., Sommer, H., and Kronke, M. (2004) Antibiotic-induced persistence of cytotoxic *Staphylococcus aureus* in non-phagocytic cells. *J. Antimicrob. Chemother.* **53**, 167–173.
102. Shopsin, B., Zhao, X., Kreiswirth, B. N., Tillotson, G. S., and Drlica, K. (2004) Are the new quinolones appropriate treatment for community-acquired methicillin-resistant *Staphylococcus aureus*? *Int. J. Antimicrob. Agents* **24**, 32–34.
103. Iyer, S. and Jones, D. H. (2004) Community-acquired methicillin-resistant skin infections: a retrospective analysis of clinical presentation and treatment of a local outbreak. *J. Am. Acad Dermatol.* **50**, 854–858.
104. Grim, S. A., Rapp, R. P., Martin, C. A., and Evans, M. E. (2005) Trimethoprim-sulfamethoxazole as a viable treatment option for infections caused by methicillin-resistant *Staphylococcus aureus*. *Pharmacotherapy* **25**, 253–264.
105. Markowitz, N., Quinn, E. L., and Saravolatz, L. D. (1992) Trimethoprim-sulfamethoxazole compared with vancomycin for the treatment of *Staphylococcus aureus* infection. *Ann. Intern. Med.* **117**, 390–398.

106. de Górgolas, M., Avilés, P., Vedejo, C., and Fernandez Guerrero, M. L. (1995) Treatment of experimental endocarditis due to methicillin-susceptible or methicillin-resistant *Staphylococcus aureus* with trimethoprim-sulfamethoxazole and antibiotics that inhibit cell wall synthesis. *Antimicrob. Agents Chemother.* **39,** 953–957.
107. Shams, W. E., McCormick, M., Rapp, R. P., and Evans, M. E. (2005) The use of trimethoprim-sulfamethoxazole for serious MRSA infections. *Infect. Med.* **22,** 507–510.
108. Ellis, M. W. and Lewis, J. S. II. (2005) Treatment approaches for community-acquired methicillin-resistant *Staphylococcus aureus*. *Curr. Opin. Infect. Dis.* **18,** 496–501.
109. Johnson, J. R. (2003) Linezolid versus vancomycin for methicillin-resistant *Staphylococcus aureus* infections. *Clin. Infect. Dis.* **36,** 236, 237.
110. Stein, A., Bataille, J. F., Drancourt, M., Curvale, G., Argenson, J. N., Groulier, P., and Raoult, D. (1998) Ambulatory treatment of multi-drug resistant *Staphylococcus*-infected orthopedic implants with high-dose oral co-trimoxazole (trimethoprim-sulfamethoxazole). *Antimicrob. Agents Chemother.* **42,** 3086–3091.
111. Howden, B. P. and Grayson, M. L. (2006) Dumb and dumber—the potential waste of a useful antistaphylococcal agent: emerging fusidic acid resistance in *Staphylococcus aureus*. *Clin. Infect. Dis.* **42,** 394–400.
112. Nimmo, G. R., Bell, J. M., Mitchell, D., Gosbell, I. B., Pearman, J. W., and Turnidge, J. D. (2003) Antimicrobial resistance in *Staphylococcus aureus* in Australian teaching hospitals, 1989–1999. *Microb. Drug Resist.* **9,** 155–160.
113. Grif, K., Dierich, M. P., Pfaller, K., Miglioli, P. A., and Allerberger, F. (2001) In vitro activity of fosfomycin in combination with various antistaphylococcal substances. *J. Antimicrob. Chemother.* **48,** 209–217.
114. Hamilton-Miller, J. M. (1992) In vitro activity of fosfomycin against 'problem' gram-positive cocci. *Microbios* **71,** 95–103.
115. Pankey, G. A. and Sabath, L. D. (2004) Clinical relevance of bacteriostatic versus bactericidal mechanisms of action in the treatment of Gram-positive bacterial infections. *Clin. Infect. Dis.* **38,** 864–870.
116. Lee, M. C., Rios, A. M., Aten, M. F., Mejias, A., Cavuoti, D., McCracken, G. H., and Hardy, R. D. (2004) Management and outcome of children with skin and soft tissue abscesses caused by community-acquired methicillin-resistant *Staphylococcus aureus*. *Pediatr. Infect. Dis. J.* **23,** 123–127.

Index

A

Adhesion, *see* Epithelial cell
AFM, *see* Atomic force microscopy
Agar dilution, minimum inhibitory concentration determination
    end points, 42
    inoculation and incubation, 41, 42
    inoculum preparation, 41
    materials, 31
    medium, 40
    plate preparation, 41
Agar screening, methicillin resistance, 31, 42, 47, 48
Allele replacement, *see* Virulence factors
Antibacterials, MRSA control, 211, 212
Antibiotic resistance mechanisms
    β-lactams, 3, 4
    vancomycin, 4, 5
Arbitrarily primed polymerase chain reaction, epidemiology studies, 11, 12
Atomic force microscopy (AFM), vancomycin-resistant *Staphylococcus aureus*, 5

B

BacLite® Rapid MRSA screening, 219
BBL Crystal™ MRSA ID system, 217, 218
Biofilm
    assays
        catheter-based model
            implantation, 139, 140, 142, 143
            infection assessment, 140, 141, 143
            inocula preparation, 137, 139, 142
        flow cell assay
            flow establishment, 133–135, 142
            inoculation, 135–137, 142
            instrumentation, 132
            plasma coating, 131, 133, 142
        materials, 129, 130, 141, 142
        microtiter plate assay, 131, 142
    formation phases, 128, 129
    staphylococcal gene expression, 129
bl2seq, sequence alignment, 154, 157, 163
BLAST, gene finding and annotation, 154, 157, 163
Broth dilution, minimum inhibitory concentration determination
    inoculum preparation, 39
    macrodilution, 38, 39
    materials, 31
    medium, 37, 38
    microdilution, 38–40
    oxacillin preparation, 37
    principles, 36, 37

C

Case studies
    clinical–molecular feature relationships, 25–27
    hypervirulent methicillin-sensitive *Staphylococcus aureus*, 24, 25
    Native American community-acquired MRSA, 23, 24
    overview, 21, 22
    transcontinental transmission of hospital-acquired MRSA, 22, 23
    transitional MRSA, 24
    treatment of MRSA, 246–249
Catheter biofilm, *see* Biofilm
ccr genes, *see* Staphylococcal cassette chromosome *mec* analysis
Ceftobiprole, MRSA management, 13, 240
Clindamycin
    MRSA management, 242
    resistance
        detection, 8, 9
        mechanisms, 230
ClustalW, multiple sequence alignment and phylogenetic tree construction, 156, 159, 161, 162, 166
Community-acquired MRSA infection

case studies, *see* Case studies
clinical significance, 2
definition, 228
Comparative genomic analysis, MRSA
    gene finding and annotation, 157
    materials, 154, 156, 147
    maximum unique matches among multiple sequences, 157, 159
    overview, 153, 154
    pathway analysis, 156, 157, 163–166
    phylogenetic tree construction, 159, 162
    sequence alignment, 157, 161
    whole-genome alignment, 159
Control and prevention, MRSA
    control
        approaches, 211, 212
        rationale, 210, 211
    detection and diagnostics, 217–219
    prevention
        antibiotic use prudence, 215
        community health habits, 216
        disinfection, 213, 214
        hand hygiene, 213
        overview, 212, 213
        personal protective equipment, 213, 214
        screening and source isolation, 214, 215
        staff education, 215, 216
    risk factor avoidance, 217
    treatment, 216, 217
    vaccine development, 219, 220
Cubicin, *see* Daptomycin

D

Dalbavancin, MRSA management, 239, 240
Dalfopristin, *see* Quinipristin-Dalfopristin
Dalvabancin, MRSA management, 13
Daptomycin (Cubicin), MRSA management, 236, 237
Diabetic foot infection, linezolid management, 233
Disk diffusion, minimum inhibitory concentration determination
    agar preparation, 32
    disk application to agar plates, 33, 42
    disk storage, 32
    inoculum preparation, 32, 33, 42
    interpretation, 33, 43, 45
    materials, 30
    plate inoculation, 33
    turbidity standard, 33
DNA microarray
    complementary DNA
        agarose gel electrophoresis, 174
        first-strand cDNA synthesis, 173
        fragmentation, 173, 174, 176
        purification, 173, 176
        terminal labeling and efficiency analysis, 174
    epidemiology studies, 12
    hybridization and analysis, 174
    materials, 170, 171
    open reading frame detection, 190
    principles, 169, 170
    RNA extraction
        phenol extraction, 172, 173, 176
        Wizard® system, 171, 172
Doxycycline, MRSA management, 243

E

eBURST, multilocus sequence typing and strain comparison, 81, 82
Environmental surveillance, MRSA
    challenges, 202, 203
    culture
        broth enrichment, 205, 207
        direct culture, 205, 207
        materials, 204
        overview, 203
    infection sources, 202
    polymerase chain reaction detection, 203, 204, 206
    sampling
        air
            active sampling, 206
            passive sampling, 206
        contact plates, 205, 207
        electrostatic cloth, 206, 207
        materials, 204, 205
        overview, 203
        swabs, 205, 206
    survival duration, 201, 210
Epidemiology, molecular techniques for study
    arbitrarily primed polymerase chain reaction, 11, 12
    DNA microarray, 12

Index

multilocus sequence typing, 11
overview, 10, 11
pulsed-field gel electrophoresis, 11
spa typing, 12
Epithelial cell
    Staphylococcus aureus adhesion
        assay, 146, 148–151
        signaling, 145
    Staphylococcus aureus invasion
        assay, 146–148, 150, 151
        overview, 146
Etest, minimum inhibitory concentration determination
    agar plate inoculation, 35
    inoculum preparation, 35
    interpretation, 36, 45
    materials, 31
    medium, 35
    principles, 34, 35, 42, 45
    strip application to plates, 35, 36
    strip storage, 35
Evolution, MRSA, 6, 7

F

FASTA, multilocus sequence typing and strain comparison, 82
Fosfomycin, MRSA management, 246
Fusidic acid, MRSA management, 245

G

GATA program, sequence alignment, 154, 157, 158, 163, 166
Gatifloxacin, MRSA management, 244
Gene disruption, *see* Virulence factors
Gene expression analysis, *see* DNA microarray; Reverse transcription-polymerase chain reaction
Glimmer, gene finding and annotation, 154, 157

H

Hand hygiene, MRSA prevention, 213

I

IDI-MRSA™ test, 218
Internal transcribed spacer-polymerase chain reaction (ITS-PCR)
    cell lysis, 53
    DNA extraction, 53, 54, 56
    materials, 52, 53, 56

microchip gel electrophoresis, 54–56
multiplex polymerase chain reaction, 54
principles of MRSA identification, 51, 52
Invasion, *see* Epithelial cell
ITS-PCR, *see* Internal transcribed spacer-polymerase chain reaction

K

KEGG, pathway analysis, 156, 157, 163–166

L

Latex agglutination MRSA-Screen, 218
Linezolid (Zyvox)
    adverse effects, 235
    costs, 235, 236, 242, 243
    MRSA management, 13, 231–236
    resistance concerns, 236
    vancomycin comparison trials, 233, 234

M

Mass spectrometry, proteomics, 184, 187–189
mecA, *see* Staphylococcal cassette chromosome *mec* analysis
Methicillin resistance
    detection
        agar screening, 31, 42, 47, 48
        minimum inhibitory concentration
            agar dilution, 40–42
            broth dilution, 36–40
            disk diffusion, 32, 33
            Etest, 34–36
            materials, 30–32
            overview, 8, 29, 30
        rapid detection, 10
    mechanisms, 3, 4, 87, 88
    SCC*mec* analysis, *see* Staphylococcal cassette chromosome *mec* analysis
MIC, *see* Minimum inhibitory concentration
Minimum inhibitory concentration (MIC)
    agar dilution
        end points, 42
        inoculation and incubation, 41, 42
        inoculum preparation, 41
        materials, 31
        medium, 40
        plate preparation, 41

broth dilution
    inoculum preparation, 39
    macrodilution, 38, 39
    materials, 31
    medium, 37, 38
    microdilution, 38–40
    oxacillin preparation, 37
    principles, 36, 37
disk diffusion
    agar preparation, 32
    disk application to agar plates, 33, 42
    disk storage, 32
    inoculum preparation, 32, 33, 42
    interpretation, 33, 43, 45
    materials, 30
    plate inoculation, 33
    turbidity standard, 33
Etest
    agar plate inoculation, 35
    inoculum preparation, 35
    interpretation, 36, 45
    materials, 31
    medium, 35
    principles, 34, 35, 42, 45
    strip application to plates, 35, 36
    strip storage, 35
Minocycline, MRSA management, 243
MLST, *see* Multilocus sequence typing
Moxifloxacin, MRSA management, 244
MRSA Evigene™, 218
Multi-LAGAN, whole genome alignment, 156, 159, 161, 163
Multilocus sequence typing (MLST)
    accuracy and quality control, 73
    amplification product purification, 78, 79, 82
    clinical–molecular feature relationships in MRSA case studies, 25–27
    data quality assessment, 80
    direct analysis, 75
    epidemiology studies, 11
    genomic DNA isolation, 77, 78
    interpretation
        allele number assignment, 80, 81
        overview, 72, 73
        strain comparison

eBURST, 81, 82
FASTA, 82
materials, 76, 77, 82
polymerase chain reaction, 78, 82
resequencing arrays, 75
scheme for *Staphylococcus aureus*
    overview, 71, 72
    worldwide coordination, 73, 74
sequencing, 79, 80, 84
subculture of MRSA, 77, 82
MUMmer, maximum unique matches among multiple sequences, 154, 156, 157, 159, 163, 166

N

Nosocomial MRSA infection
    case studies, *see* Case studies
    clinical significance, 2

O

Origins, MRSA, 6, 7
Oritavancin, MRSA management, 13, 239
Osteomyelitis, linezolid management, 233

P

Panton-Valentine leukociden (PVL)
    prognostic value, 232
    virulence factor, 3
PBP, *see* Penicillin-binding protein
PCR, *see* Polymerase chain reaction
Penicillin-binding protein (PBP), MRSA expression, 3, 4
PFGE, *see* Pulsed-field gel electrophoresis
Phylogenetic tree, *see* Comparative genomic analysis
Polymerase chain reaction (PCR), *see also* Arbitrarily primed polymerase chain reaction; Internal transcribed spacer-polymerase chain reaction; Reverse transcription-polymerase chain reaction
    environmental surveillance, 203, 204, 206
    multilocus sequence typing, 78, 82
    plasmid construction for allele replacement, 105
    staphylococcal cassette chromosome *mec* analysis, 93, 94

Index

staphylococcal superantigen gene amplification, 119, 120, 123, 124
Prevention, *see* Control and prevention, MRSA
Proteomics, MRSA
    bacteriophage studies, 193
    cell culture and lysis, 181
    clinical implications, 193, 194
    extraction of membrane proteins, 181, 185, 186, 194, 195
    mass spectrometry, 184, 187–189
    materials, 181–184, 194, 195
    overview, 179–181
    polyacrylamide gel electrophoresis
        denaturing gel electrophoresis, 182, 195
        in-gel trypsinization, 183, 186, 187, 195, 196
        staining, 182, 183, 195
        two-dimensional gel electrophoresis, 182, 194, 195
    resistance studies, 192
Pulsed-field gel electrophoresis (PFGE)
    agarose plug preparation, 62
    cell suspension preparation, 63, 68
    clinical–molecular feature relationships in MRSA case studies, 25–27
    electrophoresis, 65, 66, 68
    epidemiology studies, 11
    gel casting and loading, 64, 68
    imaging and analysis, 66, 68
    materials, 60–62, 66, 68
    plug transfer, lysis, and washing, 63, 68
    principles, 59, 60
    restriction digestion, 63–65, 68
    staining, 66
    subculture of MRSA isolates, 62
PVL, *see* Panton-Valentine leukociden

Q

Quinipristin-Dalfopristin (Synercid), MRSA management, 238

R

RBSfinder, gene finding and annotation, 154, 157

Reverse transcription-polymerase chain reaction (RT-PCR)
    amplification reactions, 175–177
    complementary DNA
        agarose gel electrophoresis, 174
        first-strand cDNA synthesis, 173, 175
        fragmentation, 173, 174, 176
        purification, 173, 176
        terminal labeling and efficiency analysis, 174
    materials, 170, 171
    principles, 169, 170
    RNA extraction
        phenol extraction, 172, 173, 176
        Wizard® system, 171, 172
Rifampin, MRSA management, 243, 244
RT-PCR, *see* Reverse transcription-polymerase chain reaction

S

SAgs, *see* Staphylococcal superantigens
SCC*mec*, *see* Staphylococcal cassette chromosome *mec* analysis
spa typing, epidemiology studies, 12
Staphylococcal cassette chromosome *mec* analysis (SCC*mec*)
    assignment
        ccr, 96, 97, 100
        J regions, 97
        mec, 94, 95, 100
        SCC*mec* elements and nomenclature, 97–100, 228
    gnomic DNA extraction, 94
    materials, 89–93, 100
    overview, 87–89
    polymerase chain reaction, 93, 94
Staphylococcal superantigens (SAgs)
    immune response, 115
    molecular analysis
        cell culture, 120, 124
        materials, 115–119, 123
        polymerase chain reaction, 119, 120, 123, 124
        Western blot, 120–122, 124
    types, 114
Streptogramin pristinamycin, MRSA management, 245, 246

Sulfamethoxazole, *see* Trimethoprim-sulfamethoxazole
Superantigens, *see* Staphylococcal superantigens
Surveillance, *see* Environmental surveillance, MRSA
Synercid, *see* Quinipristin-Dalfopristin

T

Telavancin, MRSA management, 240, 241
Tigacyl, *see* Tigecycline
Tigecycline (Tigacyl), MRSA management, 13, 238, 239
TMP-SMX, *see* Trimethoprim-sulfamethoxazole
Toxic shock syndrome (TSS), clinical features, 114, 115
Toxic shock syndrome toxin-1 (TSST-1), virulence factor, 3, 114
Treatment, MRSA
    antibiotic therapy indications, 246
    bactericidal versus bacteriostatic agents, 246
    case studies, 246–249
    ceftobiprole, 240
    clindamycin, 242
    cost concerns, 241–243
    dalbavancin, 239, 240
    daptomycin, 236, 237
    doxycycline, 243
    fosfomycin, 246
    fusidic acid, 245
    gatifloxacin, 244
    linezolid, 231–236
    minocycline, 243
    moxifloxacin, 244
    oral versus intravenous therapy, 241–245
    oritavancin, 239
    overview, 216, 217, 228, 229, 249, 250
    rifampin, 243, 244
    streptogramin pristinamycin, 245, 246
    Synercid, 238
    telavancin, 240, 241
    tigecycline, 13, 238, 239
    trimethoprim-sulfamethoxazole, 244, 245
    vancomycin, 230, 231

TreeView, multiple sequence alignment and phylogenetic tree construction, 156, 159, 162, 166
Trimethoprim-sulfamethoxazole (TMP-SMX), MRSA management, 244, 245
TSS, *see* Toxic shock syndrome
TSST-1, *see* Toxic shock syndrome toxin-1
Two-dimensional polyacrylamide gel electrophoresis, *see* Proteomics

V

Vaccine, development for MRSA, 194, 219, 220
Vancomycin
    MRSA management, 230, 231
    resistance
        detection, 9
        heteroresistance, 234, 235
        isolates, 228, 230
        mechanisms, 4, 5, 230
VAP, *see* Ventilator-associated pneumonia
Velogene™ Rapid MRSA identification assay, 217, 218
Ventilator-associated pneumonia (VAP), linezolid management, 232
Virulence factors, *see also* Staphylococcal superantigens
    MRSA, 3
    targeted gene disruption
        allele replacement by temperature shifting with enrichment, 108, 109, 111
        cloning and electroporation, 107
        materials, 104, 105
        overview, 103, 104, 110
        phage stock preparation and transduction, 107, 108, 110, 111
        polymerase chain reaction for plasmid construction, 105
        verification, 109, 110

W

Western blot, staphylococcal superantigens, 120–122, 124

Z

Zyvox, *see* Linezolid

Printed in the United States
154204LV00001B/44/P